Lattice Gauge Theory '86

NATO ASI Series
Advanced Science Institutes Series

A series presenting the results of activities sponsored by the NATO Science Committee, which aims at the dissemination of advanced scientific and technological knowledge, with a view to strengthening links between scientific communities.

The series is published by an international board of publishers in conjunction with the NATO Scientific Affairs Division

A	**Life Sciences**	Plenum Publishing Corporation
B	**Physics**	New York and London
C	**Mathematical and Physical Sciences**	D. Reidel Publishing Company Dordrecht, Boston, and Lancaster
D	**Behavioral and Social Sciences**	Martinus Nijhoff Publishers
E	**Engineering and Materials Sciences**	The Hague, Boston, Dordrecht, and Lancaster
F	**Computer and Systems Sciences**	Springer-Verlag
G	**Ecological Sciences**	Berlin, Heidelberg, New York. London,
H	**Cell Biology**	Paris, and Tokyo

Recent Volumes in this Series

Series B: Physics

Lattice Gauge Theory '86

Edited by

Helmut Satz

Universität Bielefeld
Bielefeld, Federal Republic of Germany
and Brookhaven National Laboratory
Upton, New York

Isabel Harrity
Jean Potvin

Brookhaven National Laboratory
Upton, New York

Plenum Press
New York and London
Published in cooperation with NATO Scientific Affairs Division

Proceedings of a NATO Advanced Research Workshop entitled
Lattice Gauge Theory 1986,
held September 15–19, 1986,
at Brookhaven National Laboratory, Upton, New York

Library of Congress Cataloging in Publication Data

NATO Advanced Research Workshop on Lattice Gauge Theory (1986: Upton, N.Y.)
    Lattice gauge theory '86.

(NATO ASI series. Series B, Physics; v. 159)
    "Proceedings of a NATO Advanced Research Workshop on Lattice Gauge
Theory, 1986, held September 15–19, 1986, at Brookhaven National Laboratory,
Upton, New York"—T.p. verso.
    "Published in cooperation with NATO Scientific Affairs Division."
    Bibliography: p.
    Includes index.
    1. Gauge fields (Physics)—Data processing—Congresses. 2. Lattice theory
—Data processing—Congresses. 3. Quantum chromodynamics—Data pro-
cessing—Congresses. 4. Particles (Nuclear physics)—Data process-
ing—Congresses. I. Satz, Helmut. II. Harrity, Isabel. III. Potvin, Jean. IV. North
Atlantic Treaty Organization. Scientific Affairs Division. V. Title. VI. Series.
QC793.3.F5N36   1986   530.1′42   87-12276
ISBN-13: 978-1-4612-9062-9   e-ISBN-13: 978-1-4613-1909-2
DOI: 10.1007/ 978-1-4613-1909-2

© 1987 Plenum Press, New York
Softcover reprint of the hardcover 1st edition 1987
A Division of Plenum Publishing Corporation
233 Spring Street, New York, N.Y. 10013

PREFACE

This volume contains the Proceedings of the International Workshop "Lattice Gauge Theory 1986", held at Brookhaven National Laboratory, September 15 - 19, 1986. The meeting was the sequel to the one held at Wuppertal in 1985, the Proceedings of which have appeared in the same Plenum series.

During the past few years, a considerable number of meetings on lattice gauge theory have been held, on both sides of the Atlantic. With our workshop, through early planning and coordination with other prospective organizers, we tried to channel this activity into one major yearly meeting. For 1986, these efforts were successful, and it is our hope that a pattern has been set for the coming years. One result, however, was that the number of participants considerably exceeded that normally found at NATO Advanced Research Workshops. This year, a "nucleus" of NATO-supported experts induced a large number of further interested specialists to obtain their own funds – thus greatly amplifying the impact of the event. The topics covered at the workshop ranged from hadron spectra to strong interaction thermodynamics; they included spontaneous symmetry breaking and Higgs models, renormalization group methods, as well as many contributions on various possible schemes for the simulation of dynamical quarks. First systematic applications of finite size scaling to lattice gauge theory were discussed, and the approach to the continuum limit was considered in detail. The workshop also provided the first major status report in the application of the Langevin algorithm in lattice QCD. Last, but not least, it gave a survey of hardware developments, both on the commercial level and in dedicated institutional facilities.

Much progress and much further promise has come out of lattice gauge theory over the past year. The field continues to expand and attract further disciples – certainly in no small part because of the increased availability of supercomputers. Nevertheless, all that has been achieved so far is just a beginning – as was emphasized at the meeting by Nobel Laureate K. Wilson, who started it all and continues to remind us how much further we have to go to reach the standards of condensed matter physics.

A last, but very pleasant task is to express our sincere appreciation for help and support. The sponsorship of the workshop by NATO was decisive in giving the meeting the international scope we had aimed for. The smooth and efficient services of BNL made running it seem like a routine affair, and our Conference Secretary, C. Sheppard, even made sure that no one had to eat lobster at the conference clambake unless he or she really wanted to. Finally, many thanks are due S. Sanielevici (BNL) for help in proofreading the manuscripts; and Timothy Chiu, Eric Myers and Frank Paige (BNL) who wrote the T$_{\!E}$X macros for this Proceedings. They provided invaluable guidance and assistance throughout.

Helmut Satz

Conference Organizers Michael Creutz, Sid Kahana, Helmut Satz and
Claudio Rebbi with Ken Wilson (seated right).

CONTENTS

LATTICE QCD AT FINITE DENSITY

I.M. Barbour

Department of Physics and Astronomy
University of Glasgow
Glasgow G12 8QQ, U.K.

INTRODUCTION

Lattice simulations of QCD at non-zero temperature in the 'quenched' approximation and with dynamical fermions have successfully probed the deconfinement and the chiral symmetry restoration transitions at zero chemical potential. We report here on some studies of the chiral symmetry restoration transition at non-zero density in the unquenched theory for SU(2) on small lattices and comment on possible mechanisms controlling this phase transition in SU(3).

We include[1,2] a chemical potential in our simulations by multiplying links in the positive imaginary time direction by e^{μ} and links in the opposite direction by $e^{-\mu}$ where μ is the chemical potential in lattice units.

Studies of lattice QCD using the above formalism together with Kogut-Susskind staggered quarks in the quenched approximation gave results[3] that did not fit with expectations. Mean-field as well as quenched Monte-Carlo calculations indicate that QCD at zero temperature and finite density has either a lowest baryon excitation level with mass m_{baryon} proportional to m_{π} or a state of quark matter which has a bulk energy per baryon corresponding to this mass.

For the gauge group SU(2) the result can be explained when the degeneracy of the baryons and mesons is taken into account together with the possibility of the realignment of the vacuum in the presence of baryonic matter[4]. For SU(3) the result is totally unexpected implying a chiral symmetry restoration

1

transition at $\mu_C \simeq \frac{1}{2}m_\pi$ rather than $^m p/3$. Gibbs[5] has shown, via studies of the one dimensional U(1) model at finite density, that there is good reason for suspecting that the quenched approximation is invalid for $\mu \neq 0$: the transition observed at $\mu_C \simeq \frac{1}{2}m_\pi$ could be spurious.

It is clearly necessary to extend calculations of QCD at finite density to beyond the quenched approximation. Hopefully in the full theory the spurious transition will be smoothed and only the real transition at $^M N/3$ will remain.

Calculations beyond the quenched model via numerical simulations for Wilson fermions[6] using a hopping parameter expansion in SU(3) and for staggered (Kogut-Susskind) fermions using pseudofermion techniques[7] in SU(2) and complex Langevin[8] for SU(3) essentially confirm the quenched results for SU(2) but do not yet resolve the contradictions in SU(3). We describe below an exact algorithm for studying the numerical simulation of the theory on small lattices and present initial results for the gauge group SU(2).

THE UPDATING PROCEDURE (SU(3))

The method used is a generalization of that used to investigate QCD at zero chemical potential[9]. A chequer-board lattice in terms of elementary hypercubes is used. Block Lanczos inversion gives the 48 x 48 block of the inverse of the fermion matrix M necessary for updating any of the 32 links on a hypercube. We go round each hypercube ~4 times hitting each link 10 times on each visit. The new 48 x 48 block is obtained after updating a link by means of rank annihilation.

The procedure is generalized to handle the non-hermitian fermion matrix which arises for non-zero chemical potential. In addition, we use Lanczos inversion to calculate the 48 x 48 block of M^{-2} on each hypercube. This then enables us to calculate the difference

$$\text{Tr } (M + \Delta)^{-1} - \text{Tr}(M).$$

Lanczos diagonalization at the beginning of each sweep gives us all the eigenvalues of M and hence the initial values for $|M|$ and Tr M^{-1}. The above algorithms then enable us to monitor $|M|$ and Tr M^{-1} throughout each sweep.

Note that for SU(2) the fundamental representation is pseudoreal and hence $|M|$ and Tr M^{-1} are real and positive. For SU(3) the determinant and trace can be complex. In general det $M = ||M|| e^{i\Phi}$ where $\Phi = 0$ for SU(2). Then

$$\langle \bar\psi\psi \rangle = \frac{\langle \text{Tr } M^{-1} e^{i\Phi} \rangle}{\langle e^{i\Phi} \rangle} \tag{1}$$

2

if $||M||$ is used as a contribution to the effective action in a Metropolis procedure. An equivalent expression for the condensate (in the limit of averaging over an infinite number of configurations) for SU(2) is

$$\langle \bar{\psi}\psi \rangle = \frac{\langle 1 \rangle}{\langle (Tr\ M^{-1})^{-1} \rangle} \qquad (2)$$

if $Tr\ M^{-1} \det M$ is used in the effective action. This representation for the condensate has a driving force which should have a weaker repulsion of the small eigenvalues compared to that of Eqn (1).

The above procedure should allow us to measure μ_c for chiral symmetry restoration and also importance sampling via the choice of driving force over a finite number of configurations.

RESULTS: SU(2)

A. Fermionic Weight $|M|$

With this choice of weight the condensate is given by

$$\langle Tr\ M^{-1} \rangle / \langle 1 \rangle$$

and is the analogue of the electric field in 2-dimensional electrostatics arising from unit positive charges at (x_i, y_i) where the i-th eigenvalue of the fermion matrix $\lambda_i = x_i + iy_i + m_q$ and the electric field $\underline{E} \equiv \langle \bar{\psi}\psi \rangle$ is measured at $(m_q, 0)$. An immediate consequence of this analogy (via Gauss's Theorem) is that $\langle \bar{\psi}\psi \rangle$ is non-zero in the limit of zero quark mass if (and only if) there is an excess of small eigenvalues on the imaginary axis relative to the neighbouring region.

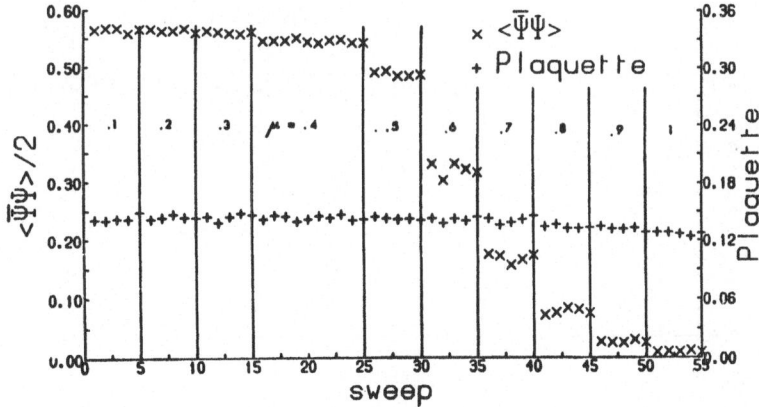

Fig. 1   The history of the condensate and average plaquette, on a 4^4 lattice at $\beta = 0.5$ and $m_q = 0.2$ for 4 flavours, as the chemical potential is increased from 0 to 1.0.

3

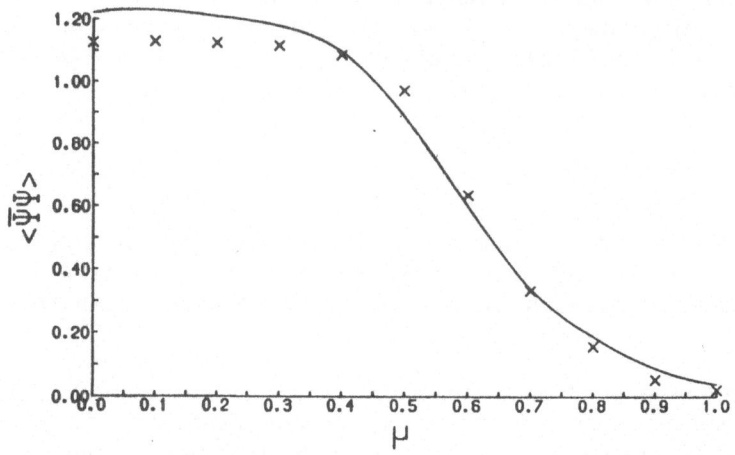

Fig. 2   Comparison of the condensate as measured above with the strong coupling prediction of Dagotto, Moreo and Wolff[10].

Results in the strong coupling region (β = 0.5) for 4-flavours and quark mass of 0.2 in lattice units are shown in Figs. 1 and 2. Fig. 1 shows the behaviour of the condensate and average plaquette (as the chemical potential μ is increased) as a function of sweep number.   On a 4^4 lattice the updating procedure appears to give an immediate reaction to the change in density.   Fig. 2 shows a comparison of the numerical data with the strong coupling prediction of Dagotto Moreo and Wolff[10].   Similar agreement has been obtained with the pseudofermion calculation of ref (7).

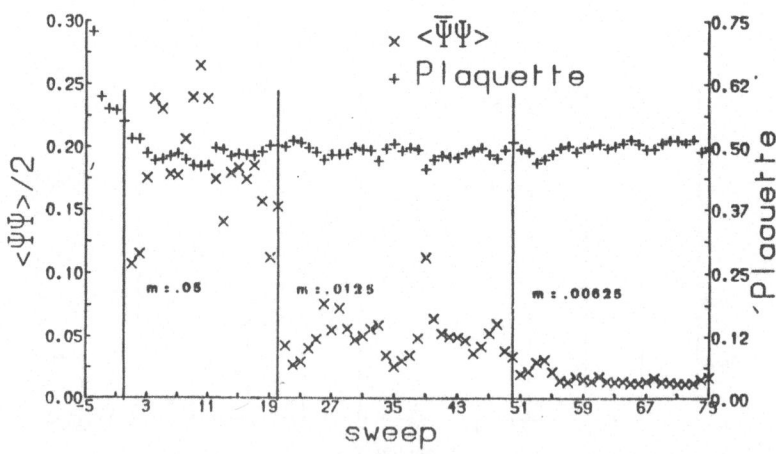

Fig. 3   The history of the condensate and average plaquette, on a 4^4 lattice at β = 1.7 and μ = 0.1 for three values of the quark mass and 4 flavours.

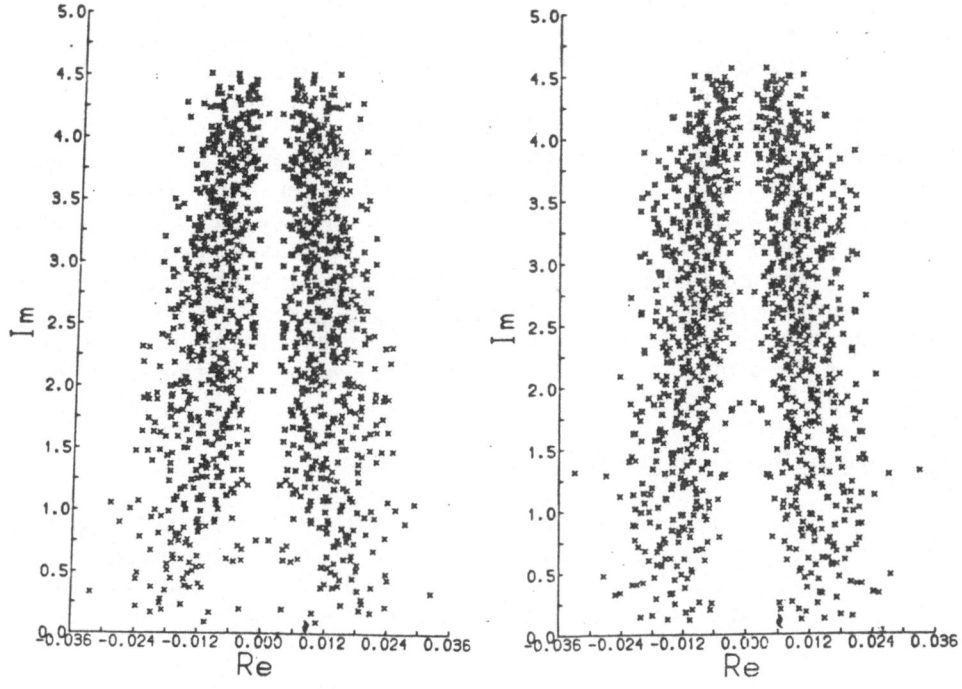

Fig. 4a

Fig. 4b

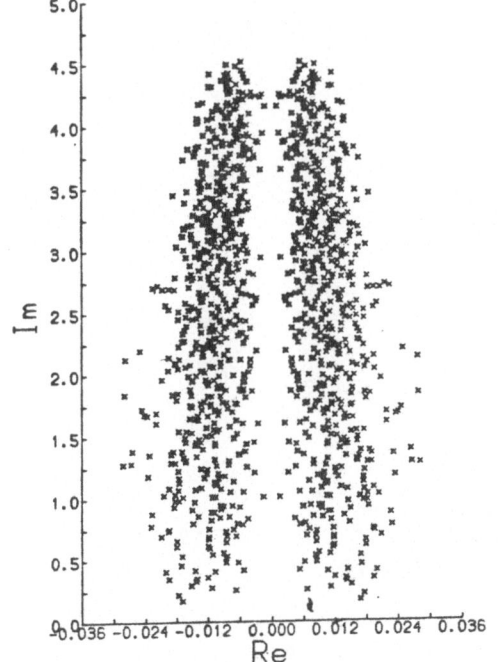

Fig. 4c

Fig. 4 The eigenvalue
distributions for the
fermion matrix for 5
superimposed
configurations.
(a) at $m_q = 0.05$,
(b) at $m_q = 0.0125$ and
(c) at $m_q = 0.00625$
on a 4^4 lattice at
$\beta = 1.7$. Note the
eigenvalues shown are
for $2(M - m_q)$.

Results in the intermediate coupling region (β = 1.7) for 4 flavours and chemical potential μ = 0.1 are summarized in Fig. 3 for three values of quark mass, m_q = 0.05, 0.0125, 0.00625. The region of quark mass explored is determined primarily by the width of the 'strip' of eigenvalues of the fermion matrix (at m_q = 0). This is insensitive to the dynamical mass. The distribution of eigenvalues at the three mass values for 4 sweeps in each case is shown in Fig. 4. Two points emerge. On this small lattice there is no sign of the small eigenvalues being driven to the imaginary axis as the dynamical mass decreases and enters the strip. Indeed there is a positive signal that there is a gap along the imaginary axis implying that $\lim_{m \to 0} \langle \bar\psi\psi \rangle$ = 0 as in the quenched theory. Also, albeit with limited statistics, we see a signal of fermionic repulsion of the zero-modes as the mass enters the strip.

B.   Fermionic Weight Tr M^{-1} det M.

We repeated the above calculation with different driving force to investigate importance sampling with limited statistics. The results are summarized in Figs. 5 and 6. The eigenvalue distributions are very similar to those of Fig. 4

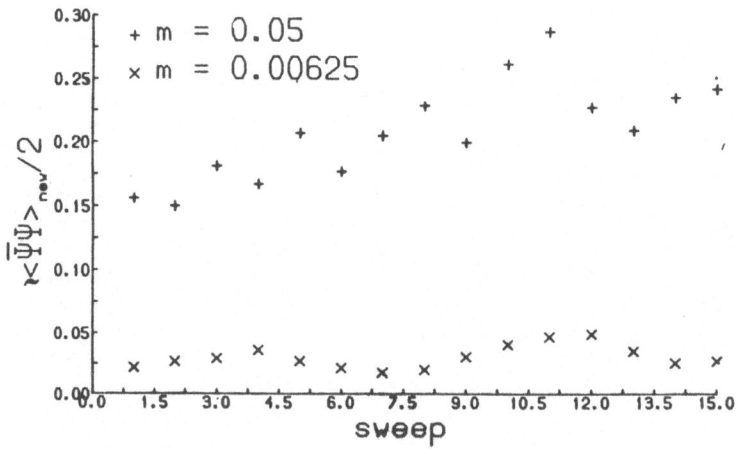

Fig. 5   The history of the condensate on a 4^4 lattice at β = 1.7 and μ = 0.1 for two values of the quark mass. Note the different fermionic weight compared to Fig. 3.

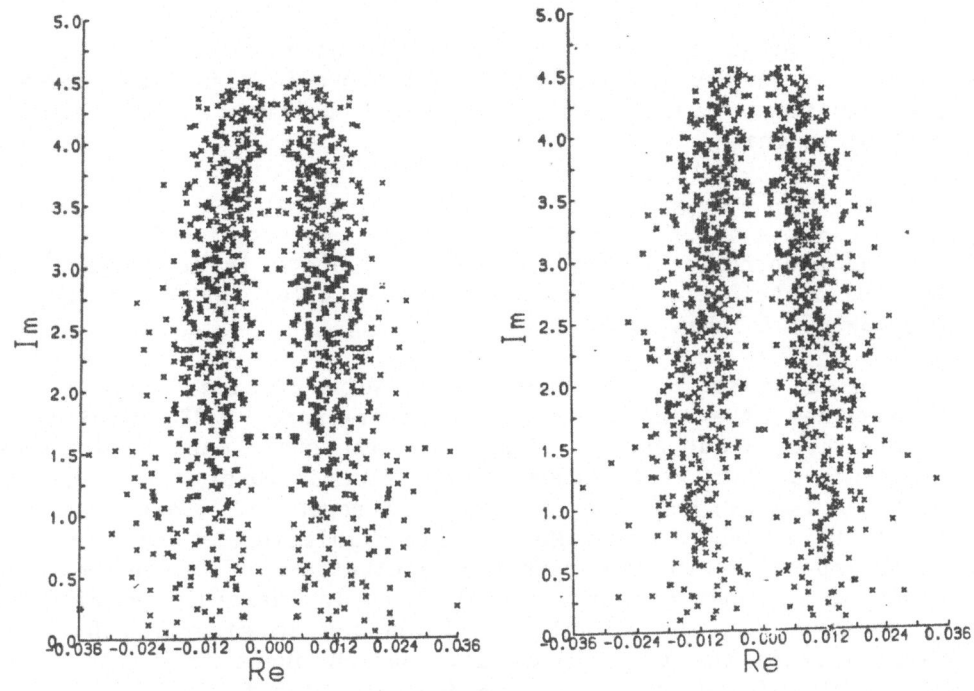

Fig. 6a   Fig. 6b

Fig. 6 The eigenvalue distributions for 4 superimposed
    configurations at (a) m_q = 0.05 and (b) at m_q = 0.00625 to be
    compared with Fig. 4a and 4c with the different fermionic
    weight.

    However the Table below, summarizing the value for the
condensate obtained in each simulation, is consistent with this
weight enhancing the density of small eigenvalues.

$$\langle \bar{\psi}\psi \rangle$$

m	|M|	Tr M^{-1} |M|
.05	0.169 ± .033	0.234 ± .027
.0125	0.049 ± .018	
.00625	0.014 ± .002	0.0315 ± .010

    In conclusion we find for SU(2) at finite density, the chiral
condensate transition is consistent with $\mu_c \sim m_\pi$. We emphasise
however that the chiral symmetry may be broken (even though
$\langle \bar{\psi}\psi \rangle$ = 0) via the mechanism proposed by Dagotto, Karsch and
Moreo[4].

COMMENTS: SU(3)

With the exception of SU(2) the fermion determinant will in general be complex for $\mu \neq 0$. The Metropolis algorithm relies on configurations being generated via a positive weight, such as |det M| so that

$$\langle \bar{\psi}\psi \rangle = \frac{\langle \text{Tr } M^{-1} e^{i\phi} \rangle}{\langle e^{i\phi} \rangle}$$

The electrostatic analogy discussed above is no longer valid but the distribution of eigenvalues is similar (but with only the ± symmetry) to that of SU(2).

For a given μ and quark mass well outside the strip of eigenvalues the phase of the determinant will have small fluctuations about zero. However as the quark mass is decreased and enters the strip then these fluctuations will increase. Gibbs[5] has conjectured that in this region of quark mass ($\mu \geqslant m_\pi/2$) it is possible that Monte-Carlo simulations break down since it is impossible to cope with the fluctuations of the phase. The onset of a possible phase transition could therefore be difficult to measure, but the actual transition may be signalled by a gap developing in the eigenvalue distribution about the real axis (as at the finite temperature phase transition for $\mu = 0$). These ideas will be developed further by Philip Gibbs in a subsequent talk.

ACKNOWLEDGEMENTS

I would like to thank my collaborators, Clive Baillie, Ken Bowler, Philip Gibbs and Muhamed Rafique, for their contributions to the above. I am also grateful to Tassos Vladikas, Frithjof Karsch and Bill Wyld for useful discussions.

REFERENCES

1. J. Kogut, H. Matsuoka, M. Stone, H.W. Wyld, S. Shenker, J. Shigemitsu and D.K. Sinclair, Nucl. Phys. B225 (FS9) (1983) 93.
2. P. Hasenfratz and F. Karsch, Phys. Lett. 125B (1983) 308. H. Matsuoka and M. Stone, Phys. Lett. 136B (1984) 204.
3. I. Barbour, N. Behilil, E. Dagotto, F. Karsch, A. Moreo, M. Stone and H.W. Wyld, Nucl. Phys. B275 (FS17) (1986) 296.
4. E. Dagotto, F. Karsch and A. Moreo, Phys. Lett. 169B (1986) 349.
5. P. Gibbs, 'Lattice Monte Carlo Simulations of QCD at Finite Baryonic Density', Glasgow preprint.
6. B. Berg, J. Engels, E. Kehl, B. Waltl, H. Satz, Bielefeld preprint BI-TP 86/05.

7. E. Dagotto, F. Karsch and A. Moreo, Phys. Lett. 169B (1986)
        421.
8. F. Karsch and H.W. Wyld, private communication.
9. I. Barbour, <u>in</u>: Lattice Gauge Theory, a Challenge in
        Large-Scale Computing, B. Bunk, K.H. Mütter and K.
        Schilling, eds., Plenum Press, New York (1986).
10. E. Dagotto, A. Moreo and U. Wolff, 'Study of Lattice SU(N) QCD
        at Finite Baryon Density', Illinois preprint,
        ILL-(337)-86-12 (1986).

FINITE SIZE SCALING AND LATTICE GAUGE THEORY

B.A. Berg

Department of Physics and
Supercomputer Computations Research Institute
The Florida State University
Tallahassee, FL 32306,

ABSTRACT

Finite size (Fisher) scaling is investigated for four dimensional SU(2) and SU(3) lattice gauge theories without quarks. It allows to disentangle violations of (asymptotic) scaling and finite volume corrections. Mass spectrum, string tension, deconfinement temperature and lattice β-function are considered. For appropriate volumes, Monte Carlo investigations seem to be able to control the **finite volume** continuum limit. Contact is made with Lüscher's small volume expansion and possibly also with the asymptotic large volume behaviour.

This talk [1] will be published elsewhere. Parts of it rely heavily on Refs. [2-4].

ACKNOWLEDGMENTS

My lecture would have been impossible without continuous collaboration with Alain Billoire. Our SU(3) results rely on work with Claus Vohwinkel [2,4,5] and part of our SU(2) data were obtained in collaboration with Enzo Marinari [2]. Further, I would like to thank Gyan Bhanot, Khalil Bitar, Dennis Duke, Anna and Peter Hasenfratz, Tony Kennedy, Roman Salvador nand Peter Weisz for useful discussions, help or correspondence. This work received support by the Department of Energy under cooperative agreement number DE-FC05-85ER25000 and from the Florida State University through use of its Cyber 205.

REFERENCES

1.   B. Berg, Preprint, FSU-SCRI-86-89.
2.   Work in preparation.
3.   B. Berg and A. Billoire, Phys. Lett. 166B (1986)203 and erratum (to be published).
4.   B. Berg, A. Billoire and C. Vohwinkel, Phys. Rev. Lett. 57 (1986)400.
5.   B. Berg, A. Billoire and C. Vohwinkel, Preprint, FSU-SCRI-86-70.

RECENT RESULTS FROM THE UNIVERSITY OF CALIFORNIA

WEAK MATRIX ELEMENT CALCULATION[*]

C. Bernard, T. Draper,[†] G. Hockney, and A. Soni

University of California, Los Angeles, CA 90024
University of California, Irvine, CA 92717

INTRODUCTION

For several years now we have been making a major effort to calculate hadronic matrix elements of weak operators by lattice Monte Carlo methods. Some of the basic techniques were described at the Argonne, Tallahassee, and Wuppertal Conferences;[1-3] results of a "first generation" calculation were published last year.[4] Since that time we have accumulated considerably more data: we have greatly increased the number of configurations on our original small lattice ($6^3 \times 17$ at $\beta = 5.7$) and have gone on to examine two larger lattice sizes: $10^3 \times 20$ at $\beta = 5.7$ to look at finite size effects and $12^3 \times 33$ at $\beta = 6.1$ to look at scaling. Here, we present the new results from the small lattice and the preliminary results from the larger lattices (where we are continuing to accumulate statistics). We treat primarily the $K^0-\bar{K}^0$ mixing and $K \to \pi$ $\Delta I = 3/2$ amplitudes, where the theoretical situation is more or less clear; however, our current approach to the problems of calculating the $K \to \pi$ $\Delta I = 1/2$ amplitude is also briefly described, and is used to obtain some very preliminary numerical results.

As discussed in Ref. 1, contributions to the matrix elements of the 4-quark weak operators of interest can be divided into two classes: "figure-eight graphs" in which all quark fields contained in the weak operator contract with the interpolating fields of the external mesons, and "eye graphs" in which two

---

[*]Presented by C. Bernard.
[†]Present address: TRIUMF, 4004 Wesbrook Mall, Vancouver, B.C. V6T2A3, Canada.

quark fields in the operator contract with each other and there is a spectator quark connecting the external mesons. Eye graphs are the more difficult to calculate. Numerically, the problem is that the straightforward application of the techniques used for hadron mass calculations require much too much computing time. In fact, the first published Monte Carlo calculations of weak matrix elements[5] simply left out the eye graphs entirely. A method which sidesteps the numerical difficulties is discussed in Ref. 1 (see also Ref. 6 for a very similar approach); a more convenient and versatile version is described in Ref. 3.

Eye graphs also present a theoretical difficulty which arises from mixing with lower dimensional operators. Such mixing is impossible in figure-eight graphs since the flavor quantum numbers of all four quarks in the operator must appear in the external states. The first attempt to deal with the mixing problem for eye graphs was within the context of chiral perturbation theory.[7] We did not, however, immediately realize that Ref. 7 could not be used directly for a lattice calculation with Wilson fermions. The reason is that the breaking of chiral symmetry by the lattice regulator introduces, effectively, an additional operator which does not appear in the continuum.[8] (On the other hand, with staggered fermions the techniques of Ref. 7 are directly applicable to the mixing problem and seem, in fact, to be the method of choice.[9]) L. Maiani et al.[10] have extended the chiral perturbation theory approach to the case of Wilson fermions. For reasons explained below, however, we now have doubts whether chiral perturbation theory is appropriate at all for meson masses in the range of the physical kaon. Near the end of this talk, we propose an alternative approach to the eye graph mixing problem and apply it to the calculation of the $\Delta I = 1/2$ amplitude. This new approach has some advantages over the method of Ref. 10, but also some disadvantages, and it is not yet clear which, if either, will produce reliable answers in practice. We turn first, however to some technical details.

TECHNICAL DETAILS

Our method for extracting the physical amplitudes from the lattice calculation is straightforward and applies in its basic form to both figure-eight and eye graphs. We compute the lattice Green's functions

$$G_{A\theta B}(t,t') \equiv \sum_{\vec{x},\vec{x}'} <0,|\chi_A(t,\vec{x})\ \theta(0,\vec{0})\ \chi_B^{\dagger}(t',\vec{x}')|0> \qquad (1)$$

where $\theta(0,\vec{0})$ is the 4-quark weak operator evaluated at the origin, χ_A, χ_B are interpolating fields for the mesons, and $t > 0$, $t' < 0$. The sum over the spatial variables $\vec{x}$ and $\vec{x}'$ is performed to ensure that the intermediate states will have zero momentum. For figure-eight graphs, the right-hand side of (1) may be easily calculated from the quark propagators starting at the origin; those propagators are obtained by standard methods. (We use an over-relaxed Gauss-Seidel algorithm.) One then simply averages over gauge configurations to find $G_{A\theta B}$. The configurations may be either quenched or unquenched. The results presented here use only quenched configurations, but we plan, in the near future, to include dynamical fermions by using unquenched configurations generated by Ph. de Forcrand and I. O. Stametescu.

To extract matrix elements from $G_{A\theta B}$ one also needs the meson propagators:

$$G_A(t,0) \equiv \sum_{\vec{x}} <0|\chi_A(t,\vec{x})\ \chi_A^\dagger(0,\vec{0})|0> \to Z_A\ e^{-m_A t} \qquad (2)$$

where we show the large t behavior, with m_A the mass of the lowest meson in the A channel and Z_A a constant. A similar expression holds for $G_B(0,t')$. G_A and G_B are obtained from the same quark propagators as $G_{A\theta B}$. We define

$$\tilde{G}_{A\theta B}(t,t') \equiv \frac{G_{A\theta B}(t,t')}{G_A(t,0)\ G_B(0,t')} \ . \qquad (3)$$

For t and -t' large enough one then has

$$\tilde{G}_{A\theta B}(t,t') \to \frac{1}{2\sqrt{Z_A Z_B m_A m_B}}\ <A|\theta|B> \qquad (4)$$

where $\sqrt{Z_A Z_B}$ is defined to have the same phase as Z_A and Z_B, which are obtained from Eq. (2).

We thus have the matrix element of interest, $<A|\theta|B>$. The method of extraction is extremely simple and has the additional advantage that there is an internal check: $\tilde{G}_{A\theta B}(t,t')$ should become independent of t and t' for t and -t' "large enough." We found that this is indeed the case. $\tilde{G}_{A\theta B}(t,t')$ varied by at most ~ 10% (and usually considerably less than that) for the values of t,t' for which $<A|\theta|B>$ was extracted. The only exceptions were in cases where the statistical errors, over configurations, indicated that $<A|\theta|B>$ was in fact consistent with zero. This is as expected; non-constancy for t and -t' large is simply another signal that the noise is dominating.

In practice, t and -t' "large enough" meant values of 4 to 6 on the $\beta = 5.7$ lattices and 7 to 11 on the $\beta = 6.1$ lattice. For these time values, not only $\tilde{G}_{A\theta B}$ but also the "effective mass"[11] of the mesons was approximately constant. In order to accommodate the correspondingly large values of t-t' and to prevent quark propagation "around the back" on a periodic lattice, we resorted to aperiodic boundary conditions (typically "free flat") in the time direction for our quarks.[11] This meant that we had to keep the fields away from the boundaries: the distance away was at least 3 lattice spacings on the small β = 5.7 lattices, at least 4 spacings on the large β = 5.7 lattices, and at least 6 spacings on the β = 6.1 lattices. We chose the "safe" distance by moving the boundary farther away and requiring that the pion propagator be changed by less than 1%. (Matrix elements were slightly more sensitive, ~ 2-3%.) To "soften" the boundary conditions in this way (and to allow for the largest of the t-t' values used) we extended by replication the periodic gauge lattices in the time direction before calculating the quark propagators. These extensions are as follows: $6^3 \times 10$ gauge (β = 5.7) → $6^3 \times 17$ quark; 10^4 gauge (β = 5.7) → $10^3 \times 20$ quark; $12^3 \times 16$ gauge (β = 6.1) → $12^3 \times 33$ quark. In specifying the size of the quark lattices we do not include the sites whose values are fixed by the boundary conditions.

We averaged over 204 configurations for the $6^3 \times 17$ lattices, with hopping parameter k = .094, .155, .162, .164, and .165. For the $10^3 \times 20$ lattices we averaged over 18 configurations, with k = .094, .155, and .164. In each set the configurations were separated by 1000 pseudo heat bath[12] passes; there were five independent sets making up the 204 configurations (3 cold, 2 hot starts) and 1 set for the 18. All the above had β = 5.7. At β = 6.1, we had 22 $12^3 \times 33$ configurations (generated by S. Otto) which were separated by 100 pseudo-heat-bath passes, and we have been generating our own configurations, separated by 2000 passes. Only 6 of these new configurations have been included so far (at k = .151); 6 of Otto's (separated by 400 passes) are used at k = .151 and .130, and all 22 of his are used at k = .153. We have been generating new configurations at β = 6.1 because there seem to be correlations when the configurations are separated by only 100 passes, with a correlation length of approximately 400 passes. No correlations were observed in our β = 5.7 samples.

Our statistical errors were in general computed using the "jackknife" method,[6] which was also useful in finding correlations between configurations. However, we have not yet fully implemented jackknifing in the final answers, so in a few instances (where there were large errors in more than one contributing factor) we determined the errors by dividing the configurations into two subgroups and multiplying the deviations of the subgroups from the average by $\sqrt{2}$.

Finally, we remark that we have observed the presence of a few very unusual configurations in our largest data sample (β = 5.7, $6^3 \times 17$). These configurations had a pion mass which was strongly suppressed over several time slices, resulting in a pion propagator which was many standard deviations larger than average in the region after the suppression. The effect was more pronounced at higher k values. At least in their rough features, these configurations seem to correspond to the "exceptional configurations" discussed by K. Mütter.[13] In line with his suggestion that fermion zero modes may be responsible, a search for a correlation between the presence of instantons and the "exceptionality" of a configuration has recently been made.[14] Though it appears that such a correlation is absent, it may well be that zero modes with other origins are responsible. This question is under investigation.[14]

RESULTS FROM FIGURE EIGHT GRAPHS

Here we examine the $K^0-\bar{K}^0$ mixing amplitude, the $\Delta I = 3/2$ $K \to \pi$ transition, and the figure eight contribution to the $\Delta I = 1/2$ $K \to \pi$ transition. In the standard model, the CP violating $K^0-\bar{K}^0$ transition comes from the the matrix element of the operator

$$H_{LL}^{\Delta S=2} = \bar{s}\gamma_\mu(1-\gamma_5)d\ \bar{s}\gamma^\mu(1-\gamma_5)\ d\ . \tag{5}$$

We also looked at the $K^0-\bar{K}^0$ matrix element of

$$H_{LR}^{\Delta S=2} = \bar{s}\gamma_\mu(1-\gamma_5)d\ \bar{s}\gamma^\mu(1+\gamma_5)\ d\ . \tag{6}$$

This left-right operator appears in various extensions of the standard model, and also provides a useful check on the calculation, since it fluctuates less and has different chiral behavior than $H_{LL}^{\Delta S=2}$. We show, in Figs. 1 and 2 the behavior of these matrix elements as a function of meson masses squared. Both points with degenerate and with non-degenerate hopping parameter values are shown. To convert the matrix elements and masses to physical units we have used $a^{-1} = 1.00$ GeV at $\beta = 5.7$ and $a^{-1} = 1.96$ GeV at $\beta = 6.1$; these come from a scaling (not asymptotic scaling) interpretation of string tension and potential calculations.[15] Of course, there are systematic errors introduced in this way (one certainly does not expect scaling to be exact at these β values) but one must make some such assumption in order to compare results at different β's. We also show, in Figs. 3 and 4, B_{LL} and B_{LR}, which are the amplitudes of Figs. 1 and 2 divided by their "vacuum saturation" values as computed on the lattice. The pseudoscalar decay constant which appears in vacuum saturation is computed directly from the vacuum-meson matrix element of the axial current and not from its divergence.

We have the following comments on the results:

1. The matrix elements look physical. In the LR case there is a rather non-trivial scaling test (even though the errors are large) because the matrix elements go like a^{-4} and a differs by almost a factor of 2 between $\beta = 5.7$ and 6.1. Computing B_{LR} reduces the errors and some scaling violation may be visible; in any case B_{LR} is close to 1. In the LL case, the errors in the

amplitude are still too large to say more than that the result is
consistent with scaling; B_{LL} actually has larger errors than the
matrix element itself, which means that the amplitude and its
vacuum saturated value have rather independent fluctuations. In
both LL and LR, the finite size effects appear under control at
the large masses but may be showing up at the smaller masses.
Even there, the finite size effects do not appear enormous, at
least compared to the statistical errors. All the results are
consistent with our preliminary study.[4]

2. The LL matrix element appears to go to zero (and perhaps
cross zero) at values of the meson mass close to the physical
region ($m^2 \sim .25$ GeV^2). If this is the case, one may at best
have a bound on the amplitude and may at worst (if zero is crossed
steeply) find it extremely difficult to make any prediction.
Preliminary results for a lighter mass on $\beta = 5.7$ $10^3 \times 20$ lattices seem to show a considerably gentler zero crossing (but a
crossing, nonetheless) than for the smaller lattices.

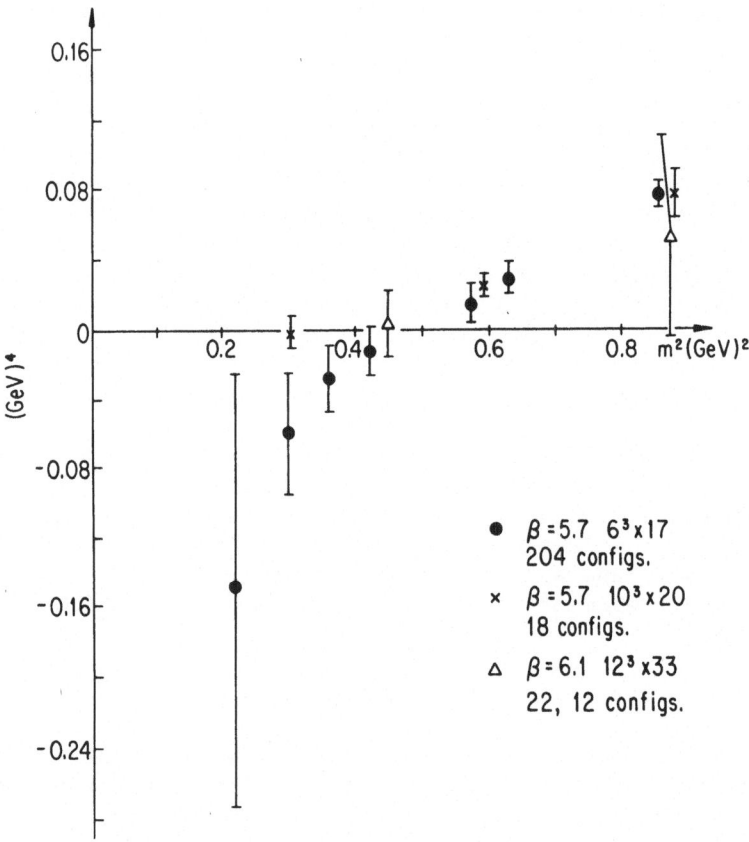

Fig. 1. The K^0-$\bar{K}^0$ matrix elements of $H_{LL}^{\Delta S=2}$ vs. meson mass
squared.

3. The LR behavior is roughly consistent with that predicted by lowest order chiral perturbation theory ($\rightarrow$ constant as $m^2 \rightarrow 0$), but the LL is not (vanishing linearly with m^2 is the lowest order CPTh result). For the LL amplitude this appears to disagree with results of Maiani and Martinelli which were presented at Berkeley[16]--they saw behavior consistent with lowest order CPTh up to masses ~ .9 GeV or more. There seems to be a numerical disagreement since we agree on the procedure to be used. (The weak coupling corrections which we have calculated[4,17] differ slightly from those computed by Martinelli,[18] but we do not expect the difference to have a significant effect; this will be checked.) Some other possible sources for the disagreement are:

(a) The way the data is analyzed. They use a more complicated method to extract the amplitudes.

(b) The coupling at which the calculations are done. Most of our runs are at a stronger coupling (β = 5.7) than their β = 6.0. Our β = 6.1 data here has large errors and it is hard to say much

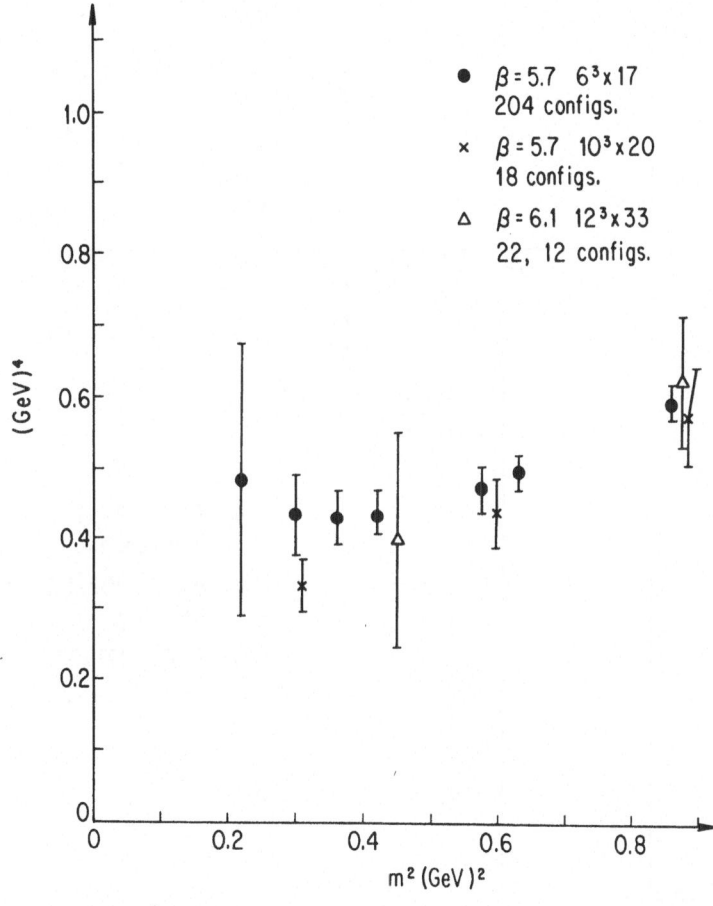

Fig. 2. The K^0-$\bar{K}^0$ matrix elements of $H_{LR}^{\Delta S=2}$ vs. meson mass squared.

at the moment. However, for the $\Delta\underline{I} = 3/2$ $K \to \pi$ amplitude, which has similar behavior to the LL K^O-$\bar{K}^O$ mixing, there is some evidence for consistency between our $\beta = 5.7$ and $\beta = 6.1$ results --see below. We plan to increase the statistics on the $\beta = 6.1$ lattices in the near future.

(c) The statistics. The existence of "exceptional" configurations means that small data samples may give misleading results or misleading error estimates, as we have learned in going from 15 to 204 configurations on our small lattice.

Our results pass several consistency checks. For large meson masses, the B parameters should be close to 1 (i.e. vacuum saturation should be good). We find, at $\beta = 5.7$, $k = .094$, that $B_{LL} = .86 \pm .02$ and $B_{LR} = .97 \pm .02$; and at $\beta = 6.1$, $k = .130$ that $B_{LL} = .86 \pm .08$ and $B_{LR} = .99 \pm .02$. (The meson masses in both these cases are $\simeq 3$ GeV, and are not shown in Figs. 3 and 4.) Further, we find that our results are unchanged if we Fierz

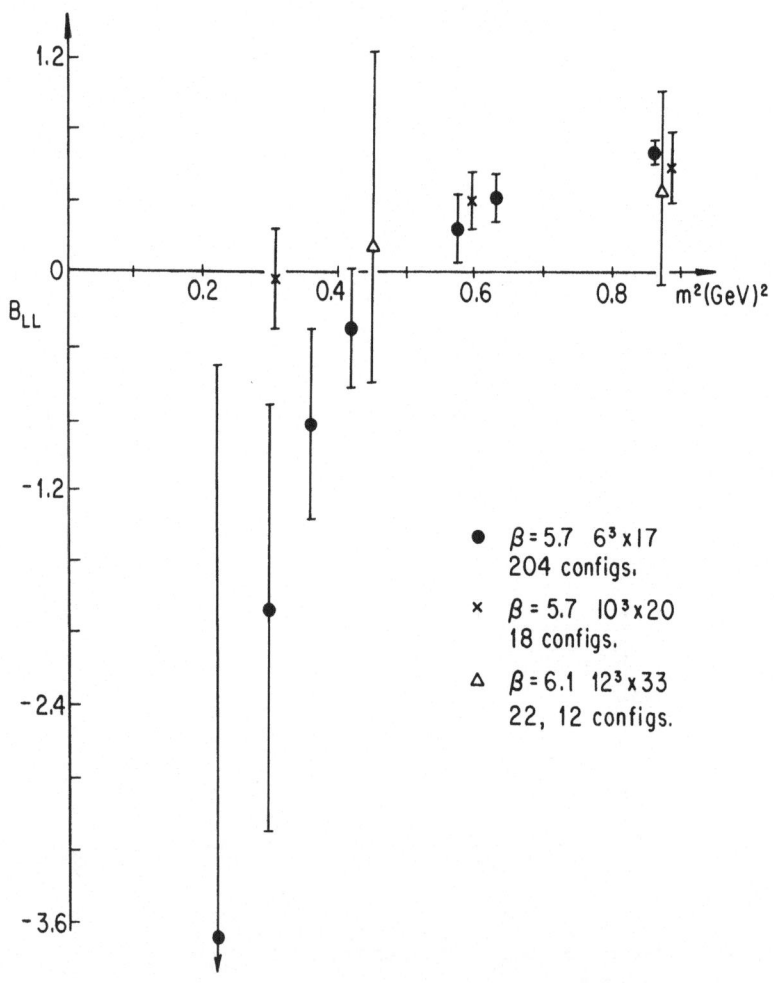

Fig. 3. B_{LL} vs. meson mass squared.

20

the weak operators before doing the calculation. This is a non-trivial check of the part of the program which puts quark propagators together into matrix elements. Finally, we note that we have used three different lattice generation routines (a pseudo-heat-bath program written by us, Otto's program, and a Metropolis program for the "first generation" results), and two different matrix inversion routines which were checked against each other (the earlier one was also checked against a program of Gupta and Patel). We think therefore that an error in these routines is rather unlikely.

Before going on to the eye graphs, we present, in Fig. 5, results for the $K \to \pi$, $\Delta I = 3/2$ amplitude and the figure-eight contribution to the $K \to \pi$, $\Delta I = 1/2$ amplitude. Two comments are in order:

1. The two amplitudes have roughly the same behavior, which in turn is roughly the same as that of the LL K-$\bar{K}^0$ amplitude. (This is perhaps not surprising since all three are LL figure-eight matrix elements.) There is thus no evidence for a $\Delta I = 1/2$ rule coming from figure eights alone: eye graphs seem to be required. In addition, near the actual K mass our $\Delta I = 3/2$ amplitude of the same order of magnitude as the "physical"

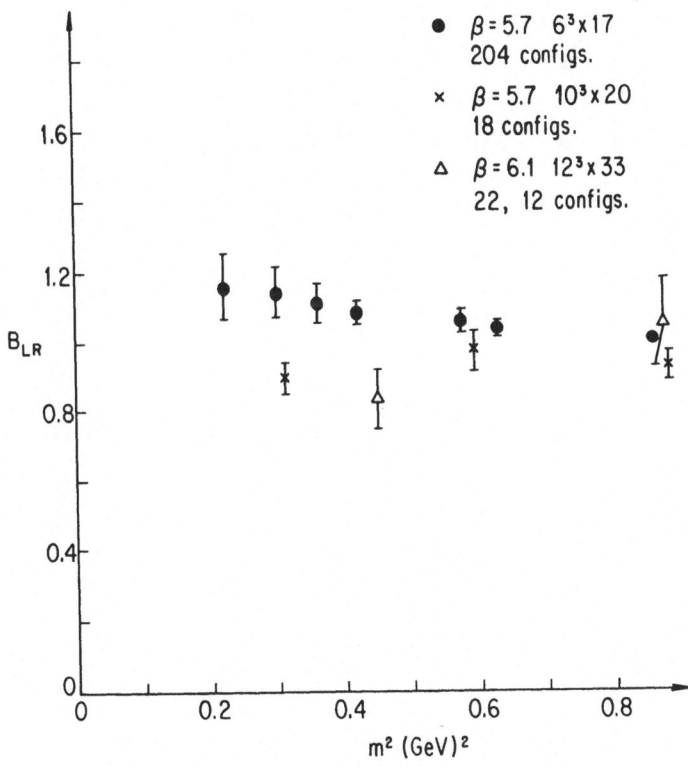

Fig. 4. B_{LR} vs. meson mass squared.

amplitude, as shown in Fig. 5. (The absolute sign of the experimental amplitude is not determined.) "Physical" is put in quotes here because while chiral perturbation theory is needed to define a "physical" $K \to \pi$ amplitude, the mass dependence of our amplitudes clearly violates lowest order CPTh. Still the order of magnitude agreement may perhaps be taken as an indication that we are in the "correct ballpark."

2. There is some evidence for scaling in these amplitudes. As before, despite fairly large errors, this is non-trivial since the matrix elements go like a^{-4}. We are therefore encouraged in the belief that results are physically meaningful, depsite the disagreement with the lowest order CPTh prediction.

THE EYE GRAPHS

As discussed in the Introduction, the eye graphs present a theoretical difficulty because of mixings with lower dimensional

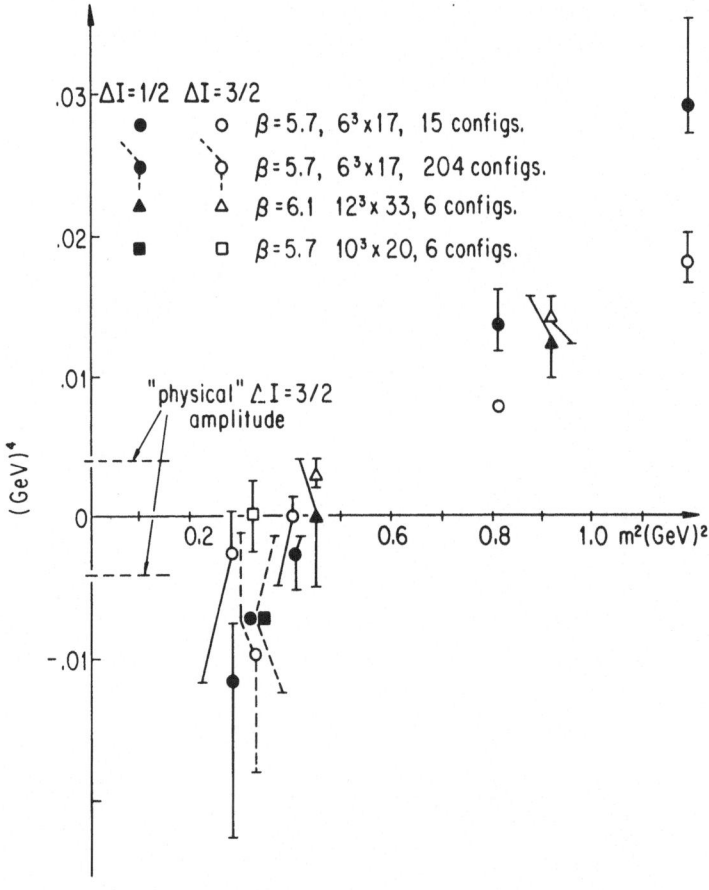

Fig. 5.   The $\Delta I = 3/2$ $K \to \pi$ amplitude and the figure-eight contribution to the $\Delta I = 1/2$ $K \to \pi$ amplitude.

operators. For example, through a graph such as Fig. 6, a LL 4-quark operator can mix with the operator $\bar{s}d$. (For LL operators, the one-loop graph and all corrections to the quark loop alone vanish.[3]) This would imply that the operator $(\delta/a^3)\,\bar{s}d$, where δ is a constant, needs to be subtracted in order to define a renormalized 4-quark operator. (The powers of a come from dimensional counting; for a GIM subtracted operator, the appropriate term would be $(\delta'/a^2)(m_c-m_u)\,\bar{s}d$.) Now, we see no reason why coefficients like δ would not have a well convergent perturbative expansion for weak enough coupling. The problem is that as a → 0, the subtraction gets more and more delicate while the coupling, g(a), vanishes only logarithmically. This means that exponentially large numbers of terms in the perturbative expansion of the coefficients would be needed as the a → 0 limit is approached. However, numerical calculations are always done at finite a, and it is therefore possible (but not required) that there exists a "window," i.e. a range of couplings for which the perturbative expansion is convergent but a is not so small that the cancellation is prohibitively delicate. We have chosen to do the perturbative calculations in the hope that a window can be found; it would show up as a range of g for which the subtracted matrix elements scale.

The alternative approach using lowest order chiral perturbation theory for Wilson fermions has also been developed.[10] This method has the advantage that it is completely nonperturbative in g. However, our $\Delta I = 3/2$ and $K^0-\bar{K}^0$ results make us extremely wary of relying on CPTh in the region of the K mass. In addition, we note that the CPTh approach can be reliable at most within a "window" also: as the subtractions become more and more delicate for small a, any approximate scheme must eventually fail, and CPTh is perforce approximate since one can never work at vanishing quark mass.

We have calculated the coefficient of $\bar{s}d$ for LL operators up

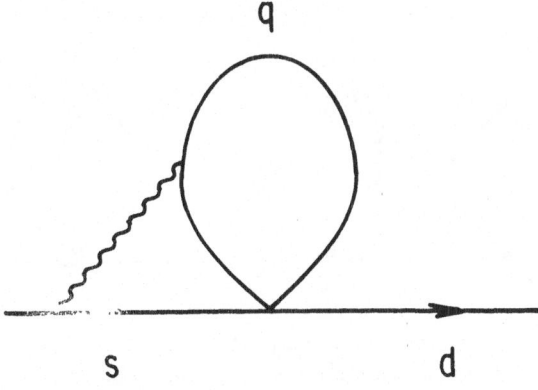

Fig. 6. A "direct" graph contributing to the mixing with $\bar{s}d$ of a LL 4-quark operator.

through two loops through Fig. 6 (and its reflection) and through the "indirect" route of Fig. 7 (the original LL operator mixes at one loop with operators of different chiral structure, which in turn can have one loop mixings with $\bar{s}d$). In the direct two-loop contributions we have so far gone only to leading non-trivial (i.e. linear) order in the "small" quark masses m_s, m_d, m_u. These linear terms are indeed small but we plan eventually to include all contributions. In the indirect case, one can replace the quark loop in the last diagram of Fig. 7 by the lattice value of $\langle\bar{\psi}\psi\rangle$ and thereby include a--hopefully important--part of the higher order contributions. For the operator $\bar{s}\sigma_{\mu\nu}F^{\mu\nu}d$ (the only other relevant operator for the two-meson on-shell matrix element) the one-loop correction to the LL operators also vanishes; at two loops, we have so far calculated the indirect but not the direct contribution (there are many diagrams). The indirect contribution is small.[19]

At the moment all these perturbative calculations still need to be checked; we have, though, used them to take a preliminary look at the eye contribution to the $\Delta I = 1/2$, $K \to \pi$ amplitude with the charm quark left in. The results are shown in Fig. 8, along with the $\Delta I = 3/2$ amplitude for comparison. We have the following comments:

1. The errors are in general very large, as should perhaps be expected for a "twice-subtracted" amplitude; subtracted once by the lower dimension operators and once by the GIM mechanism. The only $\Delta I = 1/2$ amplitudes with small errors come from our "first-generation" data for 15 configurations; when these are replaced by 204 configurations the errors <u>increase</u> substantially. This shows the dangers of using small numbers of configurations, especially for such subtracted amplitudes; it seems likely that such statistical problems will persist no matter how the subtraction is performed.

2. There is certainly no evidence for (or against) the

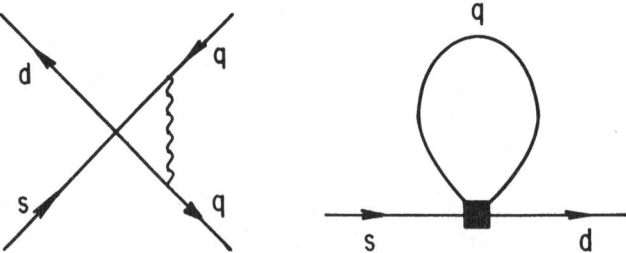

Fig. 7. The "indirect" mixing of a LL 4-quark operator with $\bar{s}d$. The first graph mixes the LL operator with a 4-quark operator of different chiral structure, which in turn can mix with $\bar{s}d$ at one loop (second graph).

presence of a "window" where perturbative subtraction works. Considerably more data is needed, especially for β = 6.1. The 204 β = 5.7 configurations should also be analyzed at additional mass values. At present, the most that can be said is that the eye contributions have the potential to be quite large, and they cover a range of values comparable to the "physical" ΔI = 1/2 amplitude. ("Physical " is in quotes here for the same reason as in the ΔI = 3/2 case.)

CONCLUSIONS

For the figure-eight amplitudes (B, ΔI = 3/2, etc.) a reasonably consistent picture exists for various lattice sizes and spacings. For large meson mass they have the expected size and sign, but for m ~ .6-.7 GeV, the amplitudes approach zero and may even change sign. This may make predictions for B difficult. This behavior appears to disagree with the results of Maiani and

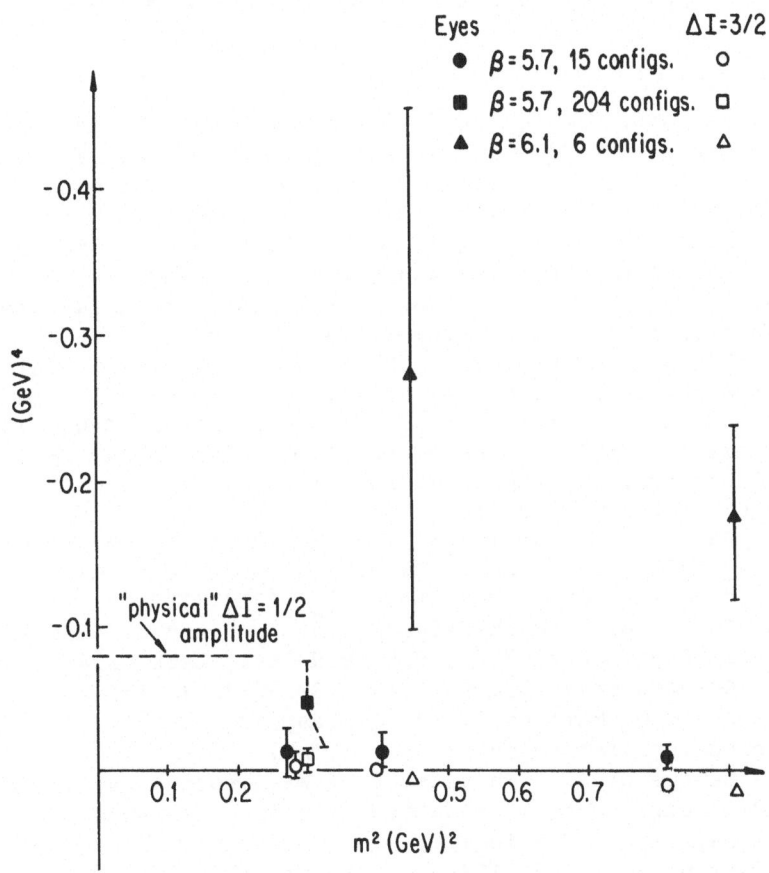

Fig. 8.   The eye contribution to the ΔI = 1/2 K $\rightarrow$ π amplitude, perturbatively subtracted, and the ΔI = 3/2 K $\rightarrow$ π amplitude.

Martinelli and with lowest order chiral perturbation theory. The figure eights alone do not come close to the physical $\Delta I = 1/2$ amplitude.

In the case of the eye graphs, it is hard to say much at this point. They seem big, big enough perhaps to explain the $\Delta I = 1/2$ rule, but there are large cancellations and hence very large errors. Much more numerical and analytical work needs to be done before one can prove or disprove the existence of a reliable "window" for perturbative subtraction.

We would like to thank J. Jennings and H. Tsuchida for computations which contributed to this work, S. Otto for generating configurations for our use, and H. D. Politzer and M. B. Wise for helpful discussions. The research of C.B., T.D., and G.H. was supported in part by the NSF; the research of A.S. was supported in part by the DOE O.J.I. Program. The computing was done at the San Diego Supercomputer Center and on the MFE Network.

REFERENCES

1. C. Bernard, "Lattice Calculation of Hadronic Weak Matrix Elements: The $\Delta I = 1/2$ Rule," in C. Zachos, et al., eds., *Gauge Theory on a Lattice: 1984*, National Technical Information Service, Springfield, VA, 1984.
2. C. Bernard, T. Draper, H. D. Politzer, and A. Soni, "Lattice Calculation of Weak Matrix Elements: A Progress Report," in D. W. Duke and J. F. Owens, eds., *Advances in Lattice Gauge Theory*, World Scientific, Singapore, 1985.
3. C. Bernard, T. Draper, G. Hockney, and A. Soni, "Calculation of Weak Matrix Elements: Some Technical Aspects," in B. Bunk, K. H. Mütter and K. Schilling, eds., *Lattice Gauge Theory: A Challenge in Large Scale Computing*, Plenum, New York, 1986.
4. C. Bernard, T. Draper, G. Hockney, and A. Soni, Phys. Rev. Lett. 55, 2770 (1985).
5. N. Cabibbo, G. Martinelli, and R. Petronzio, Nucl. Phys. B244, 381 (1984); R. C. Brower, M. B. Gavela, R. Gupta, and G. Maturana, Phys. Rev. Lett. 53, 1318 (1984).
6. S. Gottlieb, P. MacKenzie, H. Thacker, and D. Weingarten, Nucl. Phys. B263, 704 (1986) and Phys. Lett. 134B (1984).
7. C. Bernard et al., Phys. Rev. D32, 2343 (1985).
8. We thank G. Martinelli for correspondence which contributed to our understanding of this point.
9. S. Sharpe et al., University of Washington Preprint UW/400 48-11 P6, and talk presented at this conference.
10. L. Maiani, G. Martinelli, G. Rossi, and M. Testa, CERN Preprint TH. 4447 (1986); M. Bochicchio et al., Nucl. Phys. B262, 331 (1985).
11. C. Bernard, T. Draper, and K. Olynyk, Phys. Rev. D27, 227 (1983).

12. N. Cabibbo and E. Marinari, Phys. Lett. 119B, 387 (1982);
    A. Kennedy and B. Pendleton, Phys. Lett. 156B, 393 (1985).
13. K. H. Mütter, talk presented at this conference.
14. J. Jennings, UCLA Preprint, in preparation.
15. D. Barkai, K. J. M. Moriarty, and C. Rebbi, Phys. Rev. D30,
    1293 (1984); S. Otto (private communication).
16. L. Maiani and G. Martinelli, talk presented at the 23rd
    International Conference on High Energy Physics, Berkeley,
    1986.
17. C. Bernard, T. Draper, and A. Soni, UCLA Preprint, in
    preparation.
18. G. Martinelli, Phys. Lett. 141B, 395 (1984).
19. H. Tsuchida, UCLA Preprint, in preparation.

FINITE DENSITY AGGREGATION

Gyan Bhanot

Physics Department
Supercomputer Computations Research Institute
Florida State University
Tallahassee, FL 32306

ABSTRACT

I review some recent work on Finite Density Aggregation done in collaboration with Anna Hasenfratz and Mihaly Horanyi (Ref. 1).

I would like to review some work I did recently on Finite Density Aggregation in three dimensions in collaboration with Anna Hasenfratz and Mihaly Horanyi[1]. Previous work by Voss[2] was done in two dimensions and our conclusions are different from those of this work. Specifically, we find that for a range of densities of the aggregating particles, the aggregates have a characteristic Hausdorf dimension less than the embedding dimension (three in our case). In Ref. 2, the conclusion was that in all cases, the fractal dimension of the aggregate was equal to the embedding dimension.

The motivation to study such aggregates comes from many sources.

1) Industrial: Certain coagulated aerosols, many colloids, chemical precipitates and Gels are characterized by a wispy appearance and a size much larger than the range of forces binding them[3]. The weight of these objects also scales with their size in an anomalous way pointing to the possibility that these objects are fractal.

2) Astrophysical: Samples of dust collected in the upper atmosphere look like fractals. These might have their origin in the interplanetary

void or in cometary atmospheres. These dust flecks form in a low density, low gravity environment and can probably be modeled by finite density, diffusion limited aggregation.

3) Field Theory: Recently, Parisi and Zhang[4] have shown that certain kinds of Eden models are related to Reggeon field theory. If a general method to write down the field theory action corresponding to a growth process can be found, it would open up a new way to study field theories.

Since this is a field theory conference, let me explain the last connection in some more detail. For a real explanation, the reader is referred to Ref. 4.

The original Eden model is obtained as follows: One starts with a seed particle at the origin. At each time step, a new particle is added to the boundary of the growing cluster. Consider the ensemble of all Eden clusters and define a sequence of density correlation functions:

$$\rho_1(x_1; n) = < \rho(x_1) >_n \qquad [1a]$$

$$\rho_2(x_1, x_2; n) = < \rho(x_1)\rho(x_2) >_n \qquad [1b]$$

etc..... In general,

$$\rho_i(x_1, x_2, ..., x_i; n) = < \rho(x_1)\rho(x_2)...\rho(x_i) >_n \qquad [1c]$$

where $\rho(x)$ is zero (site unoccupied) or one (site occupied) at time step n. The average $< .. >_n$ is over the ensemble of clusters grown up to time step n. Parisi and Zhang[4] showed that with some modifications in the Eden growth process, the ρ_i's satisfy a set of coupled differential equations that are equivalent to the Schwinger-Dyson equations of the theory defined by the action:

$$S = \int dx dt [\phi^* \delta_t \phi + \alpha \nabla \phi^* \nabla \phi - \phi^* \phi - ig\phi^*(\phi^* + \phi)\phi] \qquad [2]$$

The precise connection between the Green's functions of this theory and the growth process is:

$$\rho_i(x_1, x_2, ..., x_i; n) = \frac{1}{(ig)^{i-1}} << \phi(x_1, t)\phi(x_2, t)...\phi(x_i, t)\phi^*(0,0) >> \qquad [3]$$

with $t = \ln(n)$ and the average $<< ... >>$ is meant in the usual field theory sense. This connection prompts the following possible analogies, the first two of which are implicit in Ref.4 and the last one is a conjecture that we will shortly test.

1) Parameters such as $g_0^2, m^2, \lambda..$ etc. in the Lagrangian of field theories (ie. coupling constants, masses etc.) are related to the growth kinematics, the embedding dimension and so on.

2) Critical exponents of the field theory, the anomalous dimensions of field operators etc. are related to the various fractal dimensions of the aggregates.

3) Relevant perturbations in the field theory such as the number of dimensions, the number of components of the field and the symmetry of the order parameter might be related to other aspects of the growth. I would conjecture that the density of the aggregating particles is such a relevant perturbation. Indeed, as I will show below, the Hausdorf dimension of the fractal varies continuously with the density and above a certain critical value, it saturates to the embedding dimension. This is similar to the critical exponent of the field theory attaining a fixed (mean field) value for dimensions bigger than the upper critical dimension.

Let me now describe the simulation we did[1]. We start with a seed particle at the origin of a three dimensional lattice. Surrounding this seed particle is a bath of free particles which are non-interacting and bosonic (so two of them can occupy the same site). The density of free particles (called walkers), is kept fixed in a shell whose inner radius is kept a distance 10 away from the longest limb of the growing cluster. At each time step, the walkers move randomly in the $\pm x, \pm y,$ or $\pm z$ direction. Whenever a walker reaches the cluster, he sticks to it. Particles leaving the outer shell are replaced uniformly over the outer shell in each time step. The code is vectorized for a CYBER-205. Because of the way the code was written, it shuts down when the number of walkers exceeds 2^{16} or when any arm of the cluster is more than 62 lattice units away from the original seed particle. Because of these limitations, our cluster sizes were limited to between $10,000$ and $20,000$ particles. However, these are not inherent limitations and we plan in the future to eliminate them in order to grow much larger clusters.

Once the clusters are grown, we need to find their Hausdorf dimension D_H. This of course assumes that the clusters are large enough that they are already in the scaling regime. One way to find D_H is to measure the gyroradius R_g of the cluster as a function of its mass M. R_g is given by,

$$R_g^2(M) = \frac{1}{M^2} \sum {}_{i,j}^{M} (r_i - r_j)^2 \qquad [4]$$

where, r_i is the position of the i th particle and i, j run over all the particles in the cluster. It is obvious that for large M,

$$R_g \sim M^{1/D_H}. \qquad [5]$$

Another way to measure D_H is to measure the mass M_s in a shell at a distance R from the seed. Clearly, for large enough M,

$$M_s \sim R^{D_H - 1} \qquad [6]$$

Using these two methods to analyze our clusters, we get the dependence of D_H on ρ, the density of the aggregating particles. This relationship is shown in Figure 1. From this figure it is evident that there is a phase transition in the Hausdorf dimension around $\rho_c = 0.5$. For $\rho < \rho_c$ the Hausdorf dimension is less than the embedding dimension (three). Above ρ_c, the Hausdorf dimension becomes equal to the embedding dimension.

Note that if the Hausdorf dimension is smaller than the embedding dimension then the density of particles in the cluster (the number of particles in a shell of unit thickness divided by the number of lattice points in the shell) goes to zero as we go far from the center of the cluster. An objection to our result can be that this is difficult to understand because the density of walkers is asymptotically constant. What seems to happen is that for a low density of walkers, the cluster creates in its neighborhood a region of almost zero density. The cluster then effectively grows in this region of very low density. We have checked that the density profile of the walkers is indeed of this kind.

Consider a model of gelation. Here one starts many particles off and allows them to stick whenever two of them come together. The stuck particles stick to more particles/clusters and so on. If the density of walkers is low enough, one would tend to grow a large number of clusters, each with a negligible fraction of the total mass. These clusters would be fractal. On the other hand, if the density were high enough, one would tend to grow a few clusters, each containing a finite fraction of the total mass. These more compact clusters would have the same Hausdorf dimension as the embedding

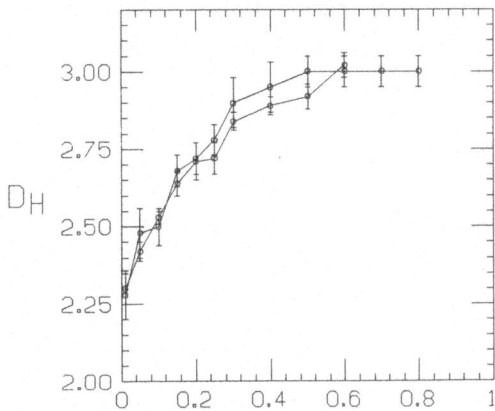

Figure 1: The Hausdorf dimension $\dot{D}_H$ as a function of ρ. The dots represent results from analyzing the gyroradius, the crosses from analyzing the mass in shells of increasing radii. We have connected the data points in the figures by straight segments to guide the eye.

space. One can also draw an analogy with the freezing of water to ice (which is lighter than the water that formed it). The way this happens is that the ice expands, pushing the water out[5].

We have also produced quite large clusters (with more than 35,000 particles) in $d = 2$ at low densities and find again a fractal structure. Our results are in contradiction to those of Ref. 2. One possibility is that our clusters are not large enough to show scaling behavior. However, as we argue in Ref. 1, our results indicate that our clusters do indeed scale.

In conclusion, I have made some analogies between field theory models and growth models motivated by the work of Parisi and Zhang and our numerical simulations. Our study seems to indicate that these analogies may lead to deep connections. Much more theoretical and numerical work needs to be done to clarify these connections.

ACKNOWLEDGMENTS

I thank Mihaly Horanyi for stimulating me to work in this field. This work was supported by the Department of Energy under cooperative agreement number DE-FC05-85ER25000.

REFERENCES

1) G. Bhanot, A. Hasenfratz and M. Horanyi, Diffusion Limited Aggregation at Constant Asymptotic Density, Florida State University Preprint FSU-SCRI-86-47, June 1986.

2) R.F. Voss, Phys. Rev. B30:334 (1984).

3) T. A. Witten and L. M. Sander, Phys. Rev. Letts. 47:1400 (1981).

   P. Meakin, Phys. Rev. A27:604 (1983); *ibid* A27:1495 (1983).

   T. A. Witten and L. M. Sander, Phys. Rev. B27:5686 (1983) and references in [3].

4) G. Parisi and Y.-C. Zhang, J. of Stat. Phys. 41:1 (1985).

5) We thank Peter Hasenfratz for pointing out this analogy.

RECENT RESULTS ON MCRG STUDIES OF THE SU(3) β-FUNCTION

AND ON QUENCHED HADRON MASSES

K.C. Bowler

Physics Department
University of Edinburgh
Edinburgh EH9 3JZ

INTRODUCTION

We report results from two projects involving members of the Edinburgh group: the extension of earlier measurements by the European lattice gauge theory collaboration [1,2] of the beta function of pure SU(3) lattice gauge theory, using Monte Carlo renormalization group methods, to β = 6.9 & 7.2, and new results on quenched hadron masses on 16^3x24 lattices at β = 6.15.

THE β-FUNCTION AT LARGE COUPLING

Monte Carlo renormalization group (MCRG) methods have been used recently to extract information about the β-function of lattice QCD from numerical simulations and to attempt to determine the regime of couplings where asymptotic scaling sets in, i.e. where physical observables scale according to the universal part of the SU(N) β-function

$$\beta(g) = - b_0 g^3 - b_1 g^5 + O(g^7) \qquad (1)$$

where

$$b_0 = 11N/48\pi^2, \quad b_1 = (34/3)\left(N/16\pi^2\right)^2$$

The MCRG methods determine not the β-function directly but the related quantity $\Delta\beta(\beta) = \beta - \beta'$, the shift in coupling which corresponds to a change in length scale by a factor b. In particular, the b = 2 optimized block spin method used in the present study and described in detail in [1], uses a 2-lattice

comparison. Starting with a 16^4 lattice permits three subsequent
block transformations. On each blocking level, Wilson loops are
measured and compared with corresponding loops measured on an
independent sequence of blocked lattices starting from an 8^4
lattice. It is of course essential that the comparison is always
made between lattices of the same size in order to minimize
finite-size effects. Our analysis at β = 6.9 and 7.2 follows the
same criteria as used in our earlier calculations and is based on
very similar statistics: 60 configurations on the 16^4 lattices at β
= 6.9 and 7.2, 100 configurations on 8^4 lattices at β= 6.3 and 6.4,
and 96 configurations on 8^4 lattices at β= 6.6 and 6.7. The 60
configurations on the large lattice were separated by 224
pseudo-heatbath sweeps at β = 6.9 and 112 at β = 7.2.

In our previous analysis we found that at a given blocking
level, results for $\Delta\beta(\beta)$ depended linearly on $1/P$, where P is the
adjustable parameter in the block spin transformation suggested by
Swendsen [3]. Thus interpolation between different P-values could
be used to determine the optimal P-value where best agreement for
the various observables which are compared could be reached after
the first blocking step. In the present analysis we used P = 25 and
30 at β = 6.9 and P = 27 and 40 at β = 7.2. For the purpose of
illustration we show in Fig. 1 the matching predictions for $\Delta\beta(\beta)$
after 2 blocking steps at β = 7.2, for the various Wilson loops.

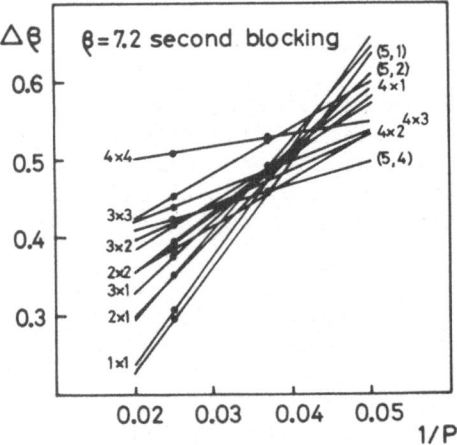

Fig. 1: matching values at β = 7.2

Further details may be found in our forthcoming paper [4]. We remark that a comparison of the present results with those of the earlier study shows that, although the P-dependence of $\Delta\beta(\beta)$ decreases as the number of blocking steps increases, the matching predictions at a given blocking level depend more strongly on P as β increases.

Our earlier results obtained with a scale factor b = 2 [1,2] were broadly consistent with other MCRG calculations using different scale factors [5] and also with scaling behaviour of physical observables like the deconfinement temperature [6] and the string tension [7] (although the 0^+ glueball may be exceptional [8]). All these calculations seem to suggest that deviations from asymptotic scaling, as described by equn.(1) are small for couplings somewhat larger than β = 6.0, although a recent analysis by Petcher [9] in which he fits a phenomenological ansatz for the β-function to the MCRG results, suggests that notwithstanding reasonable agreement between fits to the b = 2 and b = √3 data, it may be premature to conclude that the regime of asymptotic scaling has been reached. Fig. 2 shows a selection of results for $\Delta\beta(\beta)$ from earlier studies, together with our best estimates from the present study:

$$\Delta\beta(\beta=6.9) = 0.51 \pm 0.06, \qquad \Delta\beta(\beta=7.2) = 0.51 \pm 0.07$$

Thus although the matching values at β = 6.9 and 7.2 agree within errors with our earlier results at β = 6.6, they are substantially

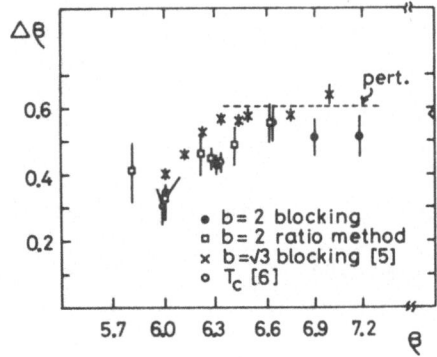

Fig. 2: results for $\Delta\beta(\beta)$

smaller than expected if asymptotic scaling were to hold. These results are also in conflict with those presented in [5] at $\beta = 7.0$ for $\sqrt{3}$ blockspin transformations. However, in all these calculations the error on $\Delta\beta(\beta)$ is large. Much higher statistics would be necessary to clarify the behaviour of the β-function at these large coupling values. Although a 16^4 lattice at $\beta = 6.9$ is already in the deconfined region this should not invalidate the MCRG method in principal, since the universal perturbative β-function is well defined and calculable even in a finite box. It is, however, likely that in practice the transient effects become more severe at larger β; an analysis of larger lattices would permit more blocking steps and hence better control over such problems.

QUENCHED HADRON MASSES

Results obtained by the Edinburgh group for quenched hadron masses using staggered fermions on 16^4 and periodically extended $16^3 \times 24$ lattices at $\beta = 5.7$ & 6.0 were reported in a recent paper [10]. Here we present new data at $\beta = 6.15$ for full $16^3 \times 24$ lattices. All these calculations used the Edinburgh DAPs. As in our earlier study we employ Dirichlet boundary conditions in the Euclidean time direction. So far we have analyzed 24 configurations imposing periodic boundary conditions in the spatial directions on the quark propagator. In order to make a direct assessment of the importance of finite-size effects, we are also repeating the analysis with antiperiodic boundary conditions in the spatial directions on the same set of configurations; to date, however, we have only completed this latter analysis on 8 configurations.

We have used a block SOR algorithm [11] to invert the fermion matrix on each configuration. If we define

$$M_k(n,0) = \sum_{a,b} |(D[\{U\}_k] + M)^{-1} {}^{ab}|^2_{n0}$$

and the configurational average

$$M(n,0) = (1/C) \sum_{k=1}^{C} M_k(n,0)$$

then the complete set of local meson time-slice propagators is:

$$M_{PS}(n_4) = \sum_{\underline{n}} M(n,0)$$

$$M_{VT}(n_4) = \sum_{\underline{n}} [(-1)^{n1} + (-1)^{n2} + (-1)^{n3}] M(n,0)$$

$$M_{PV}(n_4) = \sum_{\underline{n}} [(-1)^{n1+n2} + (-1)^{n2+n3} + (-1)^{n1+n3}] M(n,0)$$

38

$$M_{SC}(n_4) = \sum_{\underline{n}} (-1)^{n1+n2+n3} M(n,0)$$

For Dirichlet boundary conditions in the time direction, the expected behaviour of the meson time-slice propagators is

$$M(n_4) = \sum_i A_i \exp(-m_i n_4) + (-1)^{n4} \sum_i \overline{A}_i \exp(-\overline{m}_i n_4)$$

where the terms alternating in sign (the "oscillating channel") correspond to contributions from intermediate states of opposite parity to those in the first sum (the "direct channel"). A similar set of baryon propagators, based on local lattice operators, may be constructed from the inverse of the fermion matrix, as discussed in [10]. However, the symmetry first noted by Morel and Rodrigues [12] implies that, for an infinite lattice, the four baryon time-slice propagators, B_{PS}, B_{VT}, B_{PV} and B_{SC} are not independent. Consequently, there is only one distinct local baryon propagator which we take to be

$$B_{even} = -(1/8)[B_{PS} + B_{VT} - B_{PV} + B_{SC}]$$

and corresponds to taking the spatial average over a 3-dimensional

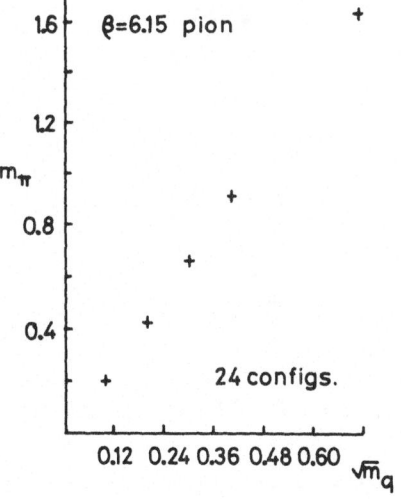

Fig. 3: pion from 24 configs. at $\beta = 6.15$

lattice of spacing 2 which includes the origin. As a check on the lattice symmetry we compare B_{even} with $B_{all} = B_{PS}$; on a finite lattice with anti- periodic boundary conditions, the symmetry is violated, so that any discrepancy between B_{even} and B_{all} may be an indication of finite-size effects.

Fig. 3 shows results for m_π as a function of $\sqrt{m_q}$ for 5 different values of the quark mass, obtained from the analysis of 24 configurations discussed above. There is some evidence for a departure from linearity at the lightest quark mass, although a quadratic fit yields results consistent with zero pion mass at $m_q=0$. Fig. 4 plots results for the rho mass as a function of m_q for two different flavour combinations associated with the direct channels in the PV and VT propagators respectively. The good agreement between the two mass estimates is convincing evidence for the restoration of flavour symmetry at this β-value. We remark in passing that a direct comparison between the pion time-slice propagator computed using anti-periodic boundary conditions for the quarks, on an ensemble of 8 configurations, and computed with periodic boundary conditions on the same set of 8 gauge configurations, shows no evidence for finite-size effects. Similar results were obtained for the rho time-slice propagator.

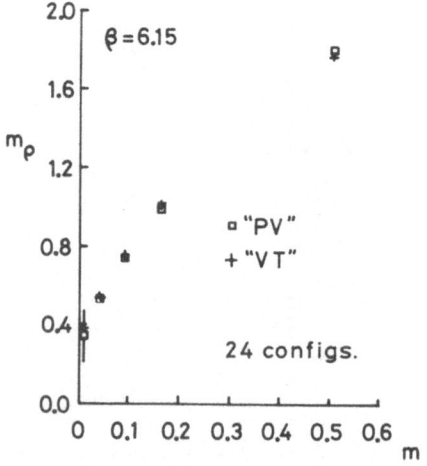

Fig. 4: flavour symmetry test for rho.

We turn now to a discussion of our results for baryon propagators. Figs. 5 & 6 show comparisons, at $m_q= 0.16$ & 0.01 respectively, of the time-slice propagators for the 'all' nucleon, again computed with periodic or anti-periodic boundary conditions in the spatial directions, on the set of 8 gauge configurations. There is a clear discrepancy, at least for values of $n_4 > 12$, with the anti-periodic propagator lying systematically below the periodic propagator, an effect which persists at the 2 intermediate quark mass values at which measurements were also made. This effect has been noted previously for free quark 'baryon' propagators by Baillie and Carpenter [13]. It is possible that this discrepancy is associated with the fact that, for anti-periodic boundary conditions, there is a non-zero minimum momentum for the quarks; we are investigating other correlation functions for baryons which might yield better signals. In any case it is clear that the difference is a finite-size effect. In [10] we reported another indication of possible finite-size effects in baryon propagators when we noted a systematic difference between B_{all} and B_{even} for the case of anti-periodic boundary conditions. We have performed a preliminary analysis on the ensemble of 8 configurations discussed above which appears to show a similar effect: there is good agreement between B_{all} and B_{even} at all quark mass values if periodic boundary conditions are used, but a discrepancy, which increases as m_q decreases, seems to be present for the anti-periodic

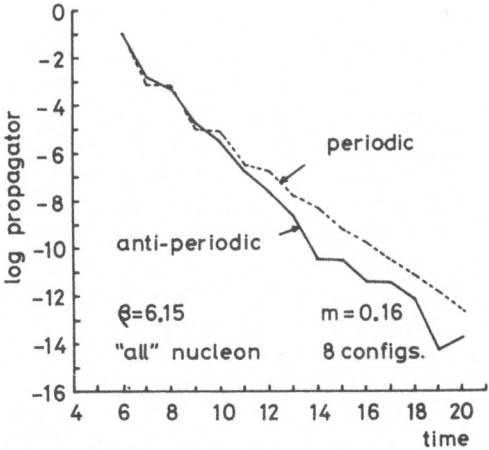

Fig. 5: comparison of boundary conditions at m = 0.16

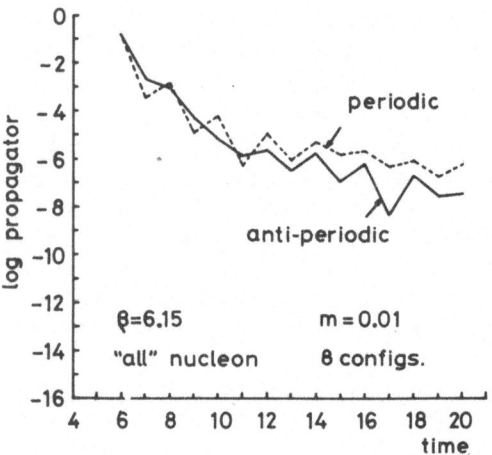

Fig. 6: comparison of boundary conditions at m = 0.01

case, although firm conclusions must await a careful analysis of the errors, to be certain that the discrepancy is statistically significant.

Fig. 7 displays a summary of the results for meson and nucleon masses at β = 6.15, based on an analysis of 24 configurations with periodic spatial boundary conditions for the quarks, and on 2-exponential fits to the time-slice propagators. As is clear from the figure, there is excellent agreement between the nucleon mass estimates obtained from B_{all} and B_{even}. In Fig. 8 we plot the mass ratio m_N/m_ρ versus the mass ratio m_π/m_ρ, quantities which should be universal in the scaling limit. These preliminary results are in marked contrast to the behaviour seen at β =5.7 and show some signs of crossing over from the heavy quark limit towards the experimental value, which is also indicated in the figure, although the cautionary remarks made above concerning finite-size effects in baryon propagators should be borne in mind. We expect to finish an analysis of 32 configurations, using both boundary conditions, shortly and results will be reported elsewhere [14].

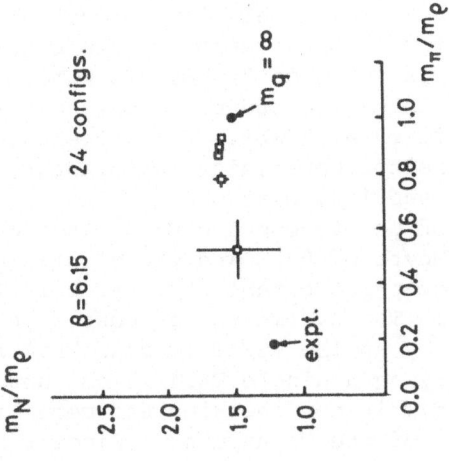

Fig. 8: mass ratios at β = 6.15

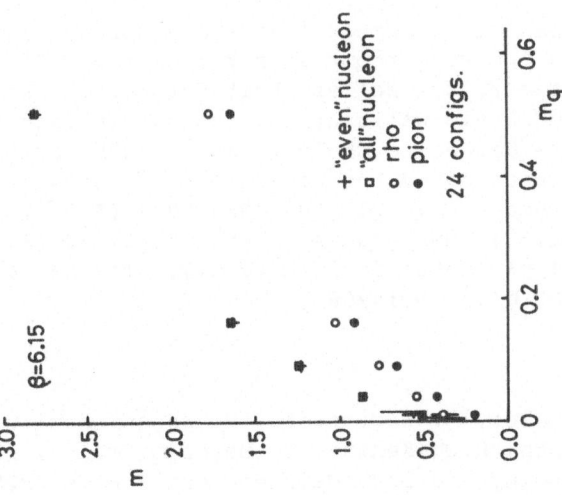

Fig. 7: hadron masses at β = 6.15

NEW DEVELOPMENTS

We complete this report on current work by the Edinburgh group with a brief description of the recently acquired Meiko Computing Surface [15], a modular reconfigurable Transputer array. The Inmos T414 Transputer, which forms the basic computing element in the Computing Surface, is a VLSI micro-processor, comprising 32-bit cpu rated at 7.5 Mips, on-chip memory of 2K bytes, 4 Gigabyte linear address space, 32 bit wide 25 MByte/s memory interface and, most importantly, 4 inter-transputer links, each with full duplex DMA transfer capability at 10Mbits/sec. The Edinburgh system consists of a 40-slot cabinet containing, at present, 1 control board, 10 compute boards and 1 display board. The control board has 1 T414 plus 3 Mbytes of parity-checked DRAM, each compute board has 4 T414's plus 1 Mbyte of DRAM and the display board, which gives the system a fast graphics capability, has 1 T414 with 3 Mbytes of display memory. The system may be hosted in 2 ways; either by a MicroVAX II or by an IBM PC/AT (Compaq 286) augmented by an Inmos B004 board carrying a single T414 which runs the Occam 2 Programming System used for writing, checking and compiling code in Occam 2, the native language of the Transputer. Since taking delivery of the system in spring '86, we have developed a number of application codes on the Surface, including a microcanonical simulation of lattice QED, including dynamical fermions, on lattices up to 14^4. We believe that the Computing Surface offers a very flexible, modular system with a clear upgrade path to super-computer performance for a fraction of the cost.

ACKNOWLEDGEMENTS

It is a pleasure to acknowledge the contributions of my collaborators in the work reported here: on the MCRG project, Anna and Peter Hasenfratz, Urs Heller, Frithjof Karsch, Richard Kenway, Stuart Pawley and David Wallace; for the hadron mass work, the latter three plus Catherine Chalmers and Duncan Roweth. The work was partially supported by the Department of Energy under cooperative agreement no. DE-FC05-85ER25000 (A.H.), Schweizerische Nationalfonds (P.H.), NSF grants NSF-PHY-82-01948 (F.K.) and PHY-82-17853, supplemented by NASA (U.H.). The Edinburgh DAP's were supported by SERC grant NG15908.

REFERENCES

1) K.C. Bowler, A. Hasenfratz, P. Hasenfratz, U. Heller, F. Karsch, R.D. Kenway, H. Meyer-Ortmanns, I. Montvay, G.S. Pawley and D.J. Wallace, Nucl. Phys. B257[FS14] (1985) 155.
2) K.C. Bowler, F. Gutbrod, P. Hasenfratz, U. Heller, F. Karsch, R. D. Kenway, I. Montvay, G.S. Pawley, J. Smit and D.J. Wallace, Phys. Lett. 163B (1985) 367.

3) R.H. Swendsen, Phys. Rev. Lett. 47 (1981) 1775 and in "Statistical and Particle Physics", K.C. Bowler and A.J. McKane, eds., SUSSP, Edinburgh (1984).

4) K.C. Bowler, A. Hasenfratz, P. Hasenfratz, U. Heller, F. Karsch, R.D. Kenway, G.S. Pawley and D.J. Wallace, Phys. Lett. 179B (1986) 375.

5) R. Gupta, G. Guralnik, A. Patel, T. Warnock and C. Zemach, Phys. Rev. Lett. 53 (1984) 352 and Phys. Lett. 161B (1985) 352.

6) S. Gottlieb, A.D. Kennedy, J. Kuti, S. Meyer, B.J. Pendleton, R. Sugar and D. Toussaint, Phys. Rev. Lett. 55 (1985) 1958; N.H. Christ and A. Terrano, Phys. Rev. Lett. 56 (1986) 111.

7) D. Barkai, K.J.M. Moriarty and C. Rebbi, Phys. Rev. D30 (1984) 1293.

8) Ph. de Forcrand, G. Schierholz, H. Schneider and M. Teper, Phys. Lett. 152B (1985) 107; A. Patel, R. Gupta, G. Guralnik, G.W. Kilcup and S.R. Sharpe, San Diego preprint UCSD–10P10-260 (1986).

9) D.N. Petcher, Nucl. Phys. B275[FS17] (1986) 241.

10) K.C. Bowler, C.B. Chalmers, R.D. Kenway, G.S. Pawley and D. Roweth, Hadron mass calculations using Susskind fermions at β = 5.7 & 6.0, Edinburgh preprint 86/369, to be published in Nucl. Phys. B.

11) C.B. Chalmers, R.D. Kenway and D. Roweth, Algorithms for calculating quark proppagators on large lattices, Edinburgh preprint 86/361 (rev), to be published in J. Comp. Phys.

12) A. Morel and J.P. Rodrigues, Nucl. Phys. B247 (1984) 44.

13) C.F. Baillie and D.B. Carpenter, Nucl. Phys B260 (1985) 103.

14) K.C. Bowler, C.B. Chalmers, R.D. Kenway and D. Roweth, in preparation.

15) K.C. Bowler and R.D. Kenway, Physics Bulletin, 37 (1986) 331.

FEASIBILITY OF LATTICE QCD SIMULATIONS FOR WEAK DECAY PROCESSES

Richard C. Brower, Roscoe C. Giles

Physics Department and
Department of Electrical, Computer and Systems Engineering
Boston University
Boston, Massachusetts 02215

K.J.M. Moriarty

Institute for Computational Studies
Department of Mathematics, Statistics,
and Computing Science
Dalhousie University
Halifax, Nova Scotia B3H 4H8, Canada

ABSTRACT

The computation of a class of processes involving a weak decay vertex and one, two, or three external hadrons is discussed. The use of an extended source propagator for the quark allows calculation of matrix elements using only one to two orders of magnitude more computer time than mass calculations. Thus in the quenched approximation these calculations appear feasible within the current generation of lattice QCD simulations. In particular, semi-leptonic decay processes for heavy flavors offers lattice QCD an opportunity to contribute to the testing of the Cabbibo-Kobayashi-Maskawa six quark mixing hypothesis in the standard electroweak model of Glashow, Salam and Weinberg.

INTRODUCTION

Many interesting phenomenological questions in weak interactions are hampered by our lack of understanding of QCD corrections. One needs to take into account hadron wavefunctions and binding effects. In principle, the lattice formulation of QCD provides us with the means to perform finite non-perturbative calculations of these strong interaction effects[1]. Thus by computing more accurate weak matrix elements, lattice QCD could provide more stringent testing of the Standard Model than has hitherto been possible, and place stronger phenomenological constraints on higher level unified theories. The question that we wish to address in these remarks is the feasibility of carrying out such a program with the current state of the art of lattice computations.

To this end we will classify the weak decay processes in the order of increasing difficulty assuming the approximations of quenched Wilson fermions and effectively local four-fermi weak interaction vertices. Our intention is introduce no further approximations for their analysis. In particular, we will not assume *a priori* PCAC or integrate out heavy quark fields. Moreover, quenching out the internal quark loops is not fundamental to our considerations. Due to the usual factorization of the problem, given a sample of thermalized lattices including fermion loop effects, one could apply all the same techniques for the valence quark propagators. This classification is essentially in the order of increasing numbers of external hadrons.

LEPTONIC DECAYS

Purely leptonic decays of mesons depend on matrix elements of the form: $< 0|J^\mu(0)|M, \vec{p} >$, where $J^\mu(x)$ is one of the weak vector or axial vector currents and $|M, \vec{p} >$ represents a meson state of momentum $\vec{p}$, created by a quark-antiquark operator. For example, a zero momentum meson decay matrix element is (figure 1a):

$$M_\Gamma(T) = \sum_{\vec{x}} < 0|\bar{\Psi}(\vec{x}; -T)\Gamma\Psi(\vec{x}; -T)J_\mu(0)|0 > \qquad (1)$$

where Γ gives the flavor spin combination and the sum over spatial positions guarantees $\vec{p} = 0$. The amplitude $M_\Gamma(T)$ must be averaged over gauge field configurations. Processes of this class have been studied by several groups, notably by Maiani and Martinelli[2] in the context of their lattice current algebra work[3].

The calculation of the above matrix element for any combination of weak currents and Γ requires knowledge of the Dirac propagator from the origin to each point in the space-time lattice. To estimate the calculational magnitude of these problems, we refer to the number of "inversions" involve in finding a solutions to the quark propagator from a fixed source,

$$\sum_{j,b,x'} D_{ia,jb}(x, x')\Psi_{jb}(x') = j_{ia}(x) \qquad (2)$$

where $D_{ia,jb}(x, x')$ is the lattice Dirac operator for a single Wilson fermion flavor, and $j_{jb}(x')$ is a source with spin index j and color index b. One such inversion corresponds to a single Gauss-Seidel or conjugate-gradient operation. Calculation of the whole propagator from the origin $\left(D_{ia,jb}(x,0) = D^{-1}_{ia,jb}(x,0)\right)$ requires 12 such "inversions", one for each possible flavor-spin delta function source $j_{jb}(x') = \delta_{j,j_0}\delta_{b,b_0}\delta(x')$. The calculations of these leptonic decays can be done as a by-product of the quenched mass calculations since the inverse data required is the same.

The extraction of the contribution of the desired meson state $|M, \vec{0} >$ is done as usual by fitting the T dependence of the gauge averaged $M_\Gamma(T)$ to the appropriate sum of exponentials[4]:

$$\ll M_\Gamma(T) \gg = \sum_\alpha < 0|\bar{\Psi}\Psi|\alpha > e^{-m_\alpha T} < \alpha|J_\mu(0)|0 > \qquad (3)$$

where the sum is over all possible intermediate states α with the right quantum numbers.

Normalized current matrix elements require wavefunction normalization constants $< 0|\bar{\Psi}\Psi|\alpha >$ which can be extracted (somewhat less reliably) from the fit to the T dependence of the meson propagator itself. It is interesting to note that ratios of matrix elements involving the lowest mass state in a given color-spin channel can be estimated simply by measuring masses and the above matrix elements for a single sufficiently large T.

Consequently, the one hadron processes described here require the same number of inversions as the mass determination. Another group of processes which are also of this order of difficulty are those with internal weak interactions and no momentum transfer (see figure 1b). Although typically these processes involve further theoretical approximations to be cast the weak vertex in local form. Examples are:

1. The $\Delta I = 1/2$ rule matrix elements after PCAC has been invoked to remove one of the two π's in $K \to \pi\pi$. The effect here is very large, but the theoretical situation is complicated by operator mixing, the accuracy of PCAC, and the important contribution of the "eye" graph or non-local penguin graph.

2. The $K^0 \bar{K}^0$ mass splitting using a local effective Hamiltonian approximation. Here, the local "blob" in figure (1b) is the effective Hamiltonian which involves the exchange of a pair of W's and the integration over c, u, and t quarks (figure 2).

SEMI-LEPTONIC DECAYS

Semi-leptonic decays and form-factors of mesons and baryons are two hadron processes. Proton decay also falls in this class. (See figure 2). These these processes have one or more "spectator" quarks which propagate between the initial and final points without touching the interaction vertex.

For example, the diagram (figure 2a) for the semi-leptonic decay of a meson has the form of the following convolution of propagators:

$$A^\alpha(T, x_f) = \sum_{\vec{x}_i} \sum_{a,b,c} \sum_{k,l,m,n} \Psi_{ij}^f D_{ja,kb}^1(x_f, 0) J_{kl}^\alpha D_{lb,mc}^2(0, x_i) \Psi_{mn}^i D_{nc,ia}^3(x_i, x_f) \qquad (4)$$

where, Ψ_{ij}'s are the meson spin wavefunctions and J_{ij}^α are the spin matrix elements of the weak current α. The propagators D^1, D^2, and D^3 are Wilson fermion propagators of (possibly) different flavors. As before, the sum over

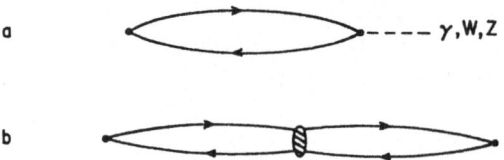

1. Class 1 diagrams. (a) meson decay, (b) meson mixing ($K^0 \bar{K}^0$ for example).

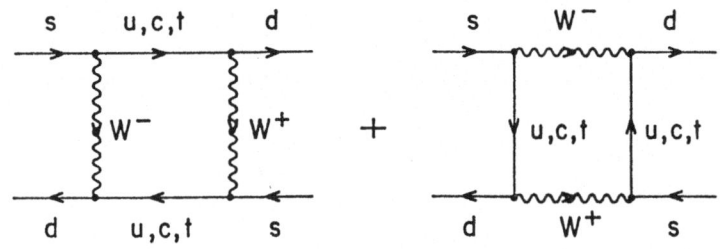

2. $K^0 \bar{K}^0$ mixing "local" kernel.

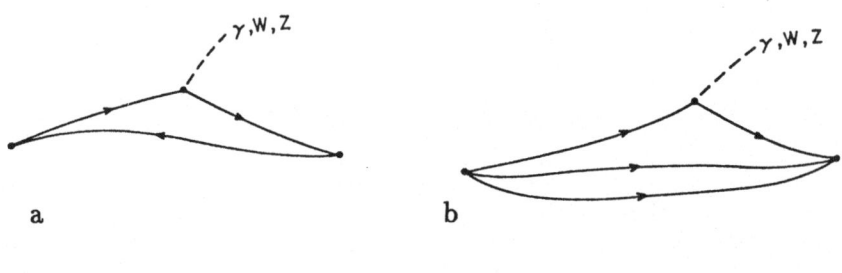

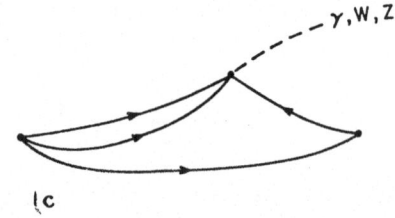

3. Semi-leptonic decay graphs: (a) Meson, (b) Baryon, (c) Proton Decay.

spatial positions of the initial meson state (with $x_i^0 = -T$) projects out the $\vec{p} = \vec{0}$ component. We have taken the weak vertex to be at the origin.

The amplitude requires knowledge both of the propagator from the origin to elsewhere in space and the propagator for the spectator quark directly from the initial to the final state. It might then appear that one needs many more inversions – a whole time plane full ($12L^3$) to perform a full convolution over $\vec{x}_i$. Fortunately, this can be avoided by using the "extended source propagator" (ESP) technique[5]. This amounts to the observation that for each color-spin component lb, the convolution:

$$\mathcal{F}_{ia}^{\dagger}(x_f) \equiv \sum_{\vec{x}_i} \sum_{c,m,n} D_{lb,mc}^2(0, x_i) \Psi_{mn}^i D_{nc,ia}^3(x_i, x_f) \tag{5}$$

is the solution of:

$$\sum_{\vec{x}_i} \sum_{c,n} D_{ia,nc}^3(x_f, x_i) \mathcal{F}_{nc}(x_i) = \sum_m \Psi_{n,m}^{i\dagger} D_{mc,lb}^{2\dagger}(x_i, 0) \tag{6}$$

The right hand side of this equation is the "extended source". Calculation of $\mathcal{F}$ requires 12 inversions (like a propagator) and amounts to doing the convolution before the inversion.

With this trick, only $3 \times 12 = 36$ inversions per initial spin state are needed even in the worst case of different mass quarks on all three legs of the diagram.

This gives the amplitude $\mathcal{A}$ connecting a $\vec{p} = \vec{0}$ meson state of definite spin created at time $-T$ to a weak decay vertex at $x = 0$ with observation of the final state $q\bar{q}$ pair at any point x_f in space-time. This data allows for momentum and energy analysis on the final state. In order to extract the contribution of a pure initial state meson, the calculation must be repeated for a variety of T values or, with less reliability, the lowest state in the $\vec{p} = \vec{0}$ channel may be assumed to dominate.

A similar method will work for the baryonic diagrams (figures 2b and 2c). In these cases, the sources are constructed from di-quarks and the baryonic spin wavefunction.

TWO BODY HADRONIC DECAYS

Two body hadronic decays involve three hadron matrix elements. These include non-leptonic decays. The diagrams of interest involve spectators which can be treated by the extended source propagator method described above. Additionally, these processes can involve spectators (figures 4a and 5) and/or exchange and annihilation (figure 4b) channels. Also the "eye" graphs (figure 6) of the $K \rightarrow \pi\pi$ amplitude which is of interest for $\Delta I = 1/2$ falls into this class when the assumption of PCAC is not used.

Typical processes in this class usually involve more inversions than class 2, but they are also approachable with a single extended source propagator for one of the external hadrons. In addition, these processes require

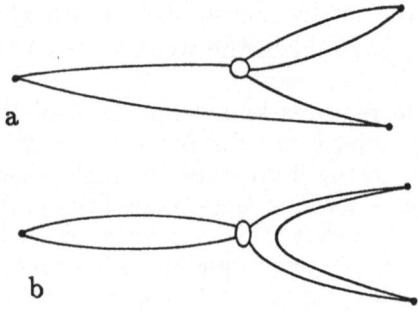

4. Non-leptonic two body decay graphs: (a) spectator, (b) exchange and annihilation.

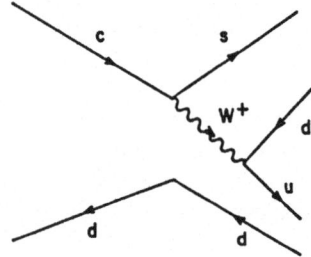

5. Spectator graph for $D^+ \to K^0 \pi^+$.

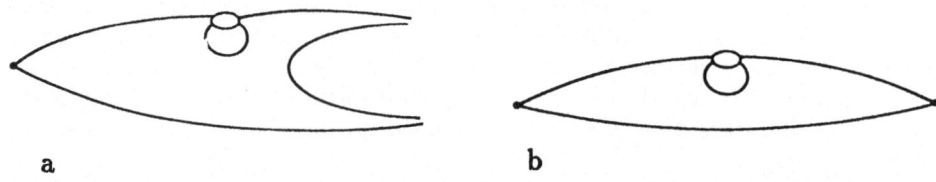

6. Eye graphs for $K \to \pi\pi$: (a) Class 3 graph, (b) PCAC approximation (class 2).

extraction of hadronic states on all three legs. Therefore they are expected to have larger errors from finite size effects and from statistics. Nonetheless, for specific channels either the number of inversions can be comparable to the semi-leptonic decays, or the experimental results better determined such that the purely hadronic channel is preferable. Much detailed phenomenology remains to be done to clearly identify the best channels. The totality of computationally feasible decay channel is somewhat daunting.

CONCLUSIONS

Many more detailed computational aspects need to be considered to seriously assess the practicality of these computations. However our rather trivial topological classification based on the valence quark propagators provides a rough initial criterion. Other equally vital questions of concern to us are: better quark inversion techniques such as fourier acceleration, improved axial and vector current insertions[3], and improved actions for Wilson fermions[6]. To address these issues, we are planning to perform an illustrative computation on one amplitude that plays an important part in determining the KM matrix[7]. Most likely a heavy quark decay process such as a B→D transition. We will use a rather small lattice to get a quick assessment of the calculational difficulty. Obviously a full scale program to do this kind of weak matrix element phenomenology requires a well organized project of several years supported by a class IV supercomputer and a group of phenomenological and computational physicists. Hopefully this effort can provide really useful numbers for the experimental test of weak interactions prior to the long awaited multi Gigaflop computer revolution.

REFERENCES

1. N. Cabbibo, G. Martinelli, and R. Petronzio, Nucl. Phys. B244 (1984) 381. R.C. Brower, M.B. Gavela, R. Gupta, and G. Maturana, Phys. Rev. Lett. 53 (1984) 1318. C. Bernard, T. Draper, G. Hockney, A.M. Rushton, and A. Soni, Phys. Rev. Lett. 55 (1985) 2770.

2. L. Maiani and G. Martinelli, CERN Preprint CERN-TH. 4467/86.

3. M. Bochicchio, L. Maiani, G. Martinelli, G.C. Rossi and M. Testa, Nucl. Phys. B262 (1985), p 331.

4. The contribution of spurious time wrapped states has been omitted for convenience of exposition.

5. This technique was used independently by authors of reference 2.

6. C. Rebbi, Boston University Preprint BUHEP-86-6, 1986

7. S. Stone, Cornell LNS Preprint CLNS-86/753, 1986

THE COLUMBIA SUPERCOMPUTER PROJECT: PHYSICS RESULTS

PRESENT STATUS AND FUTURE PLANS

N.H. Christ

Department of Physics
Columbia University
New York, NY 10027

This talk is divided into two parts. In the first part, our present results on the deconfining phase transition for pure SU(3) theory will be discussed. The second part contains a brief description of our computers: the architecture of the three machines now in varying stages of operation/construction, their performance and the expected schedual for their completion.

Our first 16-node machine has been operating since April 1985 and has been used to study the deconfining phase transition on lattices of size $16^3 \times N_t$ with N_t now ranging from 6 to 14 [1]. Our results for β_c on lattices with $N_t =10$, 12 and 14 are summarized and compared with similar calculations by the group of Gottlieb[2] et al (analysed using our angular peaking criterion) in the Table. The more difficult $N_t =12$ and 14 appear in good agreement but the smaller $N_t =10$ values of β_c significantly disagree. We have now run further on this size lattice using a new heat-bath program written by H. Q. Ding and, starting with a hot start, see a confined result for $\beta=6.1$ consistent with Gottlieb et al. However, we have no reason to distrust our earlier cold-start result for that value of β which contained 17,000 sweeps and appeared deconfined showing one jump in the Z_3 phase. We will study this question further!

Most of our running in the last five months has focused on the calculation of the latent heat of the deconfining phase transion. We have looked at lattices with $N_t =6$, 8 and 10 in an effort to understand the

TABLE

Nt	Columbia 10/85	Columbia 9/86	Gottlieb et. al.
10	6.08 +/-0.027	?	6.14 +/-0.01
12	6.261 +/-0.020	6.267 +/-0.012	6.28 +/-0.01
14	6.355 +/-0.026	6.383 +/-0.010	6.37 +/-0.01

systematics of this quite ellusive quantity[3]. Fig. 1
shows the internal energy as a function of β on the Nt=6
lattice. The line shown is a least squares fit to the
β=5.95, 6.0 and 6.1 points while the vertical step is
drawn at the point where the deconfinement fraction equals
0.5. Note that the step in the energy appears to occur at
a somewhat larger value of beta. (We see a similar
discrepancy between these two measures of Tc on the Nt=8
and 10 lattices also.) The value of the latent heat which
we deduce from height of the vertical step is
E/T^4=3.23 +/-0.55 . This agrees well with the result of
Karsch and Petronzio[4] of 2.7 +/-0.7 from a run with less
than one tenth the statistics presented here.

A similar calculation of the latent on larger
lattices (especially with Nt=10 and greater) is made
quite difficult by the quartic divergence of the single
plaquette expectation value: the signal to noise ratio in
the energy measurement described above is expected to
decrease as $1/Nt^4$. One possible remeady is to allow the
loops used to determine the energy to grow with the
lattice, always keeping the spacial and temporal extent of
the loops much less than Nt. For the long-distance modes
which participate in the deconfining phase transition,
such larger loops should be physically equivalent to the
standard 1x1 loops. However, one expects the signal,
which is proportional to the square of the flux through
the loops, to grow as the $area^2$ of the loop while the
noise remains essentially constant.

Fig. 2 shows the average of time-space minus space-
space loops for a 16^3x8 lattice as a function of the
$area^2$ of the loops at β=6.15. One sees the desired linear
dependence on $area^2$ for loops as large as 1x2. We are
currently trying to systematically incorporate the larger
loops into a measurement of the latent heat on Nt=8 and 10
lattices. It is also interesting to note that this
difference of time-space and space-space loops provides an
improved signal for the deconfining phase transition,
being approxmately zero for β < Tc and significantly
larger than that from the 1x1 loops for β > Tc.

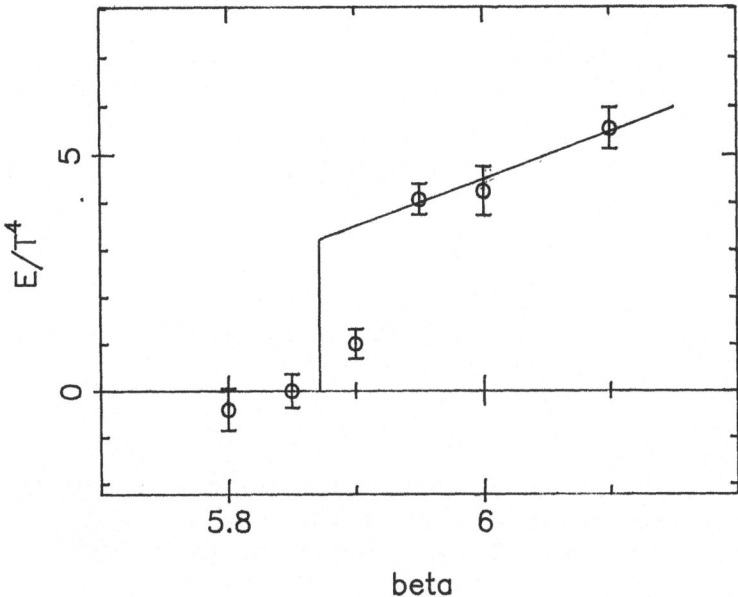

Fig. 1 The internal energy plotted as a function of β for a $16^3 \times 6$ lattice. The ordinate is the average of the quantity

$$\beta *[\ tr\{ \ U_{ts} \ \} \ - \ tr\{ \ U_{ss} \ \} \ ]*N_t^4/3$$

where U_{ts} and U_{ss} are the sum of products of the link matrices of 1x1 time-space and space-space Wilson loops.

Let us now turn to the second half of the talk: a brief description of the computers presently operating or under construction at Columbia. The results outlined here where obtained on our 16-node machine[5]. This parallel computer is connected as a 4x4 mesh of processors with the topology of a two-dimensional torus. The machine has a peak speed of 256 Mflops and 16 Mbytes of memory distributed as 1 Mbytes per node. Each node is controlled by an Intel 80286/80287 and has 128 Kbytes of fast data cache accessed by a pipelined, microcode controlled vector processor. The general architecture of a single node is shown in Fig. 3. This basic scheme is common to all three versions of our machine. It is the module contained within the dotted lines that is improved in our 64- and 256-node versions. The architecture of the 22-bit vector processor used on the 16-node machine is shown in Fig. 4.

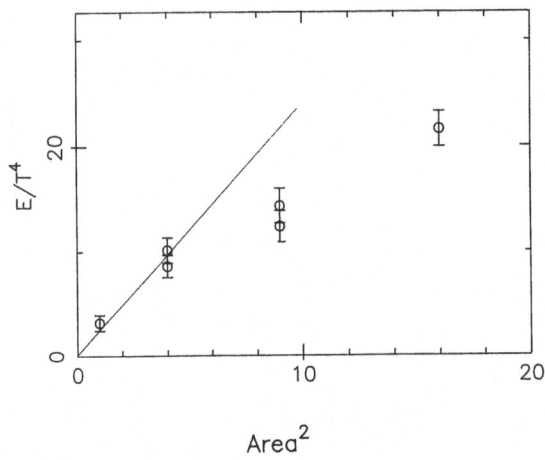

Fig. 2 The difference of the time-space and space-space
Wilson loops plotted versus their area² for a
16³x8 lattice at β=6.15. The line is drawn by eye.

This 16-node machine has been operating since April
1985, except for a 5 month period last winter when minor
changes were made to improve its performance and accuracy.
It now runs continuously with a failure on average every
two weeks caused by a soft error in the static memory.
Such an error aborts the job which must then be restarted
from the last check point, a loss of a maximum of 5 hours.
Every two months we have a more major failure do to a
faulty component or power supply. Such faults have
requried less than one day to diagnose and correct.

The next phase in our project is a 64-node machine,
interconnected as a 8x8 torus. It is identical to the 16-
node machine except for the 80286 clock (which has been
increased from 6 to 8 MHz) and the vector processor. The
64-node vector processor is shown in Fig. 5. It uses
Weitek 1033 and 1232 chips and performs IEEE 32-bit
floating point arithimetic. The arrangement is quite
similar to that of the 16-node machine except for
additional input and output buffering and an external
accumulation path for the floating point adder. Each node
of this machine will have 2 Mbytes of memory for a total
of 128 Mbytes. Five nodes are presently working with the

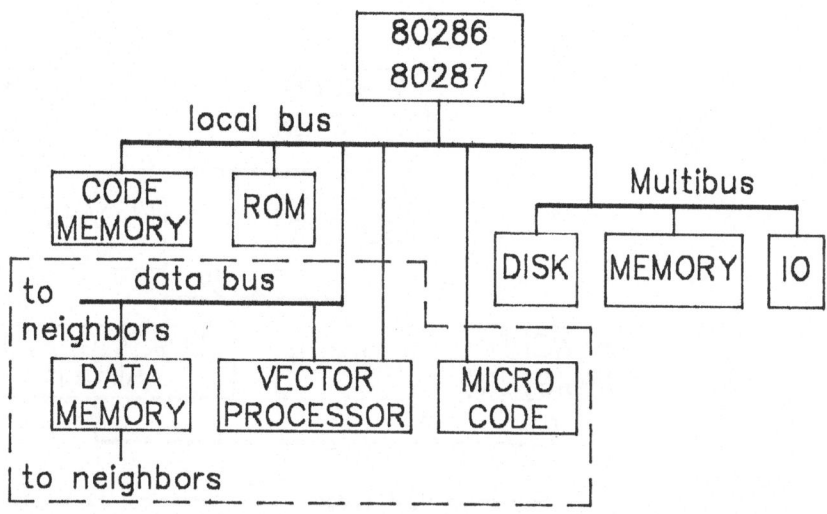

Fig. 3 The block diagram for a single node. The data bus is two independent 16-bit busses on the 16 and 64-node machines and a single 32-bit bus on the 256-node machine.

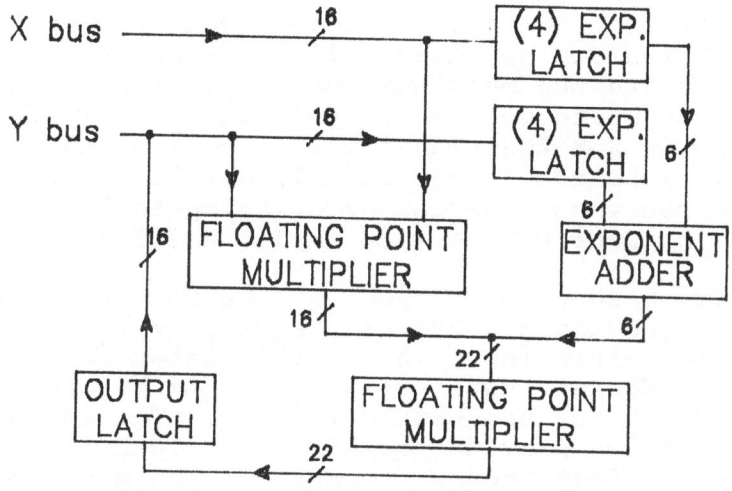

Fig. 4 The 22-bit vector processor in the 16-node machine.

remaining now being wire-wrapped. We expect to have the full 64-node machine operating before the end of the year.

The final phase of our project is a "256-node" machine which can be arranged in a 16x16 or 12x24 mesh. Although very similar to the two previous machines some improvements will be made: i) All the static memory chips will be upgraded from 16 Kbit to 64 Kbit chips with provision for 256 Kbit chips in the data and code memory when available. ii) The two memory banks and pair of 16-bit data busses have been replaced by a single memory

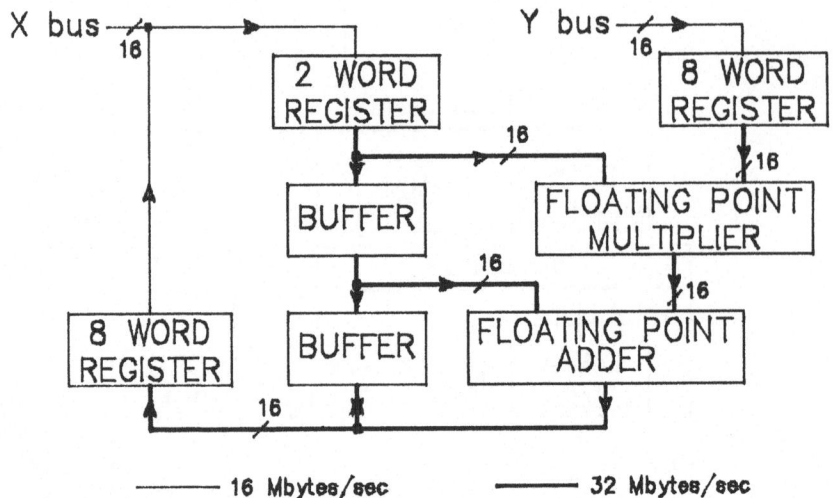

Fig. 5   The 32-bit vector processor for the 64-node machine.

bank, 32 bits wide, connected to a single 32-bit data bus.
This increases the interprocessor communication bandwidth
to 32 Mbytes/sec per node.   iii)   An LBX bus interface has
been added to the Multibus connector on each node.   This
will allow us to double the packing density of boards and
increase by 25% the transfer rate between the processor
and memory boards.   iv)   Finally we will significantly
upgrade the vector processor changing to Weitek 3332 chips
and adding four Weitek 1066 register files.

Fig. 6 shows the layout of the new vector processor.
Each 3332 contains a 32-word register file, floating point
adder and multiplier with flexible interconnection.   Each
runs at 16Mhz giving a peak rate of 64 Mflops for the pair
or 16 Gflops for the entire machine.   The added external
register files provide the high bandwidth cache necessary
feed such a fast vector processor.   The microcode is now
160 bits wide but still clocked at 8Mhz.   Double
instructions are provided in each cycle to support the
higher clock speed of the 3332's.   Each 3332 executes the
same microcode instructions but operates on different
data.   Thus a single line of microcode specifies four
vector processor cycles so we have, in total, increased
our microcode storage 16-fold.   The design of this machine
is nearly complete with prototype construction to begin
shortly.   Actual construction of the machine should start
late this Spring and be complete in one or two years.

Next let us discuss the methods used to program these
machines.   On the highest level one programs in Fortran or
PL/M to control the 80286.   Typically this code chooses

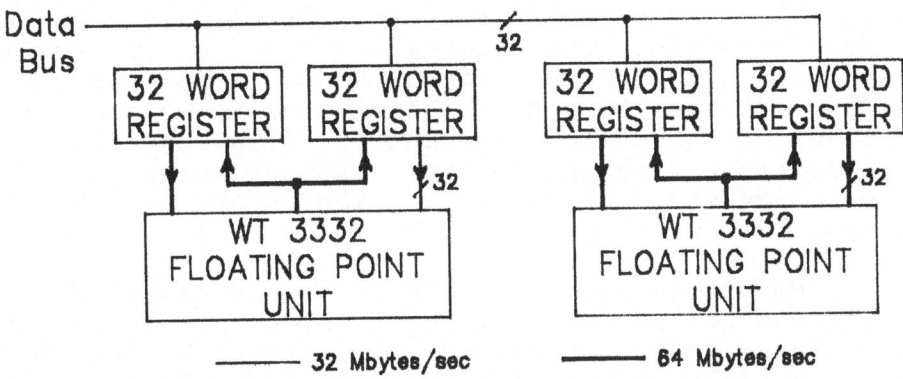

Fig. 6 The architecture of the 64 Mflop vector processor in the 256-node machine.

the order of execution of lower level modules, performs data transfers, lattice initialization etc. In fact all but the most computationally demanding parts of the program can be written as if one was programming a PC. Since the vector processor runs independently of the 80286 its microcoded programs can be viewed as subroutines to be called by the 80286. These subroutines which are written in a home-grown assembly language will multiply pairs of link matrices, compute the sine and cosine of a string of ten numbers or, for example, generate a series of Fibbonaci random numbers.

The speed of these vector processor operations is so great that only the most minimal 80286 activity can be allowed to prepare for them. This is accomplished by providing a two-layer software interface between the high-level 80286 programs and the vector processor routines. The high-level program calls a carefully written assembly language program (the microcode dispatcher) which reads a data file generated before assembly by a C program. This data file is a list of the microcode routines with their required arguments in the intended order of execution. The microcode dispatcher reads this list, moves the supplied arguments to the vector processor registers, waits for completion of the current vector processor routine and then starts execution of the next routine.

In the Cabibbo-Marinari heat bath program running on the 16-node machine each processor updates one 4x4 x-y plane at a time. This updating is described by a single data file which lists the sequence of vector processor matrix multiplies necessary to compute the environment of each link in the x-y plane. Also present in this list are instructions to jump to the updating routine when each subsequent environment has been calculated.

Finally, let us consider the machine's actual performance. The program now running on the 16-node machine uses a 3-subgroup Cabibbo-Marinari heat bath algorithm. The complete updating program, including all data transfers, link renormalizations, etc., requires 1.8 msec per link on a single node. (The complete machine takes 140 usec to update a link because we presently run it at 80% of its design frequency.) Although this is a respectable link update time, it corresponds to a sustained speed of only about 10% of the peak 256 Mflop capability of the machine.

We are currently writing a second class of programs intended to increase this efficiency by a factor of 3 to 5. These include a 64-node version of the Cabibbo-Marinari algorithm and a Langevin simulation of fermion loops written for the 16-node machine. Our basic programing strategy is not changed. However, we now vectorize over a number of links. This allows significantly more efficient non-matrix floating point operations and also compensates for the shorter matrix operations present in the Fermionic calculations.

Although we are pleased to present significant physics results from our 16-node machine, it should be emphasized that 50% of our efforts are still directed toward computer construction. Perhaps 25% of us are working to improve the software environment, while the remaining 25% of our time is devoted to physics. As work progresses on the 256-node machine we will give a larger emphasis to software design and finally to physics.

Acknowledgement. The physics results and computer design and construction described here are the joint work of a group of theoretical physicists at Columbia: K.M. Barad, F.R. Brown, F.P. Butler, Y.F. Deng, H.Q. Ding, M.S. Gao, P.F. Hsieh, S.P. Sun, L.I. Unger and T.J. Woch in addition to myself and A.E. Terrano who is now at Ruttgers.

REFERENCES

1. N. H. Christ and A. E. Terrano, Phys Rev Lett $\underline{56}$, p111 (1986).
2. S. A. Gottlieb, J. Kuti, D. Toussaint, A. D. Kennedy, S. Meyer, B. J. Pendleton, and R. L. Sugar, Phys Rev L $\underline{55}$, p1958 (1985).
3. H. Q. Ding, Columbia University PhD Dissertation, in preparation.
4. F. Karsch and R. Petronzio, Phys Lett, $\underline{139B}$, p403 (1984).
5. N. H. Christ and A. E. Terrano, Byte Magazine, Vol 11, no 4 p145, April 1986.

FOURIER ACCELERATION AND LATTICE

GAUGE THEORIES

C.T.H. Davies

Department of Physics and Astronomy,
University of Glasgow, Glasgow G12 8QQ, U.K.

Abstract

We discuss the use of the technique of Fourier acceleration to avoid critical slowing down in a wide range of numerical problems. Results are presented on the application of this technique to three different algorithms useful for the numerical simulation of lattice gauge theories: gauge fixing to Landau gauge, updating the gauge field, and calculating the fermion propagator. We estimate that our simulation algorithm for QCD including dynamical quarks will be 10 times faster when Fourier acceleration is used, at current values of lattice spacing and volume.

Introduction

The work I shall report on here was done at Cornell University in collaboration with George Batrouni, Garry Katz, Andreas Kronfeld, Peter Lepage, Pietro Rossi, Ben Svetitsky and Ken Wilson[1,2]. We have been concentrating on the development of efficient algorithms for the simulation of lattice field theories, the goal being a future study of QCD including the full effect of quarks. One problem that we have tackled with considerable success is that of critical slowing down which until now has severely hampered the approach to the continuum limit.

Critical slowing down occurs in a wide variety of iterative numerical algorithms when the matrix which controls the change from one step of the algorithm to the next is very ill-conditioned, that is, it has a large range of eigenvalues. Consider a simple-minded relaxation scheme for solving the equation $Mx = 0$ where M is an NxN matrix.

$$x^{(n+1)} = x^{(n)} - \varepsilon M x^{(n)} \tag{1}$$

It is clear that the component of x parallel to the eigenvector of M with largest eigenvalue, λ_{max}, is changing much faster than the component along the eigenvector of smallest eigenvalue, λ_{min}. For stability of the algorithm the step-size ε must be chosen to be less than the reciprocal of λ_{max}. The time taken to solve $Mx = 0$, t, is then determined by the eigenvector corresponding to λ_{min}:

$$t \propto \frac{\lambda_{max}}{\lambda_{min}} \tag{2}$$

The algorithm will be very slow when the ratio of the largest to smallest eigenvalues, the condition number of M, is large.

For physical systems M will often be at least approximately diagonal in momentum space and will have a condition number which is an inverse momentum or mass in lattice units to some power, j. The computational cost of the iterative algorithm will grow as

$$\text{cost} \propto V \times \left(\frac{1}{ma}\right)^{j} \tag{3}$$

where a is the lattice spacing, V is the number of lattice points and m is a mass in physical units. Attempting to make contact with continuum physics by taking a to zero will result in a disastrous increase in cost. This is critical slowing down.

The first factor of eq. 3 is unavoidable but the consequences of the second factor can be alleviated by a transformation (preconditioning) of M. The idea is to multiply M by a matrix E so that the product matrix has eigenvalues which are all the same. Since the cost of a typical matrix multiplication grows like V^2, this will not generally be computationally worthwhile unless E is sparse. When M is diagonal in momentum space, however, we can exploit the power of the fast Fourier transform (FFT) to cure critical slowing down. The algorithm becomes

$$x^{(n+1)} = x^{(n)} - F^{\wedge -1} \varepsilon \frac{\lambda_{pmax}}{\lambda_{pmin}} \hat{F} M x^{(n)} \tag{4}$$

where F represents an FFT and E has only diagonal elements in momentum space which are inversely proportional to the eigenvalue at that momentum. This is called Fourier acceleration. The cost of the algorithm grows like $V \ln V$, being the cost of a fast Fourier transform.

In the interesting cases of a theory with interactions M may be only approximately diagonal in momentum space. We can still gain a large factor in algorithm speed using Fourier accceleration, however, as I shall describe in the following sections. We have successfully applied this acceleration to three different areas of the simulation algorithm for a lattice gauge theory:

(i) gauge fixing to Landau gauge;

(ii) updating the gauge fields with a Langevin equation;

(iii) calculation of the fermion propagator.

Cases (i) and (ii) represent examples of j = 2 in equation 3 and (iii) has j = 1. I shall now go on to describe these in more detail.

Landau gauge fixing

Although it is not necessary to fix the gauge for simulations of lattice gauge theories, it is often desirable. To calculate matrix elements of an extended operator (as appears for example in hadronic wavefunctions) it is often easier to completely fix the gauge than to connect all points with strings of link fields. Landau gauge, $\partial_\mu A^\mu = 0$, is a good gauge to use and, as I shall show, it is easy to implement. It has none of the problems of axial gauges such as lack of translational invariance or large field variations transverse to the gauge direction.

To fix to Landau gauge is a standard optimization problem in a space with many dimensions. We wish to maximize the sum over the lattice of the real part of the trace of the link field,

$$\sum_{\mu, x} \mathrm{Tr} \left(U_\mu(x) + U_\mu^\dagger(x) \right) \tag{5}$$

in the space of gauge-equivalent fields. We do this iteratively by successively calculating gauge transformation matrices, $G(x)$ (which sit on the sites of the lattice), as functions of the link field and then applying the transformation

$$U_\mu(x) \rightarrow G(x) \, U_\mu(x) \, G^\dagger(x + \mu) \,. \tag{6}$$

The naïve steepest descents (Jacobi) method of calculating $G(x)$ gives

$$G(x) = \exp\left[\frac{\alpha}{2} \, \partial_\mu A^\mu(x) \right] \tag{7}$$

where

$$\partial_\mu A^\mu(x) = \sum_\mu \left[U_\mu(x) - U_\mu^\dagger(x) - \{ U_\mu(x-\mu) - U_\mu^\dagger(x-\mu) \} - \mathrm{trace} \right] \tag{8}$$

The step size α is generally taken between 0.05 and 0.1.

Expanding around $U = 1$ for the Abelian theory shows that $\partial_\mu A^\mu$ decays to zero in the following manner

$$\partial_\mu A^{\mu\,(n+1)} = \partial_\mu A^{\mu\,(n)} - \alpha \partial^2 \partial_\mu A^{\mu\,(n)} \,. \tag{9}$$

Critical slowing down is evident, controlled by the minimum eigenvalue of ∂^2. Fourier acceleration of this algorithm is simple. The steepest descents method becomes an improved Newton's method in momentum space where the step size is inversely proportional to the appropriate eigenvalue of ∂^2 (p^2).

$$G(x) = \exp\left[\hat{F}^{-1} \frac{\alpha}{2} \frac{p_{max}^2}{p^2} \hat{F} \, \partial_\mu A^\mu(x) \right] \tag{10}$$

The results of applying this algorithm to configurations of SU(3) link fields for an 8^4 lattice are shown below. Seven times fewer iterations are required to reach the maximum trace link (equation 5) with Fourier acceleration than without, and this improvement factor will grow with increasing volume.

Landau gauge is known not to completely fix the gauge degrees of freedom in the continuum. This may not be a problem for the lattice theory but in order to be sure of having a completely fixed gauge we combine axial and Landau gauge. Our gauge is then defined to be axial gauge followed by n hits of the Fourier accelerated Landau gauge fixing algorithm as described above. It seems satisfactory for our purposes to use values of n around 10. This gives a trace link within 10% of its maximum value.

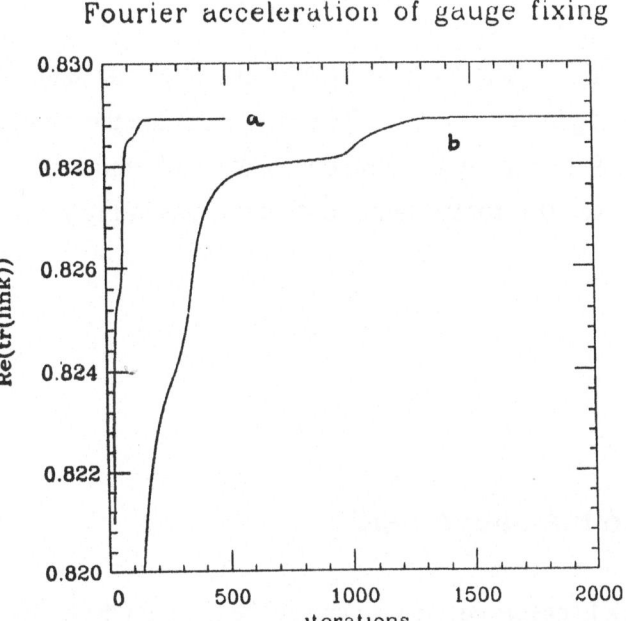

Fig.1. The growth of the average value over the lattice of the real part of the trace of the link field as the Landau gauge fixing algorithm proceeds. The curves compare the algorithm with Fourier acceleration (a), $\alpha = 0.08$, and without (b), $\alpha = 0.1$ for an 8^4 quenched SU(3) configuration at $\beta = 5.8$.

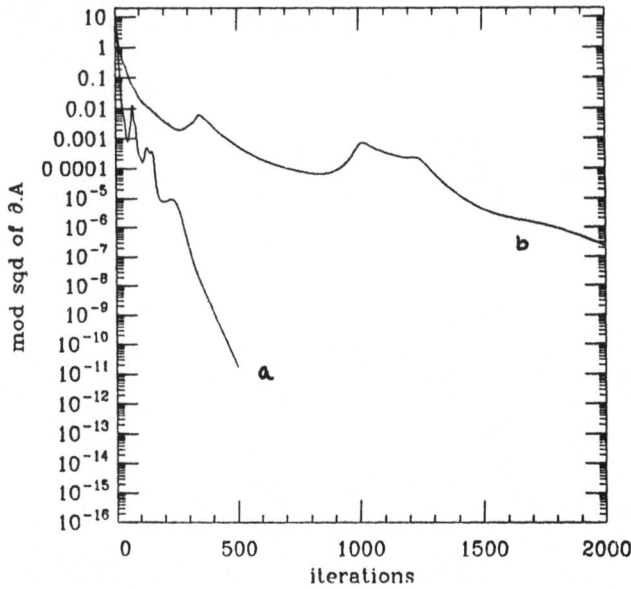

Fig.2. The decay of $|\partial_\mu A^\mu|^2$ as the Landau gauge fixing
algorithm proceeds. The curves compare the algorithm
with Fourier acceleration (a), $\alpha = 0.08$, and without (b),
$\alpha = 0.1$, for the same configuration as in Fig.1.

Updating the gauge field

We use a Langevin equation to update the fields on the lattice. The link field at
the nth step is given by

$$U^{(n+1)} = e^{-ifT} U^{(n)} \tag{11}$$

where, in the simplest algorithm,

$$f_j = \varepsilon \partial_j S + \sqrt{\varepsilon}\, \eta_j \tag{12}$$

and the T_j are the generators of the gauge group. $\partial_j S$ is the derivative of the action
with respect to the field along the group manifold. η_j is noise drawn independently

from a Gaussian distribution for each component j, for each link, at every step. The force term f_j then has two parts; one tends to minimise the action and the other provides the quantum fluctuations.

As before, expanding around $U = 1$ for the Abelian theory shows clear evidence of critical slowing down. The controlling eigenvalues are those of the operator ∂^2, causing the low momentum components of the field configurations to change very slowly.

The Fourier acceleration technique in the case of the U(1) theory is straightforward. The force term becomes

$$ f = \hat{F}^{-1} \varepsilon \frac{p^2_{max}}{p^2} \hat{F} \, \partial S + \hat{F}^{-1} \sqrt{\varepsilon \frac{p^2_{max}}{p^2} \hat{F}} \, \eta \qquad (13) $$

We have tested the efficiency of this algorithm by making measurements of the modulus squared of the force term in momentum space,

$$ F = |\tilde{f}(p)|^2 \, . \qquad (14) $$

We derive a measure of its decorrelation time in sweeps from the autocorrelation function in Langevin time , $< F(t) \, F(t + \tau) >$. In the weak coupling limit these decorrelation times can be calculated and measurements in U(1) at $\beta = 1.5$ on a 2^4 lattice confirm the results. Without acceleration the decorrelation time is inversely proportional to p^2 whereas with acceleration it is independent of p^2; critical slowing down is eliminated.

The SU(3) case is more complicated because the force term is no longer gauge-invariant and if we wish to accelerate we must fix the gauge. The updating algorithm with Fourier acceleration then consists of two parts, first an update with force term as in eq.13 followed by a complete gauge fixing as described in the previous section (axial gauge followed by Landau gauge). Unfortunately this algorithm gives an incorrect effective action because the preconditioning matrix E has become field-dependent through the gauge-fixing procedure. We correct for this by modifying the force term to contain a stochastic estimator of the derivative of E with respect to the fields, thereby cancelling the unwanted term. Tests of this modified algorithm in the weak coupling region $\beta = 10$ are shown below, where the

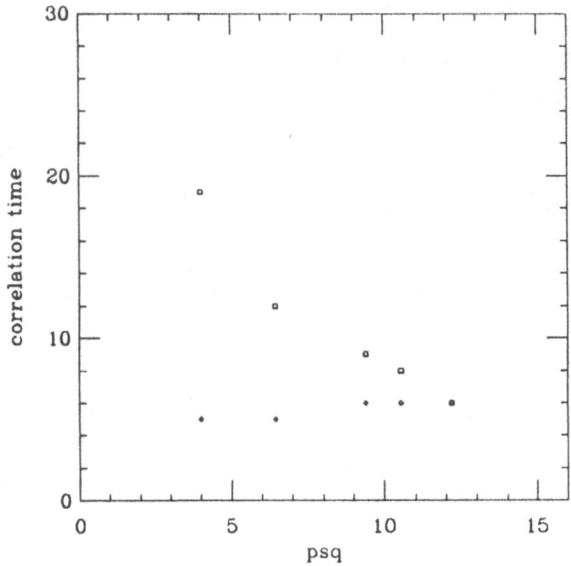

Fig.3. The correlation time in sweeps for the modulus squared of the force term in momentum space vs. p^2. Data for a 2^4 lattice, quenched QCD at $\beta = 10$, $\varepsilon = 0.01$. The squares are for the algorithm without Fourier acceleration, the diamonds with Fourier acceleration.

correlation time of the modulus squared of the force term in momentum space is plotted against p^2 (in lattice units). Gains in correlation time of a factor of 3 are obtained for this small lattice by using Fourier acceleration.

At more realistic values of the coupling, β, it is found necessary to modify the accelerating factor from p_{max}^2/p^2 to $(p_{max}^2 + m^2)/(p^2 + m^2)$. For momenta below m the interactions prevent further acceleration. At $\beta = 5.8$ we require $m^2 a^2 = 6$ but if, as we believe, m represents a physical mass then $m^2 a^2$ should fall rapidly with increasing β. For a fixed number of lattice points the gain in speed from using

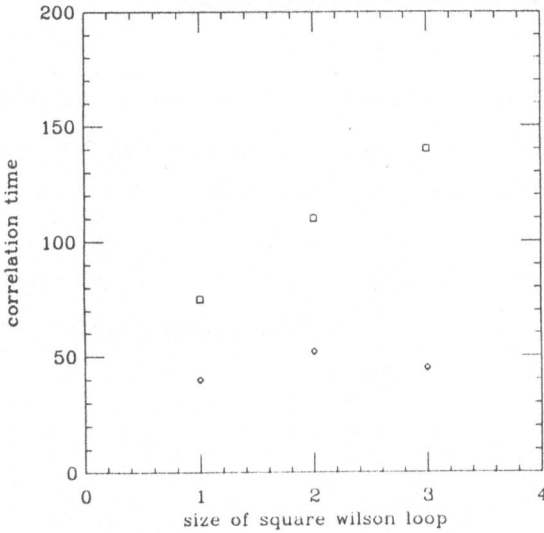

Fig.4. The correlation time in sweeps for square Wilson loops
vs. the length of the side of the loop. Data are for a 4^4 lattice
quenched QCD at $\beta = 6.2$, $\varepsilon = 0.016$. The squares represent
the algorithm without Fourier acceleration , the diamonds with
Fourier acceleration.

acceleration will then increase. The effect of changing β but maintaining a lattice of
fixed volume in physical units is not yet clear. In figure 4, I show the results for the
decorrelation time of square Wilson loops in the quenched approximation on a 4^4
lattice at $\beta = 6.2$. Without acceleration the correlation time grows with size but after
acceleration with acceleration factor $(p_{max}^2 + 4)/(p^2 + 4)$ the points have a much
flattter distribution. Gains in correlation time of at least a factor of 3 will be clearly
obtainable for the updating algorithm at current values of β and lattice volume. The
time for one iteration of the accelerated algorithm will clearly be longer than for the
unaccelerated case (perhaps by a factor of 3). The time for updating the gauge field
is still negligible compared to calculating the fermion propagator, however, so for a
simulation of full QCD this additional cost is irrelevant.

Calculation of the fermion propagator

One of the problems that has delayed the implementation of algorithms for QCD which include the full effect of quarks has been the huge computational cost of calculating the fermion propagator. The matrix M (where the fermionic contribution fo the action is $\psi M \psi$) must be inverted once every update of the gauge field and the inversion algorithm shows critical slowing down, controlled by the spectrum of M. We work with Wilson fermions and use the conjugate gradient algorithm to calculate $(\gamma_5 M)^{-1}$. The equation to be solved is

$$\left(\gamma_5 M\right)^2 \psi \; = \; \left(\gamma_5 M\right) \eta \,. \tag{15}$$

The algorithm takes a long time to converge if the parameter containing the bare quark mass, κ, is close to its critical value, κ_{crit}, where the pion mass vanishes (on averaging over configurations).

It is clear for free fermions that we could invert the matrix in one step by preconditioning $(\gamma_5 M)^2$ in a symmetric way with the square of the free fermion propagator in momentum space. This method also works well in the presence of a background gauge field when we use as preconditioning matrix the square of the free propagator with renormalised mass parameter κ_r.

I show in figure 5 how the decay of the residue $|\gamma_5 M \psi - \eta|$ is accelerated by this technique on an 8^4 lattice using a background quenched SU(3) gauge field at $\beta = 5.8$. Three to four times fewer iteratons are required with Fourier acceleration giving an overall factor in speed of 2.5-3 when an overhead of 20% for the fast Fourier transform is taken into account.

The presence of critical slowing down makes the choice of a good convergence criterion for the inversion much harder[3]. The residue itself is not sensitive to the low momentum components which are converging slowly.We recommend instead

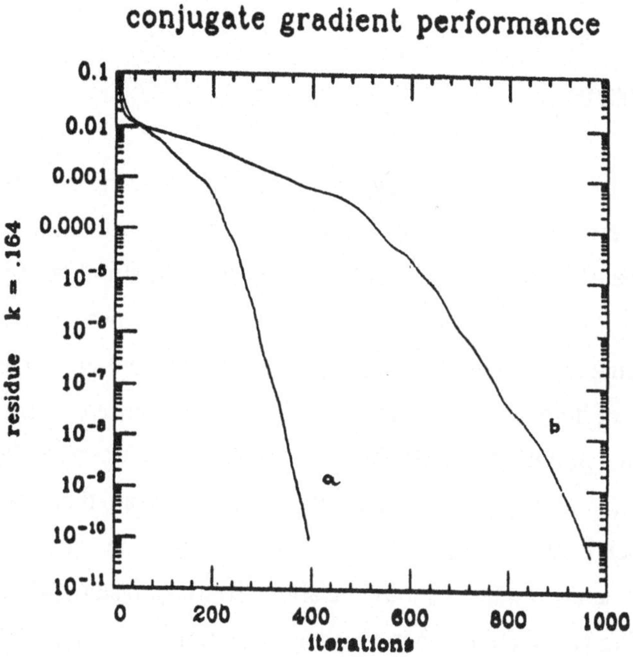

conjugate gradient performance

Fig.5. The decay of the residue $|\gamma_5 M\psi - \eta|$ during inversion with the conjugate gradient algorithm. Data are for an 8^4 quenched QCD configuration at $\beta = 5.8$, $\kappa = 0.164$ ($\kappa_{crit}=0.167$). The curves compare the algorithm with Fourier acceleration (a), $\kappa_r =0.08$, to that without acceleration (b).

using the following variable:

$$\frac{|\psi^{(n)} - \psi^{(n-1)}|}{1 - \dfrac{r^{(n)}}{r^{(n-1)}}} \qquad (16)$$

Further details will be described in a forthcoming paper.

Conclusions

In the preceding sections I have described the application of the technique of Fourier acceleration to algorithms useful for the numerical simulation of lattice gauge theories. The time for one update of an algorithm that simulates full QCD is dominated by the time required to calculate the fermion propagator. As described above this time can be cut by a factor of 2.5-3 when Fourier acceleration is used. The number of iterations of the updating algorithm required for the decorrelated measurements of operators relevant to long distance physics can also be cut using Fourier acceleration. The results described above indicate that a factor of 3.5 can also be gained here. The result is that our algorithm for the simulation of QCD including dynamical quarks should be about 10 times faster at current values of the lattice spacing and volume when Fourier acceleration is used.

References

1) G.G. Batrouni, G.R. Katz, A.S. Kronfeld, G.P. Lepage, B. Svetitsky and K. Wilson, *Phys.Rev.* D**32** (1985) 2736.
2) C. Davies, G. Batrouni, G. Katz, A. Kronfeld, P.Lepage, P. Rossi, B. Svetitsky and K. Wilson , *J.Stat.Phys.* **43** (1986) 1077.
3) G.R. Katz, PhD thesis, Cornell University (1986).

PROPERTIES OF THE GLUON PLASMA JUST ABOVE T_c

Thomas A. DeGrand

Physics Department
University of Colorado, Boulder, Colorado, 80309

Carleton E. DeTar

Physics Department
University of Utah, Salt Lake City, Utah, 84112

ABSTRACT

We describe the results of a detailed study of the properties of finite temperature SU(3) lattice gauge theory on large lattices near and above the confinement-deconfinement phase transition. We measure correlation functions of Wilson lines and local operators and observe a nonzero value for the inverse screening length in the deconfined phase. We calculate thermodynamic quantities of the plasma. We compare our results with perturbative QCD and with the model of dynamic confinement based on dimensional reduction.

INTRODUCTION

Despite the considerable theoretical effort that has been devoted to the phenomenology of the gluon plasma, very little is actually known about its properties that follow rigorously from quantum chromodynamics (QCD). It is expected from asymptotic freedom that at extremely high temperatures the plasma may be described to a good approximation as a gas of weakly interacting quarks and gluons. Numerical simulations suggest that there may be a phase transition between a low temperature hadronic gas and a high temperature plasma phase. However at temperatures achievable in heavy ion collisions, temperatures not much greater than that of the phase transitions, asymptotic freedom does not apply, and the weakly interacting gas model is surely inadequate.

Indeed, we really don't even know what the plasma is made of at the experimentally accessible temperatures–i.e., we don't know its elementary modes of excitation. Obviously, it is ultimately made of quarks and gluons, just as are the hadrons. But the excitations of the plasma are very likely not simply quasi-quarks and quasi-gluons in analogy to a weakly interacting quantum electrodynamic (QED) plasma. Without knowledge of the plasma structure, it is difficult to say with confidence what phenomena signal the formation of a plasma in a heavy ion collison.

We report here results of numerical Monte Carlo simulation of a set of quantities related to these elementary excitations, namely the static screening lengths of the SU(3) gluon plasma. These are found by measuring the decay of static correlation functions at large space-like separation in the Gibbs ensemble at a fixed temperature. The simplest such correlation involves the interaction of a pair of fixed color triplet and anti-triplet charges, i.e., Wilson lines. The associated screening length is similar to the Debye length in QED and the corresponding dynamical excitation would be the plasmon in QED. Thus studying the static screening length gives indirect information about the elementary excitations. Because the plasma is strongly interacting and perhaps even confining in a sense,[1] the analogy with Debye screening in QED is very loose. As a byproduct of these measurements, we also compute thermodynamic properties of the plasma. The detailed discussion of our work is given in Refs. (2) and (3).

OBSERVABLES

The data set consists of many runs at $N_t = 4$, 6 and 8. Typical spatial lattice sizes are 11^3 and $11^2 \times 15$. The simulations consumed about 150 hours of CDC-205 time and 120 hours of Cray time. The coupling β ranged from 5.7 to 6.2.

We have elected to use the source method[4] a source constant in imaginary time is introduced at spatial coordinate $z = 0$. Operator expectation values averaged over x, y, and z are then measured as a function of the displacement in z from the source. The observed asymptotic decay with distance gives the long range screening lengths of the system. We fit the operator expectation value at a distance z from the source to

$$< \phi\left(z\right) >= a + b\cosh\mu\left(z - L/2\right) \tag{1}$$

The parameter μ is the inverse screening length or "static screening mass," sometimes referred to here as simply "mass." We typically begin our fits to the data using (1) at a distance $z = 3$, and extend them to a distance of 6 or 7 lattice spacings from the source. We have measured expectation values of two classes of operators in the presence of a source: nonlocal operators - Polyakov loops (Wilson lines) and local operators (various closed paths of link variables). Among the local operators we have considered are 1×1, 1×2, and 2×2 x-y plaquettes, 1×1 x-t and y-t plaquettes.

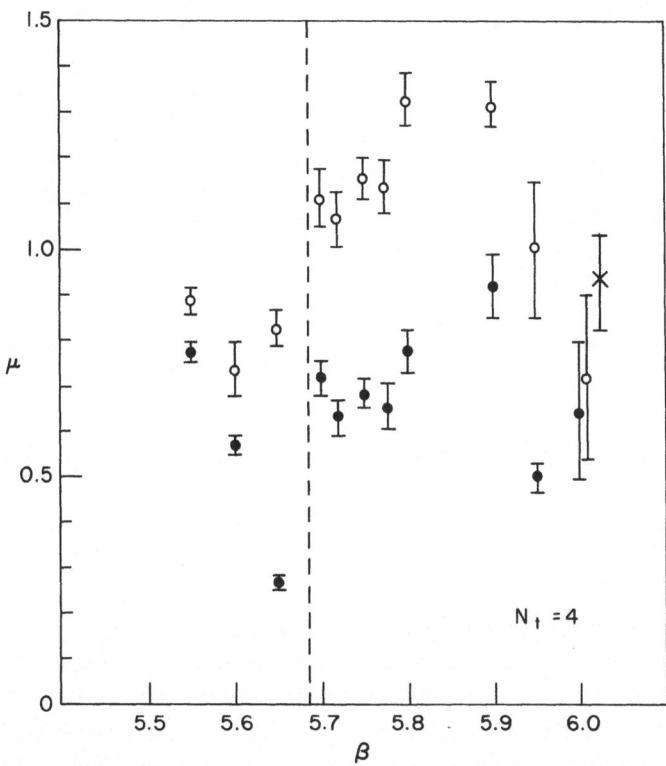

Fig. 1: Lattice masses for local operators (open circles), Wilson lines (closed circles) and adjoint Wilson lines (crosses for $N_t = 4$.

DATA ANALYSIS

There is a rather natural division of our data into correlation functions which are local and ones which are nonlocal (Wilson lines). As the quality of the two data samples is quite different, we discuss first the Wilson line measurements, and then the local operator measurements.

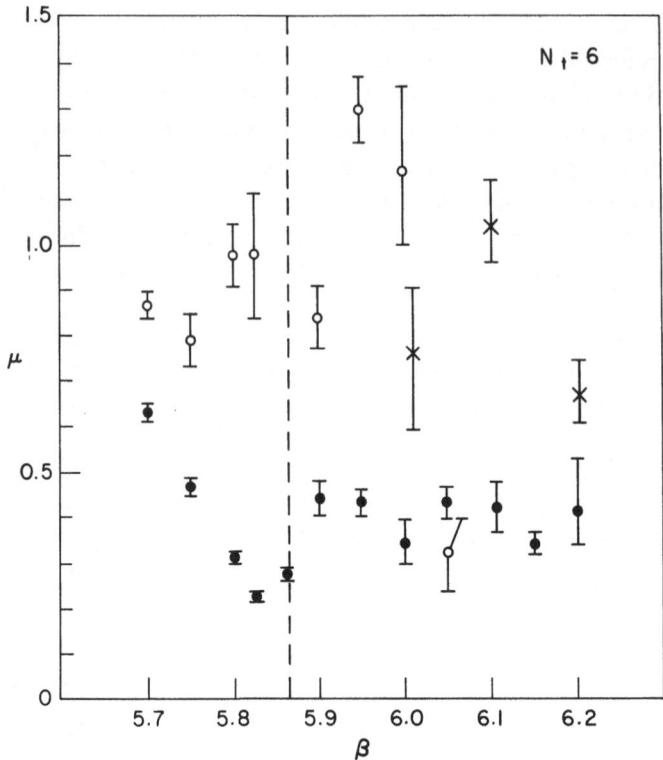

Fig. 2: Same as Fig. 1, but $N_t = 6$.

We first wish to explore the behavior of μ near the deconfinement phase transition. Since the lattice critical temperature does not show scaling, it is unjustified to make an extrapolation to the continuum limit. We elect to analyze the results of calculations at different N_t's separately. Graphs of μ vs β for the three N_t values (4,6,8) are shown in Figs. 1-3. The dotted lines show the location of the deconfinement phase transition. The open and closed circles show the masses extracted from local operators and Wilson lines respectively.

Clearly shown for each N_t is a substantial drop in μ_{WL} as β approaches β_c from below and a substantial rise just above β_c, followed by a possible slow decrease with increasing β. Whether there is a discontinuity in μ_{WL} at β_c cannot be conclusively determined from our data, although such an effect is permitted by the first-order character of the phase transition and it is suggested in the data.

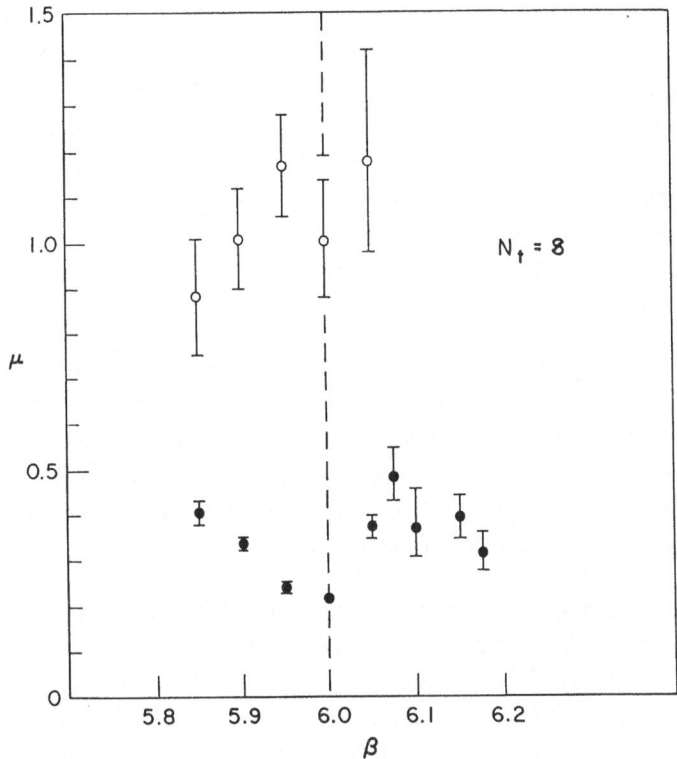

Fig. 3: Same as Fig. 1, but $N_t = 8$.

Our results from correlation functions of local operators are much less clear cut than those from Wilson lines. Consequently we will not even attempt an extrapolation to the continuum and will confine our discussion to masses measured on the lattice. Generally we find that the masses we extract from correlation functions beginning a distance z_0 from the source show much stronger z_0 dependence than masses extracted from Wilson lines. In many cases it is not possible to obtain stable masses within the constraints of our lattice size. In many cases when we do obtain a z_0-independent mass, its error bars are about twice as big as the error on a Wilson line mass from the same data sample. Also, in many cases, the value of the mass measured from local operators is larger than the mass measured from Wilson lines.

As another attempt to measure the high temperature analogs of glueballs we did a small set of measurements of expectation values of Wilson lines in the adjoint representation[5] We were inspired in our hope that they might be good antennas by the recent work of Berg and Billoire, who have claimed that adjoint Wilson lines couple strongly to the lightest glueball state[6]

The measurements are very straightforward. We have concentrated on β values greater than 6.0 since the ordinary local operators seem to work well enough at lower β, and since we are interested in the highest β possible to take a continuum limit. So we measured five points: $N_t = 4$, $\beta = 6$, $N_t = 6$, $\beta = 6.0$, 6.11, 6.2, and $N_t = 8$, $\beta = 6.05$. The last case did not give a useable signal beginning at $z_0 = 4$. The results are shown in the figures for the different N_t data (where they appear as crosses). In all cases the mass extracted from the adjoint loop is consistent with the masses gotten from local operators.

Are the adjoint lines "better" operators than the local operators? The error bars are about the same size as for the local operators, and the b parameters (see eqn. (1)) are about 10^{-4}, also like the local operators. What is significant is that we are able to measure a clean exponential at much larger β – up to $\beta = 6.2$ in one case. In that sense the adjoint lines are superior to 1×1 or 2×2 plaquettes.

The adjoint masses are larger than the fundamental masses. This is a conundrum: in perturbation theory the correlation function between adjoint loops can be mediated by exchange of two electric gluons. Why then is the mass associated with the correlation function not the Debye screening mass? Apparently what is happening here must be similar to the situation with the local operators. If below T_c an operator couples strongly to the lightest glueball, it might remember this fact for T greater than T_c. Combined with the local operators, the adjoint line provides additional evidence that there might be another correlation length in the plasma, besides the Debye screening length which governs correlation functions of fundamental lines.

THE PLASMON MASS

We next attempt to extrapolate our lattice results for β much greater than β_c to the continuum limit using the perturbative renormalization group. In the asymptotic limit the bare lattice spacing a is related to the physical temperature through

$$\frac{1}{T} = N_t a \tag{2}$$

and to the lattice cutoff Λ_L through the two-loop β-function

$$a\Lambda_L = \left(\frac{8\pi^2}{33}\beta\right)^{\frac{51}{121}} e^{-4\pi^2\beta/33}$$
$$\equiv f(\beta). \tag{3}$$

Thus a lattice of temporal extent N_t and lattice coupling β has a physical temperature

$$\frac{T}{\Lambda_L} = \frac{1}{N_t f(\beta)}. \tag{4}$$

A mass μ measured in lattice units corresponds to a physical mass m

$$\frac{m}{\Lambda_L} = \frac{\mu}{f(\beta)} \tag{5}$$

or

$$\frac{m}{T} = N_t \mu. \tag{6}$$

We face an immediate complication in our analysis in that most of our data comes from β-values where neither T_c nor the string tension scale (although $T = 0$ glueball masses might). Two groups[7,8] have recently seen indications of perturbative scaling in T_c with T_c/Λ_L about 45, on lattices of length $N_t = 10, 12,$ and 14. The critical coupling at $N_t = 10$ is at $\beta_c = 6.15$. As all of our data are taken on lattices with $N_t \leq 8$ and only two data points have $\beta > 6.15$ we are not likely to be in a scaling region.

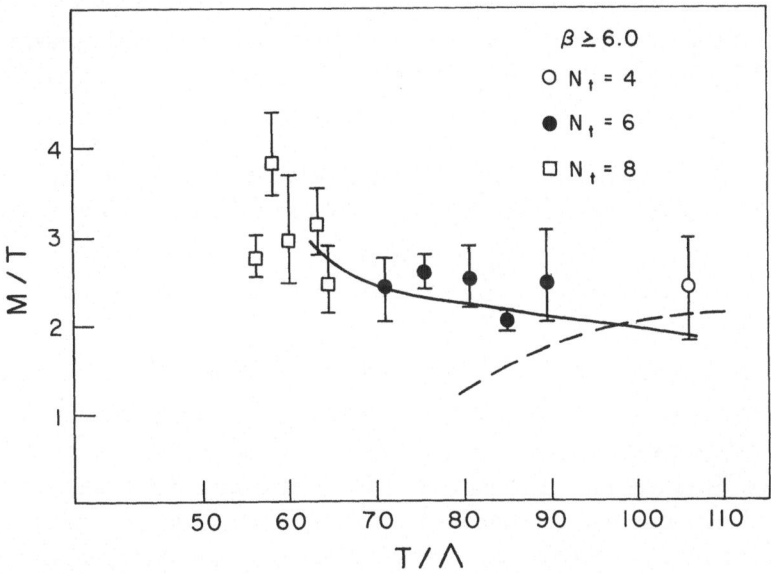

Fig. 4: The color singlet plasmon mass vs temperature.

In Fig. 4 we show the quantity m/T as a function of T/Λ_L, as extracted from measurements of the correlation function for fundamental Wilson lines. The three different temporal lattice sizes allow the data to overlap in T/Λ. The data lie in a smooth band which decreases with increasing T/Λ. In an attempt to remove the grossest scaling violations, we have cut out all points for $\beta \leq 6.0$.

In lowest order perturbation theory the plasmon or electric gluon mass is

$$m/T = g\left(T\right) \tag{7}$$

where the temperature dependent running coupling constant is

$$g\left(T\right)^2 = \frac{4\pi^2}{3}\left[\frac{11}{6}ln\left(\frac{T}{\kappa\Lambda_L}\right) + \frac{17}{22}\ln\,\ln\left(\frac{T}{\kappa\Lambda_L}\right)\right]^{-1}. \tag{8}$$

Here $\kappa = 31.3$ to make contact with the thermodynamic potential calculations of Kapusta.[9] Inspired by eqn. (7), we attempted a fit to the data of $m/T = Rg\left(T\right)$. We get a good fit with $R = 0.77 \pm 0.03$. This curve is plotted on Fig. 4 as the solid line.

How can we interpret this result? The simplest possibility is that the mass we have measured is just approximately twice the electric plasmon mass, perhaps modified by binding effects if the picture of dynamic confinement is correct. Thus we have measured an electric plasmon mass of $m \simeq 0.38Tg\left(T\right)$. This result is very different from the lowest order result $\left(R = 2.0\right)$.

We are aware of three different calculations in the literature of corrections to the lowest order formula. All are different. All are of the form

$$\frac{m_e^2}{T^2} = \frac{N}{3}g^2 + Cg^3N^{3/2} \tag{9}$$

where C is a constant. The first is due to D'Hoker:[10] $C_{DH} = 1/\left(4\pi\sqrt{3}\right)$. The second is due to Kajantie and Kapusta:[11] $C_{KK} = -1/\left(2\pi\sqrt{6}\right)$. The third is due to Furusawa and Kikkawa:[12] $C_{FK} = 1/\left(2\pi\sqrt{3}\right)$. The calculations of Kajantie and Kapusta, and Furusawa and Kikkawa, are self consistent calculations using truncated Schwinger-Dyson equations; the former in temporal axial, Coulomb, and covariant gauges, the latter in an axial gauge.

There is clearly a problem with the perturbation theory calculations! We have plotted the Kajantie-Kapusta prediction as dashed lines on Fig. 4. It is interesting that it is the only calculation which gives a lighter mass from a higher order correction, but its curvature is opposite that of the data. Its agreement with the data at greater T/Λ_L is probably fortuitious. It would appear that perturbation theory is inadequate to describe the plasmon mass.

GLUON PLASMA THERMODYNAMICS

The first computation of thermodynamic quantities for a gauge theory was done (for SU(2)) by Engels, Karsch, Montvay, and Satz.[13] Svetitsky and Fucito[14] have computed the latent heat in pure SU(3) gauge theory. More recently, Gocksch and Gavai[15] have calculated the energy density and speed of sound in pure SU(3) gauge theory on a lattice with the number of time steps N_t equal to 4. Following in this tradition, we have also calculated the energy density and pressure of pure gluonic matter and attempt to extend our results to the continuum limit. Our statistics are comparable with those of Ref. 15, while our data sample includes lattices of temporal size $N_t = 4$, 6, and 8.

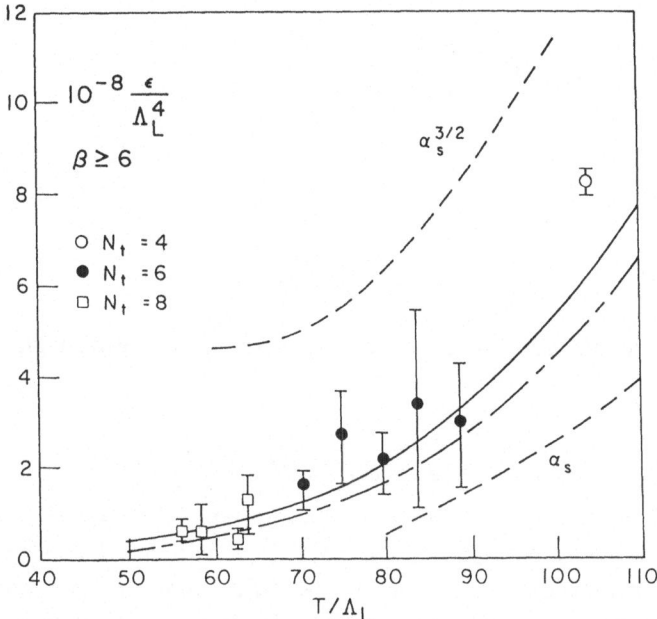

Fig. 5: The energy density of the gluon plasma. The solid line is the pure Stefan Boltzmann result for eight massless gluons. The dashed lines are perturbative predictions, and the long-and-short dashed line is from the effective model described in the text.

The energy density ϵ and pressure P are given by

$$\epsilon = \frac{3\beta}{a^4} \left[W_t - W_s + g^2 \left(C_s \left(W_s - W \right) + C_t \left(W_t - W \right) \right) \right] \qquad (10)$$

$$\epsilon - 3P = \Delta. \qquad (11)$$

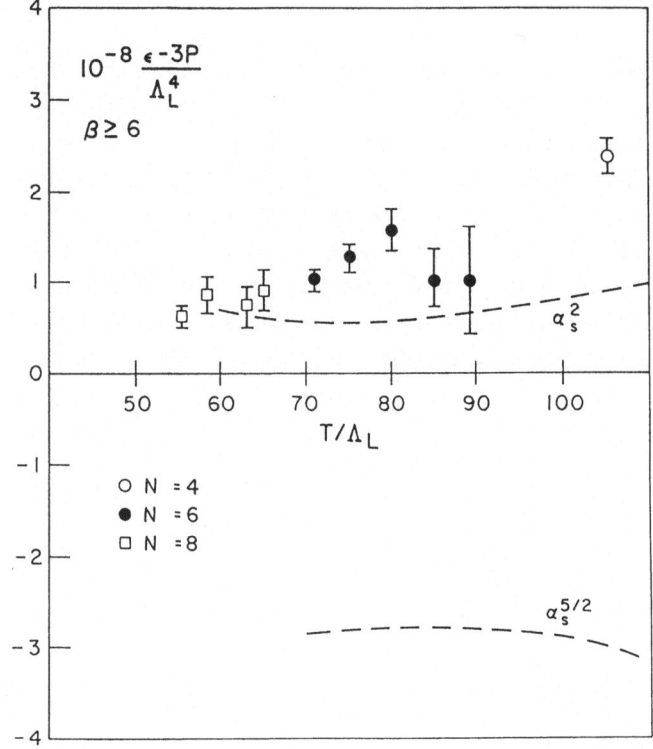

Fig. 6: The quantity $\Delta = \epsilon - 3P$ for the gluon plasma.

$$\Delta = -18a^{-3}\partial g^{-2} a \left(W_s + W_t - 2W \right). \tag{12}$$

The constants C_s and C_t were first calculated in perturbation theory by Karsch.[16] W_s and W_t are the expectation values of the space-space and space-time oriented plaquettes in the finite temperature model, while W is the plaquette on a symmetric $(T = 0)$ lattice. One must either calculate $a\partial g^{-2} a$ using perturbation theory or from Monte Carlo renormalization group methods. Opting for the first choice, we have

$$a^4 \Delta = 18 \left(\sqrt{\frac{11}{8\pi^2}} + g^2 \sqrt{\frac{51}{64\pi^4}} \right) \left(W_s + W_t - 2W \right) \tag{13}$$

The speed of sound is

$$V_s^2 = \sqrt{\frac{1}{3} \left(1 - \partial \Delta \epsilon \right)} \tag{14}$$

We also need the expectation values of plaquettes on symmetric (T=0) lattices. We did not generate those quantities as part of our simulation.

Instead, we chose to work with the high statistics plaquette expectation values of Barkai, Moriarty, and Rebbi[17] taken on a $16^3 \times 32$ lattice. These authors present data at $\beta = 5.6, 5.8, 6.0, 6.2$, and 6.4. We interpolated their data to our β values using a three point interpolation. (A similar procedure was done in Ref. 13; the authors of Ref. 15 could compare against their own $T = 0$ data.)

Much of the analysis of ϵ and Δ involves taking the continuum limit using the perturbative beta function. Unfortunately, perturbation theory is not believed to be applicable for $\beta \lesssim 6.0$; asymptotic scaling in agreement with the two-loop perturbative beta function for the deconfinement transition temperature is known to set in only at β greater than about 6.1. In an attempt to remove gross scaling violations, we plot in Figs. 5 and 6 all of our data for which $\beta \geq 6.0$. The data show a rapid increase in the energy density with temperature. For comparison, we superpose on the graph the expected energy density from a continuum gas of eight massless noninteracting vector bosons. Remarkably, the data all lie on the curve, confirming previous observations.[13,18]

We have attempted to verify this behavior by fitting the energy density to the form

$$\epsilon = A + B\frac{\pi^2}{15}T^4 \tag{15}$$

where B counts the effective number of massless vector degrees of freedom. Results of this fitting procedure are $B = 9.8 \pm 1.0$ for $N_t = 4$, while our $\beta \geq 6, N_t = 6, 8$ data give $B = 8.2 \pm 2.5$. .

The parameter Δ does not show any striking rise with temperature, unlike ϵ, suggesting that the speed of sound is approximately $\sqrt{\frac{1}{3}}$ for T/Λ_L greater than about 50.

The analysis of the pressure and energy density is much more model dependent than the analysis of other quantities on the lattice: There are delicate cancellations between the $T = 0$ and $T \neq 0$ plaquette values, and many quantities in the analysis have been calculated using perturbation theory, yet the data are mainly taken in a range of β's for which asymptotic scaling is known not to work.

We can compare our Monte Carlo results to the predictions of two models. The first model is perturbation theory. Kapusta[9] has computed the thermodynamic potential Ω to $O\left(g^3\right)$. He finds

$$\frac{\epsilon}{\Lambda_L^4} = \left(\frac{T}{\Lambda_L}\right)^4 \left(\frac{8\pi^2}{15} - 8\pi\alpha_s + 128\sqrt{\pi}\alpha_s^{3/2}\right) \qquad (16)$$

and

$$\frac{\Delta}{\Lambda_L^4} = \left(\frac{T}{\Lambda_L}\right)^4 \frac{22\alpha_s^2}{\pi} \left(\frac{8\pi}{3} - 64\sqrt{\pi\alpha_s}\right). \qquad (17)$$

The running coupling constant is $\alpha_s\left(T\right) = g^2\left(T\right)/16\pi$.

At low values of T/Λ_L perturbation theory is poorly behaved. For this reason, the much - celebrated finding that the leading order perturbative contribution, i.e. the Stefan-Boltzmann formula, agrees with the data can scarcely be taken as a confirmation of the idea that the plasma is a weakly interacting gas at these temperatures. We plot the $O\left(\alpha_s\right)$ and $O\left(\alpha_s^{3/2}\right)$ contributions to eqn. (16) on Fig. 5.

For all values of T/Λ_L appropriate to our simulation the $\alpha_s^{5/2}$ term in Δ is bigger than the α_s^2 term, and it drives Δ negative. This poor convergence property of the thermodynamic potential is well known. We compare our data with the $O\left(\alpha_s^2\right)$ and $O\left(\alpha_s^{5/2}\right)$ calculations in Fig. 6.

Our second phenomenological model is based on the picture of the plasma given in Ref. 1.

In the plasma there are at least three important length scales. First, in order of magnitude, there is the Debye screening scale

$$m_D \simeq gT \equiv \mu_D T, \qquad (18)$$

and second, the scale for magnetic confinement

$$m_M \simeq g^2 T \equiv \mu_M T. \qquad (19)$$

Third, there is an inverse mean free path m_{MF} which marks the upper momentum cutoff for the plasma to support collective hydrodynamic fluctuations. The inverse mean free path for gluons has been calculated in perturbation theory by Shuryak[19] to be

$$m_{MF} \simeq 30\alpha_s^2\left(T\right)T \equiv \mu_{MF}T. \qquad (20)$$

Since this distance is shorter than the magnetic confinement scale, the mean free path relevant to hydrodynamics should be calculated instead for the color singlet modes. The size of the color singlet modes is set by the confinement scale $1/m_M$ and, since they exist only for momenta smaller than m_M, their density, given by the Boltzmann distribution, is of order m_M^3. Thus the mean free path is of order

$$m_{MF} \simeq m_M. \tag{21}$$

For distances much shorter than $1/m_D$ or $1/m_M$ the plasma is economically described as a gas of noninteracting gluons. For distances longer than or on the order of $1/m_M$ or $1/m_D$, confining effects are important. For distances much longer than $1/m_{MF}$, hydrodynamic modes are important. Thus we suggest a crude three component model of the plasma as a gas of free high momentum gluons, low momentum noninteracting color singlet modes, and low momentum hydrodynamic phonons.[20] This model is somewhat reminiscent of the Landau theory of liquid ^{4}He.[21] Clearly there is some risk of multiple counting of degrees of freedom, since the color singlet excitations and the phonons are collective excitations of gluons. However, there is some consolation in the observation that the effect of excluding low momentum color octet gluons and replacing them by a few color singlet modes is to reduce the number of degrees of freedom compared with a pure Stefan-Boltzmann gas of gluons.

We calculate the energy density by breaking momentum space into a sequence of regions and keeping only the most important contribution in each region. For $k > m_D$ those degrees of freedom are eight noninteracting massless gluons. For $k < m_M$ they are the magnetically confined color singlet modes. For $k < m_{MF}$ the phonon also contributes to the energy density. Because all the scales in eqn. (18-21) vary linearly with the temperature (up to logarithmic corrections) each of the regions of momentum space gives a contribution to the energy density which is proportional to T^4.

We infer that μ_D (and probably μ_M) range between 1 and 3 for the temperature range of interest. We also know that m_{MF} is no bigger than $\mu_D T$. Taking $V_s = \sqrt{1/3}$, we can numerically evaluate the contributions to the internal energy. As the parameter μ rises (as the temperature falls) the phonons take a larger share of the energy density. The massive confined modes never contribute more than ten per cent of a single massless boson mode. Only for very large μ does the energy deviate appreciably from the free gluon result.

Including the variation of μ_D with T shown in Fig. 4, taking μ_D, μ_M, and μ_{MF} all equal, and counting one magnetic and one electric mode in ϵ, we get the long - and - short dashed curve shown in Fig. 5. It is entirely consistent with our $N_t = 6$ and 8 data but considerably undershoots our

$N_t = 4$ points. This simple model is much too crude to apply to the quantity Δ.

CONCLUSIONS

The gluon plasma at a temperature T slightly above T_c is much more than a free gas of gluons. Its static correlation functions are dominated by massive modes, with masses $m \simeq 2T$. Whether these modes are real color singlet bound states is not known. Continuum perturbation theory does not explain the plasmon mass which we see. It also does not explain our measurements of the energy density or the pressure of the plasma. The range of temperatures which we probe are likely to be the ones reached in ultrarelativistic nucleus-nucleus collisions; our measurements suggest that the plasma is likely to be a more complicated and more interesting medium than was originally anticipated.

ACKNOWLEDGMENTS

This work was supported by the U. S. Department of Energy (T. D.) and by the National Science Foundation (C. D.). T. D. would like to thank Keijo Kajantie and Ben Svetitsky for discussions.

REFERENCES

1. C. DeTar, Phys. Rev. D32 (1985) 276. For another viewpoint, see the contribution of J. Polonyi to these proceedings.
2. T. DeGrand and C. DeTar, Phys. Rev. D34 (1986) 2469.
3. T. DeGrand and C. DeTar, University of Colorado preprint COLO-HEP-131 (1986), Phys. Rev. D, in press.
4. Cf. Ph. de Forcrand, G. Schierholz, H. Schneider and M. Teper , Phys. Lett. 152B (1985) 107.
5. T. DeGrand, work in preparation.
6. B. Berg and A. Billoire, Phys. Lett. 166B (1986) 203; B. Berg, A. Billoire and C. Vohwinkel, Phys. Rev. Lett. 57 (1986) 400.
7. S. Gottlieb, J. Kuti, D. Toussaint, S. Meyer, B. Pendleton, R.Sugar, Phys. Rev. Lett. 55 (1985) 1958.
8. N. Christ and A. Terrano, Phys. Rev. Lett. 56 (1986) 111.
9. J. Kapusta, Nucl. Phys. B148 (1979) 461.
10. E. D'Hoker, in Proceedings of theWorkshop on Quark Matter Formation and Heavy Ion Collisions, Bielefeld, 1982.
11. K. Kajantie and J. Kapusta, Ann. Phys. (N. Y.) 160 (1985) 477.
12. T. Furusawa and K. Kikkawa, Phys. Lett. 128B (1983) 218.

13. J. Engels, F. Karsch, I. Montvay, H. Satz, Nucl. Phys. B205 (1982)545, Phys. Lett. 101B (1981) 89.
14. B. Svetitsky and F. Fucito, Phys. Lett. 131B (1983) 165.
15. A. Gocksch and R. Gavai, Phys. Rev. D33 (1986) 614.
16. F. Karsch, Nucl. Phys. B205 (1982) 285.
17. D. Barkai, K. Moriarty and C. Rebbi, Phys. Rev. D33 (1984) 2201, Phys. Lett. 115B (1982) 151.
18. I. Montvay and E. Pietarinen, Phys. Lett. 110B (1982) 148.
19. E. Shuryak, Phys. Rep. 61 (1981) 71.
20. P. Carruthers, Phys. Rev. Lett. 50 (1983) 1179.
21. See W. Yourgraw, A. van der Merwe, and G. Raw, "A Treatise on Irreversible and Statistical Thermodynamics", (McMillan, New York, 1966).

RENORMALIZATION AND CONTINUUM LIMIT

OF COMPOSITE OPERATORS IN LATTICE GAUGE THEORIES

Adriano Di Giacomo

Dip. di Fisica Univ. di Pisa Italy
I.N.F.N. Sezione di Pisa, Italy

INTRODUCTION

I will report on some problems of renormalization, which are related to the determination of the gluon condensate of dimension 4

$$G_2 = \langle 0| \frac{\alpha_s}{\pi} : G_{\mu\nu} G_{\mu\nu} : |0\rangle \qquad (1.1)$$

from Montecarlo simulations of lattice gauge theories[1].

I will first discuss the motivations of this work. Then I will review the situation of asymptotic scaling in the determination of G_2. Finally I will cover the problems of renormalization which arise when G_2 is extracted from different lattice operators.

MOTIVATIONS AND OUTLOOK

The existence of gluon condensates in the vacuum of non abelian gauge theories is a fundamental question in field theory.

Besides the phenomenological determination by SVZ sum rules, where the existence of condensates was first proposed[2], it is important to known from first principles if the condensates do exist in the ground state of the theory.

The answer from lattice is definitely affirmative, and many determinations exist in the literature, which are consistent with each other. These determinations are made for pure SU(2) or SU(3) theories with no fermions: they just answer the question of the existence of condensates. A realistic determination with the full QCD would be necessary to compare with the sum rule determination.

Problems have been raised on the very possibility of defining G_2[3][4]. In fact the definition (1.1) requires the subtraction of the perturbative value in order to isolate the non perturbative part.

It is well known that the sum of the perturbative series does not converge, even by Borel resummation, for renormalizable theories[4].

This is a desease for the definition of G_2 in the same sense that the perturbative expansion does not exist in Q.E.D. However people do compute e.g. (g-2) by perturbation theory, and the accepted philosophy is that perturbation theory is an asymptotic expansion of something which we are not able to reconstruct and to define properly.

The technique to extract G_2 from lattice simulations of the theory consists of the following steps[5)7]:
1) Determine the vacuum expectation value of any gauge invariant operator which, in the naive continuum limit a->0 is proportional to $\alpha_s G_{\mu\nu} G_{\mu\nu}$ + operators of higher dimension. E.g.

$$1 - P_{1,1} \underset{a\to 0}{\simeq} \frac{\pi^2}{24} \left(\frac{\alpha_s}{\pi} G_{\mu\nu} G_{\mu\nu} \right) a^4 + O(a^6) \qquad (2.1)$$

(By $P_{n,m}$ we denote the Wilson loop n x m)
2) Subtract the perturbative expansion

$$\langle : 1 - P_{1,1} : \rangle \equiv \langle 1 - P_{1,1} \rangle - \sum_n \frac{c_n}{\beta^n} \qquad (2.2)$$

$\beta = 2N/g_o^2$ for a group SU(N). g_o is the unrenormalized coupling constant.

At sufficiently large β the perturbative expansion, as an asymptotic expansion, will be approximated within the errors by a few terms.
3) Then

$$\langle : 1 - P_{1,1} : \rangle \underset{\beta\to\infty}{\simeq} \frac{\pi^2}{24} G_2 a^4 + O(a^6) \qquad (2.3)$$

If there is asymptotic scaling

$$a \simeq \frac{1}{\Lambda} \exp(-\beta/4Nb_o) (\beta/2Nb_o)^{b_1/2b_o^2} \qquad (2.4)$$

$$b_o = \frac{11}{3} \frac{N}{16\pi^2} \qquad b_1 = \frac{34}{3} \left(\frac{N}{16\pi^2} \right)^2 \qquad (2.5)$$

and

$$\langle : 1 - P_{1,1} : \rangle \simeq \frac{\pi^2}{24} \frac{G_2}{\Lambda^4} \exp(-\beta/Nb_o) (\beta/2Nb_o)^{2b_1/b_o^2} \qquad (2.6)$$

If the dependence on β eq.(2.6) is verified by the data, we say that there is asymptotic scaling.

Asymptotic scaling may be violated in a range of β's
(i) by the fact that the β function of the theory is not yet well approximated by its asymptotic behaviour, and therefore eq.(2.4) is not yet true.

(ii) because operators of higher dimension like those in eq.(2.1) are not yet negligible.

Violations of the kind (i) are the same for all quantities. Violations of the kind (ii) depend on the specific operator considered.

Operators for which violations of type (ii) disappear at lower values of β have a larger "continuum window" that others.

There are indeed quantities which are better than others to look at asymptotic scaling[8)9)]. G_2 is among them.

In order to get better evidence for scaling, we have tried to extract G_2 from different operators, namely

$$\left.\begin{array}{l} O_1 = 1 - P_{1,1} \\ O_2 = \frac{1}{4}(1 - P_{1,2}) \\ O_3 = \frac{1}{16}(1 - P_{2,2}) \\ O_4 = \frac{1}{4}(1 - P_{1,1}^2) \end{array}\right\} \tag{2.7}$$

$P^2_{1,1}$ is the 1x1 plaquette covered twice.

All the operators O_i in eq.(2.7) behave in the same way eq.(2.3) as the lattice spacing a->0. The numerical factors in the definitions (2.7) correspond the (area)^2 law

$$1 - P_{n,m} \underset{a \to 0}{\simeq} \frac{\pi^2}{24} \left(\frac{\alpha_S}{\pi} G_{\mu\nu} G_{\mu\nu}\right) a^4 (n \cdot m)^2 \tag{2.8}$$

and give all the O'_is the same naive normalization. However renormalization effects are expected, giving instead

$$\langle : O_i : \rangle \underset{\beta \to \infty}{\simeq} \frac{\pi^2}{24} G_2 a^4 \cdot Z_i \tag{2.9}$$

$$Z_i = 1 + \frac{\delta_i}{\beta} + \cdots \tag{2.10}$$

Z_i is a renormalization factor.

We have verifyed[1)] asymptotic scaling for all the operators $\langle : O_i : \rangle$ in the range $2.4 \leqslant \beta \leqslant 2.85$. We have then computed to one loop the factors Z_2 and Z_4, finding qualitative agreement with the observations.

MEASUREMENTS

Our lattice is 16^4. The group is full SU(2): we need measurements at large β to determine the perturbative subtraction. Moreover at our level of precision even at low β's there are systematic discrepancies between the full group and the Y(120) subgroup. The program was heat bath, with 6 trials per link. At each value of β 300 initial configurations were discarded to guarantee thermalization: the next 200 were used for measurement. A careful analysis was made of the correlations[9] to get a gaussian distribution and reliable statistical errors.

The perturbative expansion in eq.(2.2) was truncated at n=5. The coefficients C_1 C_2 are known by computation[5][6]: three more coefficients C_3 C_4 C_5, and the quantities $Z_i G_2/\Lambda^4$ were determined by best fit. Data refer to 38 values of β for each operator, 34 of which at $\beta \gtrsim 2.45$. The χ^2/N of the fit is of the order 1 for all the operators. Scaling is verified in the range $2.45 \lesssim \beta \lesssim 2.85$.
Fig.1 shows the result for O_1.
Fig.2 gives an idea of the quality of the fit.
Table I compares our result with previous determination.
We find

$$\frac{Z_2}{Z_1} = .41 \pm .05 \; ; \; \frac{Z_3}{Z_1} = .15 \pm .05 \; ; \; \frac{Z_4}{Z_1} = .35 \pm .06 \qquad (3.1)$$

$$\frac{Z_1 G_2}{\Lambda^4} = (.14 \pm .02) \cdot 10^8 \qquad (3.2)$$

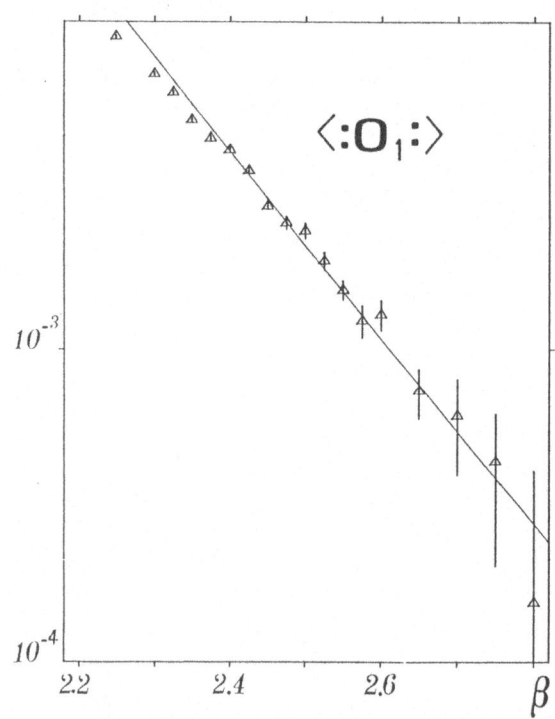

Fig.1. $\langle :O_i: \rangle$ versus β . The line in the prediction of asymptotic scaling. Only the overall normalization is determined by best fit.

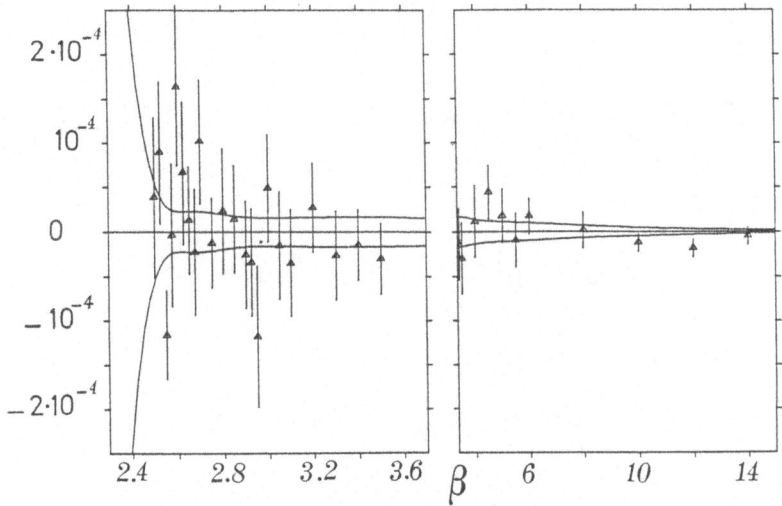

Fig.2 Deviations of the measured values $\langle O_1 \rangle$ from the optimized
formula (2.2). The error bars are experimental. The lines
are the borders of the one standard deviation of the fit.

COMPUTING THE Z_i's IN PERTURBATION THEORY

Let K be any gauge invariant operator, which goes to a constant
in the limit a->0. Then, for any of our operators O_i, we define

$$F_i = \langle O_i K \rangle \tag{4.1}$$

At large β we expect

Table I. Existing determinations of G_2. Ref.5) and 7) truncate the
perturbative expansion (2.2) at the third term. Ref. 13)
uses a completely different method.

Ref	Gauge Group	c_3	c_4	c_5	$G_2/\Lambda^4 \times 10^{-8}$	Lattice Size
5)	SU(2)	.27	0	0	$.21 \pm .03$	10^4
7)	SU(2)	.29	0	0	fraction of 1	4^4
10)	Y(120)	.16	-.35	1.85	$.14 \pm .01$	4^4
11)	Y(120)	.19	-.40	1.90	$.23 \pm .01$	8^4
12)	Y(120)	.26	-.84	2.63	$.21 \pm .01$	8^4
13)	SU(2)	---	----	----	$.09 \pm .04$	8^4
8)	SU(2)	.22	-.52	2.2	$.13 \pm .03$	8^4
8)	SU(2)	.21	-.42	1.85	$.14 \pm .02$	12^4

$$\frac{F_i}{F_j} = \frac{Z_i}{Z_j} \qquad (4.2)$$

We can write

$$F_i = \frac{1}{Z_W} \frac{d}{d\varepsilon} \iint [dU] \exp(-S_W(U) + \varepsilon O_i) K \Big|_{\varepsilon=0} \qquad (4.3)$$

$S_W(U)$ is Wilson action and

$$Z_W = \iint [dU] \exp(-S_W(U)) \qquad (4.4)$$

For all our operators O_i

$$O_i = S_W + \tilde{R}_i \qquad (4.5)$$

where $\tilde{R}_i$ is the sum of the operators of dimension higher than 4.
Then eq. (4.3) can be written

$$F_i = \frac{1}{Z_W} \frac{d}{d\varepsilon} \iint [dU] \exp(-S_\varepsilon(U)) \cdot K \Big|_{\varepsilon=0} \qquad (4.6)$$

$$S_\varepsilon(U) = (1-\varepsilon) S_W(U) - \varepsilon \tilde{R}_i \qquad (4.7)$$

In particular, for the operator O_1 $R_1=0$.
For the operator O_4 R_4 is proportional to the adjoint action.
For the operator K we have chosen the Wilson loop with 2 lines going through the lattice at constant distance. The quantity F_2 is then known from ref. (15) where it was computed for completely different purposes. The computation of F_4 is instead prohibitively difficult.
We find thus

$$\frac{Z_2}{Z_1} \simeq .6 \qquad\qquad \frac{Z_4}{Z_1} = .5 \qquad (4.8)$$

to be compared with the observed values (3.1).

CONCLUDING REMARKS

Asymptotic scaling is remarkably good for our operator. A possible explanation is that small Wilson loops are not sensitive to lattice size, i.e. to the long wavelengths which are cut off by the size of the lattice. This is evident from table I, where G_2 appears to be independent, within the errors, of the size of the lattice.

The renormalization factors Z_i agree reasonably with the computed ones already at the one loop level, even if they are sizeably different from unity.

Non trivial renormalization factors, which have to be determined by computation are a general difficulty of interpretation of lattice measurements[16].

REFERENCES

1) M.Campostrini, G.Curci, A.Di Giacomo and G.Paffuti, Zeits.fur Physik C, Particles and Fields, in Press.

2) M.A.Shifman, A.I.Vainshtein and V.I.Zakharov, Nuc.Phys. $\underline{B147}$, 385, 448, 519 (1979).

3) S.David, Nucl.Phys. $\underline{B209}$, 433 (1982).
S.David, Nucl.Phys. $\underline{B234}$, 237 (1984).

4) G.Parisi, Cargése Lectures (1977).
G.t'Hooft, Erice Lectures (1977).
B.Lautrup, Phys.Lett. $\underline{69B}$ 109 (1977).

5) A.Di Giacomo and G.Rossi, Phys.Lett. $\underline{100B}$, 481 (1981).

6) G.Curci, G.Paffuti and R.Tripiccione, Nucl.Phys. $\underline{B240}$, FS12 91 (1984).

7) T.Banks, R.Hosley, H.R.Rubinstein and V.Wolf, Nucl.Phys. $\underline{B110}$ 692 (1981).

8) A.Di Giacomo, Proceedings of International Conference on "Non-Perturbative Methods", Montpellier 1985, World Scientific.

9) B.Berg, S.Billoire, S.Mayer and C.Panagiatakapoulos, Com.Math.Phys. $\underline{97}$, 31 (1985).

10) A.Di Giacomo and G.Paffuti, Phys.Lett. $\underline{108B}$, 327 (1982).

11) M.Campostrini, A.Di Giacomo and G.Paffuti, Zeits.fur Physik $\underline{C22}$ 143 (1984).

12) K.Ishikava, G.Schierholtz, H.Schneider and M.Teper, Nucl.Phys. $\underline{B227}$, 221 (1983).

13) R.Kirschner and J.Kripfganz, Nuc.Phys. $\underline{B210}$ FS6 567 (1982).

14) K.Binder, Phase Transition and Critical Phenomena, C.Domb and M.S.Green ed., Vol.5B Academic Press N.Y. (1976).

15) G.Gurci. P.Menotti and G.Paffuti, Phys.Lett. $\underline{130B}$ (1983) 205.

16) see e.g. M.Bochicchio, L.Maiani, G.Rossi and M.Testa, Nucl.Phys. $\underline{B262}$, 331 (1985).

LANGEVIN SIMULATIONS WITH DYNAMICAL QUARK LOOPS

M. Fukugita

Research Institute for Fundamental Physics
Kyoto University
Kyoto, 606 Japan

ABSTRACT

Langevin simulation is presented for quantum chromodynamics with dynamical quark loops fully taken into account. A particular stress is made on the procedure to remove systematic errors of the simulation. Results are also given for spectroscopy of hadrons and finite temperature behaviour of quantum chromodynamics with the Wilson fermion.

INTRODUCTION

One of the final goal of lattice gauge theory is to extract physical quantities from the simulation of quantum chromodynamics (QCD) with the quark vacuum polarisation effect fully taken into account. This necessitates the evaluation of quark determinant arising from the Gaussian integral over the quark field, which is the most tedious procedure. Due to this technical reason hadron spectrum calculations made so far have ignored this dynamical quark loop effect without a proper justification.

Recent developments in fast computer technology and a significant progress made in the last few years in simulation techniques, however, have enabled us to incorporate this quark vacuum polarisation effect and to examine the validity of the quenched approximation. In this report I describe an effort towards this direction using the Langevin approach,[1,2] and present some results obtained with the use of the vector processor HITAC S-810/10(128MB) at National Laboratory for High Energy Physics (KEK) in Tsukuba.[3-6] We have studied spectroscopy of hadrons and

99

finite temperature behaviour of QCD both with the Wilson fermion and the Kogut-Susskind (staggered) fermion, but I concentrate on the simulation with the Wilson fermion in this report.

A number of methods have been proposed to incorporate the dynamical quark loops. The methods so far utilised for studying QCD on a reasonably large lattice are; (i) pseudofermion Metropolis method[7], (ii) microcanonical or molecular dynamics method[8], (iii) Langevin (stochastic quantisation) method[1,2] and (iv) a hybrid method of (ii) and (iii)[9]. All the methods in its practical application involve the source of systematic errors characteristic to them, and it is most important to know the nature of such systematic errors and to control them. In the pseudofermion Monte Carlo method, the ratio of the quark determinant is approximated by the leading term in the variation of the gauge variables, and the acceptance of the hitting and a possible violation of detailed balance may also be a source of systematic bias, especially for the system close to critical. The validity of the microcanonical method hinges on the assumption of the ergodicity: The ergodicity assumption should fail in the limit of weak coupling for the gauge system or of large quark mass for the fermion system,[10] because the system (or part of it) becomes integrable in these limits and hence by the Kolmogorov-Arnold-Moser theorem[11] invariant tori exist on the energy surface.

In the Langevin approach the systematic error arises from the discretisation of the stochastic evolution equation. The advantage with the Langevin approach, however, is that the property of such errors is relatively well-understood and they can be under control. In a hybrid approach proposed in ref.9 the stochastic noise is applied to the microcanonical system to ensure the ergodicity. In a practical application, however, the total Langevin time $\tau_L = (\Delta\tau_{mc})^2 N$ ($\Delta\tau_{mc}$=microcanonical time step, N=number of times of applied noise) is considerably smaller than unity, and the realistic case is rather far from the case for asymptotic analysis[12] where the correct limiting distribution is ensured.

FORMALISM

Langevin equation is written with the effective action

$$S_{eff}(U,Y) = S_{gauge}(U) + \Sigma Y^+ \frac{1}{D(U)} Y \qquad (1)$$

with $S_{gauge}(U)$ the pure gauge action, $D(U)$ the Dirac operator on the lattice and Y the pseudofermion variable. The Gaussian integral over the pseudofermion variable gives the correct determinant of the Dirac operator.

We formulate the Langevin equation[13] for the gauge field on the group manifold so that the unitarity constraint is automatically satisfied. The Langevin equation with a discrete time step $\Delta\tau$ is then

$$U^{(n+1)} = U^{(n)} e^{i\Delta\tau X^{(n)}} ,$$

$$X^{(n)} = -i \frac{\delta}{\delta U^{(n)}} S_{eff} + \xi^{(n)} , \qquad (2)$$

with $U^{(n)} = U(n \cdot \Delta\tau)$ and $\xi^{(n)}$ the gauge white noise of the width $(\Delta\tau)^{-1/2}$. The Langevin equation for the pseudofermionic variable may be written as

$$Y^{(n+1)} = [1-\Delta\tau B(U^{(n)})]Y^{(n)} + \Delta\tau C(U^{(n)})\eta^{(n)} , \qquad (3)$$

with B and C some functions of the gauge variable satisfying $BD+DB^\dagger = 2CC^\dagger$. Typical choices of B and C are:

(A) $\qquad B = D^{-1} , \qquad C = 1 \qquad$ (pseudofermionic)

which corresponds to the naive time discretisation[1,2] of the Langevin equation $Y=-D^{-1}Y+\eta$, and

(B) $\qquad B = \Delta\tau^{-1} , \qquad C = D^{1/2} \qquad$ (bilinear noise)

proposed in ref.2. In the latter case Y does not evolve and is written directly in terms of the white noise. As a result the Langevin equations reduce to a single equation for the gauge variable with a bilinear noise term. We call hereafter the scheme (A) and (B) pseudofermionic and bilinear noise schemes, respectively.

SYSTEMATIC ERRORS

Systematic errors of the Langevin simulation arise from discretisation of the fictitious time $d\tau\rightarrow\Delta\tau\neq0$. Finite time step evolution equation, however, is still exact as a stochastic equation, but it leads to the limiting distribution of the form

$$\rho_\infty(U,Y) = \exp[-S_{eff}+\Delta\tau S_1+(\Delta\tau)^2 S_2+\dots] , \qquad (4)$$

that differs from the desired form $\exp(-S_{eff})$ by the term $O(\Delta\tau)$.

For the pure gauge system one can develop a second-order formalism[1] in which the term proportional to $\Delta\tau$ is cancelled out. In the presence of fermion one can still remove the term $O(\Delta\tau)$ for the pseudofermionic scheme using the Runge-Kutta formalism. Using the Fokker-Planck equation one can show that the residual error is of the order of $(\Delta\tau/\lambda^2)^2$ with λ the eigenvalue of the Dirac operator $\gamma_5 D$. On the other hand, for the bilinear noise scheme the removal of $\Delta\tau/\lambda^2$ term[14] requires a widening of the effective width of the gauge white noise by a term $(\Delta\tau/\lambda^2)^2 \cdot V$ (V=volume), and the second order formalism practically does not work, unless one takes an unfeasibly small $\Delta\tau$ when V is reasonably large.[4] The residual error of the first-order formalism is $\Delta\tau/\lambda^2$.

When we treat the system with a small quark mass, λ takes a small value and for the minimum eigenvalue $\Delta\tau/\lambda^2$ may become of the order unity for the parameter range that concerns us. In such a case the second-order formalism may not be quite advantageous, and one has to devise other means to remove the error.

Let us now ask the characteristics of the systematic error. When the finite time Langevin equation is used, the long-distance modes that satisfy $\Delta\tau/\lambda^2 \gtrsim 1$ cannot be fully fledged, and the infrared modes are seriously distorted. On the other hand the short-distance mode is relatively insensitive to this cutoff and may be modified rather modestly by the fraction $\Delta\tau/<\lambda^2>$. We therefore expect a larger modification for large distance physical quantities. For the gauge sector the finite time step suppresses ultraviolet modes. This seems to be practically harmless, because the ultraviolet mode shorter than the lattice spacing is an irrelevant quantity on the lattice. Indeed, we did not find any evidence for distorsion at a short distance for pure gauge system when the second-order formalism is adopted.[1]

It is not a priori clear which fermionic scheme results in a smaller error for a given $\Delta\tau$. After several trial runs we found that the bilinear noise scheme leads to smaller errors for a light quark mass, and the error can be better tolerated. We therefore used the bilinear noise scheme in the first-order formalism for most of our simulation with partial use of the second-order formalism for the gauge sector to minimise the errors. Our Langevin equation is then[4,5]

$$X_0^a = -i\Delta\tau(\partial^a S_{gauge} - \partial^a \eta^\dagger \ell n D \cdot \eta) + (\Delta\tau)^{1/2}\xi^a$$
$$U^{(n)'} = U^{(n)} e^{i\alpha X_0}$$

$$X_1^a = -i\Delta\tau(\partial^a S'_{gauge} - \partial^a \eta'^\dagger \ell n D \cdot \eta') + (\Delta\tau)^{1/2}\xi^a$$
$$U^{(n+1)} = U^{(n)} e^{i(\beta X_0 + \gamma X_1)}$$

(5)

with $\beta = 1/2 + \Delta\tau\beta_1$, $\gamma = 1/2 + \Delta\tau\gamma_1$ $(\beta_1+\gamma_1 = C_2/12)$ and $<\eta\eta^\dagger> = <\xi\xi^\dagger> = 2$.

We expect that physical quantities behave as,

$$F = F_0 + \Delta\tau F_1$$

(6)

for a sufficiently small $\Delta\tau$. We can then separate F_0 by making the simulation with several value of $\Delta\tau$.

102

ILUCR METHOD

An important ingredient for the success of the simulation is to find an efficient algorithm for solving the linear equation $D\underline{x}=\underline{\eta}$, on which the bulk of computer time is spent (We have to solve this equation twice per gauge sweep). The standard procedure for such a problem may be to use the conjugate gradient algorithm.[15] This algorithm, however, is not fast enough for our purpose. We have overcome this problem for the case with the Wilson fermion by applying a preconditioning to the equation.[16]

Let us suppose the equation to be solved of the form

$$A\underline{x} = \underline{b} \ . \tag{7}$$

We make an incomplete LU decomposition as $A=LU+R$ with L and U a lower and an upper triangle matrices and R is the residual matrix with small entries. For our case it is easy to make such a decomposition by taking $L_{ij}=A_{ij}$ $(i>j)$, $L_{ij}=0$ $(i<j)$, $L_{ii}=1$ and $U_{ij}=A_{ij}$ $(i<j)$, $U_{ij}=0$ $(i>j)$, $U_{ii}=1$, so that the residual error matrix is of the order of K^2 (K=hopping parameter). We then solve

$$(LU)^{-1}A\underline{x} = (LU)^{-1}\underline{b} \tag{8}$$

instead of (7). The operator $(LU)^{-1}A$ is close to diagonal and a rapid convergence is ensured. As we demonstrate below we achieved with this preconditioning an improvement by a factor of 20 in computer time over the standard method.

For a solver we adopted a variant of the conjugate gradient method — conjugate residual algorithm which minimise $\|A\underline{x}-\underline{b}\|$ over the affine space $x_i + <p_i,\ldots,p_{i+k}>$ with p_i chosen to be orthogonal with respect to $A^{\dagger}A$. We used this method because of its simplicity in the algorithm, which makes it possible to save computer time by a factor of 2. An extra 30-40% improvement was also achieved in the convergence speed with the sccessive over-relaxation (SOR) type acceleration; in our case by replacing K in the LU decomposition with cK (c=1.2-1.3).

In fig.1 we compare the convergence of various method for our problem.[16] The advantage of the preconditioning is apparent. The convergence of the standard conjugate gradient (or conjugate residual) algorithm is slow in the beginning and becomes faster at several hundred iterations when the correct direction vector is found by the solver. Fig.2a(residual norm) and 2b(propagator) show a realistic example in our gauge sweeps. The convergence is quite smooth, and any desired accuracy can be achieved by continuing the iteration till an appropriate residual norm is obtained. Practically we made the iteration until $\|\underline{\eta}-D\underline{x}\| \leq 1$, which

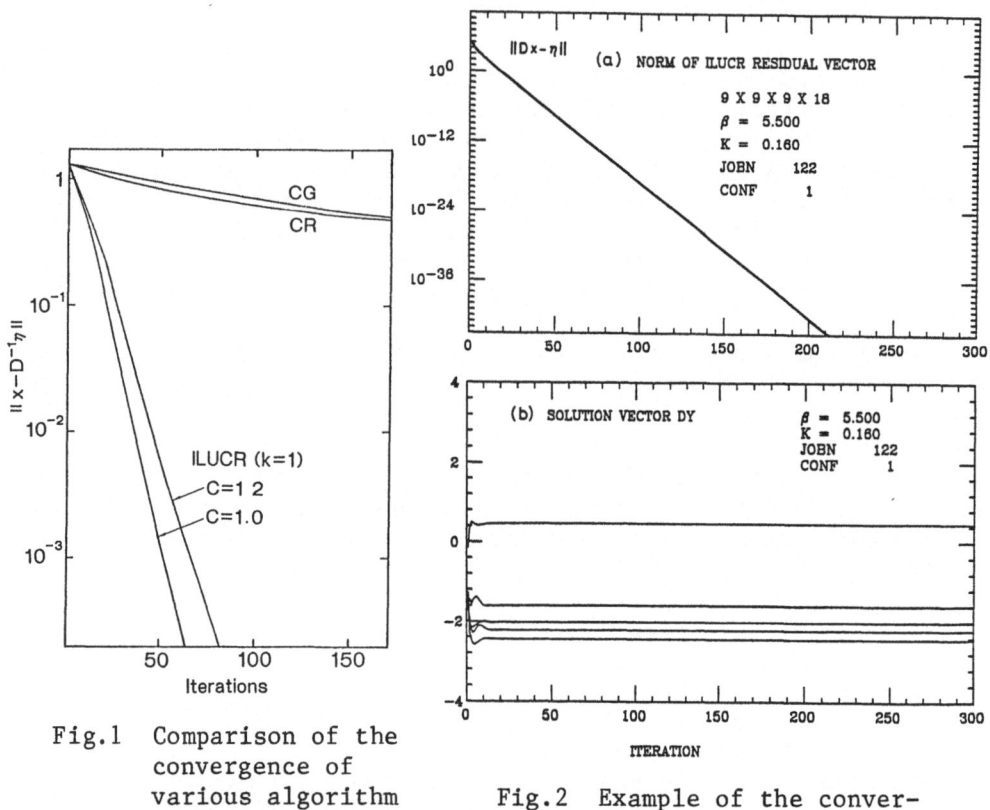

Fig.1 Comparison of the convergence of various algorithm as a function of iterations.

Fig.2 Example of the convergence for solution of Dx=η in the gauge sweep.

corresponds to 1% accuracy in each element of the vector x̲.

SIMULATION FOR SPECTROSCOPY[4,5]

Let us now describe our analysis of hadron spectroscopy with the Wilson fermion. The gauge group is SU(3) and the number of flavours N_F=2 with the same hopping parameter K. We use a $9^3 \times 18$ lattice with the periodic boundary condition. This lattice size is close to the maximum within our available central memory, if whole computation is made with the double precision accuracy. We also generated gauge configurations with the antiperiodic boundary condition imposed on the quark field in order to examine the absence of the effect from the fake loop contribution in the spatial direction.

We took a relatively small value of the gauge coupling $\beta=6/g^2$=5.5. Since the quark loop renders the gauge configuration more ordered and the lattice size effectively shrinks towards light quark masses, the value of β should be such that the finite size effect for the gauge configuration due to the fake loop

104

contribution is insignificant even then. This problem is more serious with the Wilson action than with the Kogut-Susskind action as discussed below.

We generated gauge configurations for five values of the hopping parameters K=0.14,0.15,0.155,0.16 and 0.162 (The real K_c is 0.1613. As we will see below a finite $\Delta\tau$ shifts K effectively towards a smaller value and the system with K=0.162 and with $\Delta\tau$=0.01 is still below the critical value). At the largest K, $m_\pi a \simeq 0.4$ and $m_N/m_\pi \simeq 1.7$. For each value of K, 5000 sweeps are made with $\Delta\tau$=0.01, and then 1500 and 6000 sweeps are made with $\Delta\tau$=0.02 and 0.005, respectively, starting from the last configuration of the $\Delta\tau$=0.01 runs. Fermionic quantities are calculated by solving $D \cdot \Delta = 1$ with the fixed source at every $\delta\tau = 1$. The thermalisation is checked by inspecting both Wilson loops and hadron propagators. We found that 1500 iterations are sufficient for equilibration when started with an arbitrary initial configuration, and discarded the initial 2000 iterations. For successive runs with $\Delta\tau$=0.02 and 0.005 the first one third of sweeps are discarded. We have analysed then 30-20 configurations per parameter set.

We have used the standard local operator to evaluate hadron propagators. In the presence of vacuum quark loops, hadrons other than the lowest are in general not stable, and the hadron propagator should behave as

$$G = A_i \ e^{-m_i t} + \int_{m_0}^{\infty} dm \ \sigma(m) \ e^{-mt} \ . \tag{9}$$

In the parameter region that we explored, however, the ground state hadrons are stable in each channel ($m_\rho < 2m_\pi$, $m_\Delta < m_N + m_\pi$ etc.), and the propagators may be well approximated by a pole contribution alone. In reality the quality of the fit for $t \gtrsim 5$ with the form $\exp(-m_i t)$ is generally excellent.

RESULTS FOR SPECTROSCOPY

<u>$\Delta\tau$-dependence</u>   Fig.3 gives the inverse of the minimum eigenvalue of the Dirac operator $\gamma_5 D$. We observe that $\Delta\tau/\lambda_{min}^2 \gtrsim 1$ already for K=0.155. In such a case the finite $\Delta\tau$ leads to the distorsion of the gauge configuration in which infrared modes are suppressed. This effect is apparent in fig.4, in which we show $1/\lambda_{min}$ as a function of $\Delta\tau$. For K=0.15, $1/\lambda_{min}$ does not vary with the choice of $\Delta\tau \lesssim 0.02$, but for K=0.16 $1/\lambda_{min}$ increases significantly for a smaller value of $\Delta\tau$. Infrared modes, which have small eigenvalues, are gradually fledged for smaller $\Delta\tau$. We expect that long-distance quantities are more significantly modified. This is best exemplified in the Wilson loop W(L×L) shown in fig.5. While the modification with $\Delta\tau$=0.01 is less than 2% for W(1×1) at K=0.16, it reaches almost a factor of 2 for W(4×4). Fortunately the dependence of the Wilson loop with respect to $\Delta\tau$ is almost

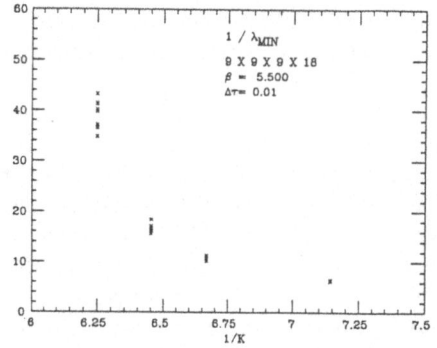

Fig.3 Inverse of the minimum eigenvalue of the Dirac operator as a function of 1/K.

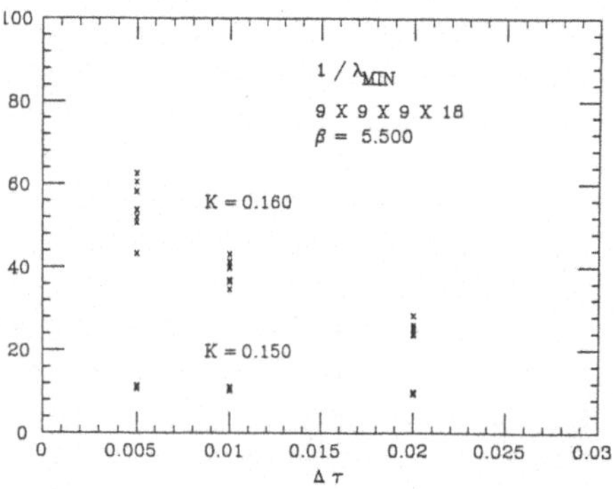

Fig.4 Inverse of the minimum eigenvalue of the Dirac operator as a function of $\Delta\tau$.

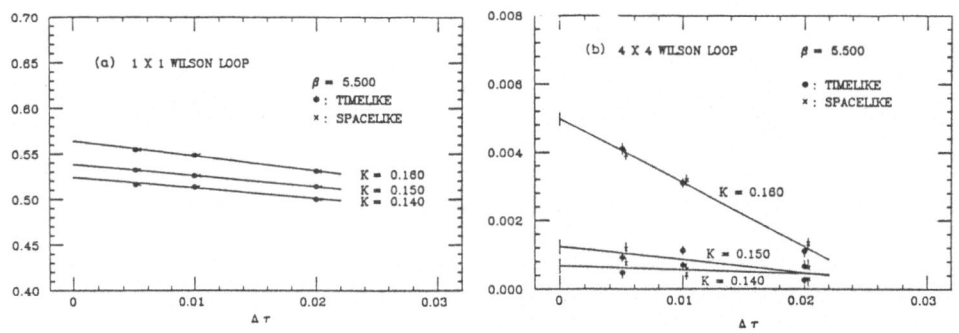

Fig.5 Wilson loop as a function of $\Delta\tau$.

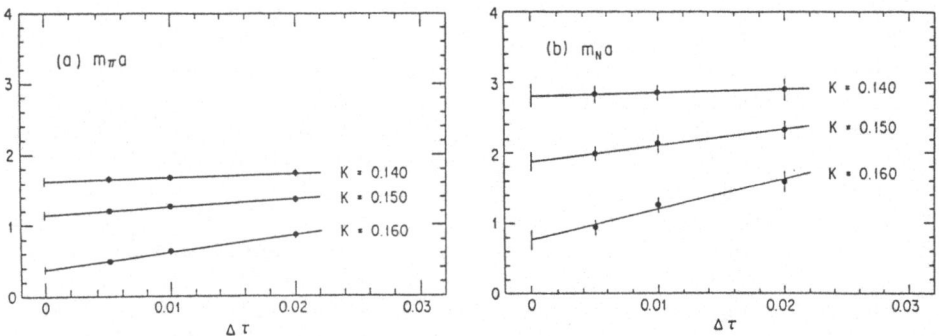

Fig.6 Hadron mass as a function of $\Delta\tau$.

linear and the extrapolation procedure well applies to find the value for $\Delta\tau=0$. Fig.6 shows the $\Delta\tau$-dependence for m_π and m_N. We see here also an approximate linearity with respect to $\Delta\tau$ that validates the extrapolation procedure to $\Delta\tau=0$.

<u>Hadron mass</u>   We plot in fig.7 m_π^2 as a function of K for a fixed $\Delta\tau$, as well as the value obtained by the $\Delta\tau\to0$ extrapolation at each K. We also added a curve for the quenched approximation for comparison. We then fit m_π^2 with the form

$$(m_\pi a)^2 = A_\pi (K^{-1} - K_c^{-1}) . \tag{10}$$

For ρ,N and Δ we only show the result for $\Delta\tau=0$, as compared with their quenched values in fig.8. The mass values are fitted with

$$m_i a = A_i (K^{-1} - K_c^{-1}) + B_i . \quad (i=\rho,N,\Delta) \tag{11}$$

Some parameters obtained in this analysis is presented in table I. The effect due to the inclusion of vacuum quark loops is quite significant as clearly seen in fig.7 and 8. The lattice spacing shrinks by 40%. A better agreement of m_N/m_ρ and m_Δ/m_N for $\Delta\tau=0$ with the experimental values ($m_N/m_\rho=1.22$ and $m_\Delta/m_N=1.31$) is caused by a steeper slope of the $\Delta\tau$ extrapolation curve for nucleon (see fig.6b). We cannot conclude, however, whether this is a real physical effect or merely represents either a size effect that the nucleon is not contained in the lattice or an artifact due to our extrapolation procedure combined with large statistical errors. In any case, an analysis at a different value of β should be made before concluding its physical significance.

<u>Physical implication of the vacuum quark loop effect</u>   It is quite conceivable that a part of vacuum polarisation effect can be absorbed into the shift of the gauge coupling β. To investigate this point we analysed an effective shift of β by matching the Wilson loop at various values of K and $\Delta\tau$ with that for the pure gauge system (see table II). The shift $\Delta\beta$ given by the Wilson loop of various size agrees within 20–30% even at K=0.16, but with a trend that a larger Wilson loop gives slightly a larger shift. (A somewhat stronger size dependence is seen with the Kogut-Susskind fermions.[17,18])

We then calculate hadron propagators in the quenched approximation at $\beta'=\beta+\Delta\beta(K,\Delta\tau)$. Here we took a $\Delta\beta$ given by the 4×4 (the largest) Wilson loop, because the hadron mass is controlled by the large-distance behaviour of propagators. In fig.9 the masses for full QCD are compared with those in the quenched approximation for the shifted β. The agreement between two cases is excellent for any K and $\Delta\tau$, and no appreciable deviation can be seen within our statistics. We have also made a similar comparison for the chiral order parameter $\langle\bar{\psi}\psi\rangle$ with

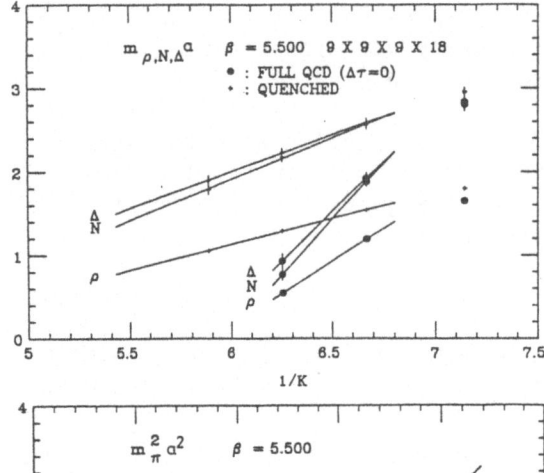

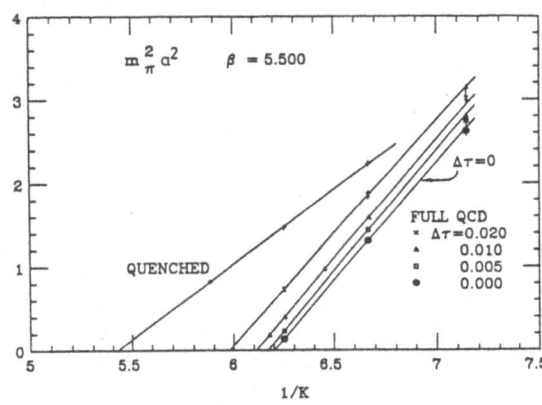

Fig.7 Pion mass squared as a function of 1/K.

Fig.8 ρ,N and Δ mass as a function of 1/K.

Table I. Summary of spectroscopic parameters.

| | quenched | full QCD | |
		$\Delta\tau = 0.01$	$\Delta\tau = 0$
K_c	0.1844 ± 0.0009	0.1637 ± 0.0005	0.1613 ± 0.0003
a^{-1}	0.99 GeV (0.20 fm)	1.31 GeV (0.15 fm)	1.63 GeV (0.12 fm)
m_N/m_ρ	1.73 ± 0.10	1.68 ± 0.15	1.36 ± 0.19
m_Δ/m_N	1.09 ± 0.04	1.10 ± 0.03	1.28 ± 0.03

Table II. $\Delta\beta$ from the Wilson loop matching.

| | $\Delta\tau = 0.01$ | | | $\Delta\tau = 0$ | | |
K =	0.14	0.15	0.16	0.14	0.15	0.16
W(1×1)	.06	.10	.20	.09	.16	.26
W(2×2)	.06	.11	.23	.09	.17	.29
W(3×3)	.06	.12	.24	.09	.17	.31
W(4×4)	--	.12	.25	--	.16	.30

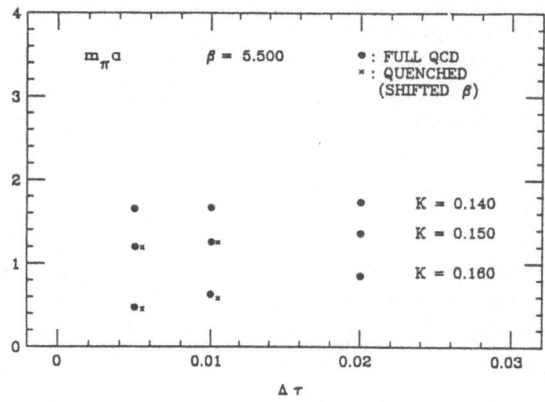

Fig.9 Hadron mass: A comparison of full QCD to the quenched case
with the shifted β.

Δβ(K,Δτ) for this case, however, from W(1×1), since $\langle\bar{\psi}\psi\rangle$ is a
local quantity. A good agreement is found between the two values.

From this analysis we conclude that a bulk of vacuum
polarisation effect can be absorbed into the shift of the gauge
coupling.[4] At a more precise level, however, the shift seems to
depend slightly on the length. In other words, the gauge
configuration with the quark vacuum polarisation is very similar
to the pure gauge configuration with a shifted effective coupling
where the value varies slightly with the length scale. A good
agreement of the full QCD and the quenched data at any Δτ also
suggests that the finite time step effect can also be absorbed
effectively into the shift of β.

It is an interesting question to ask where one can see a
clear net effect of quark vacuum loops. In our simulation we
could not explore the most interesting region of opening threshold
for ρ→2π and Δ→Nπ etc., where one may expect a net effect. For
this purpose we have to prepare a considerably larger lattice and
the simulation is much more time consuming. This requirement
necessitates a further improvement of the computing algorithm and a
sizable increase of the computing power.

Perhaps an easier way to look a net vacuum effect may be to
study the spectrum for the excited hadrons. We have calculated
the static potential for full QCD from the rectangular Wilson loop
by a straightforward application of the standard procedure.[19] The
deviation of the potential from that for the pure gauge system
is very small except for that at a large distance, where we see
some indications of the departure from the linear potential. This
implies that, while the mass of ground state hadrons which stay in
the inner part of the potential hardly get a correction from
vacuum quark loops, their effect may be seen with excited hadrons.

Pseudoscalar two-loop operator $<[\mathrm{Tr}(\gamma_5\Delta(0,0))]^2>$ leads to the difference of π and η propagators $G_{\eta-\pi}$ at t=0. We found that the value for full QCD does not differ much from that for the quenched case, and $G_{\eta-\pi}(0)/G_\pi(0) \sim 10^{-3}$ at K=0.16. The effect of vacuum quark loop is not likely to boost the pseudoscalar two loop operator.

FINITE TEMPERATURE BEHAVIOUR [6]

Studies of finite temperature behaviour of QCD made so far are mostly confined to the case with the Kogut-Susskind action,[20] which keeps a remnant of chiral symmetry. It has been suggested that chiral symmetry plays a dynamical role for the deconfining transition at a small quark mass.[21,3] It is therefore interesting to see whether the deconfining transition appears in the same way also with the Wilson action, which violates chiral symmetry explicitly.

To study a finite temperature behaviour we employed a rather small lattice of $5^3 \times 3$, because the lattice size should not be crucial for the study of the phase structure. We restrict our study to the case with $N_F=4$ to facilitate a comparison to the analysis with the Kogut-Susskind fermion.[3] We have also made a separate spectroscopic analysis to determine the critical hopping parameter K_c (see Table III). We observed the behaviour of the Polyakov line $<\Omega>$ and the gluon internal energy E_g/T^4 as a function of β at fixed values of K. A rapid crossover is seen at all the value of K that we explored (K=10^{-4},0.10-0.20) before reaching the critical line of the hopping parameter. The transition seems to become gradually less pronounced as K increases, with the transition region shifted towards smaller β. In fig.10 we plot the location of the transition region as deduced from our analysis.

We observe that the zone of transition does not meet the line of critical hopping parameter K=$K_c(\beta)$ even at $\beta=3.5$. This is quite different from what might be expected from the behaviour of the system with the Kogut-Susskind action. If the critical line

Table III. Critical hopping parameters K_c.

	$N_F = 0$	2	4	
4.0	0.226	---	0.221	$6^3 \times 12$ ($\Delta\tau=0.01$)
5.0	0.208	0.193	0.186	$6^3 \times 12$ ($\Delta\tau=0.01$)
5.5	0.1844	0.1635	---	$9^3 \times 18$ ($\Delta\tau=0.01$)

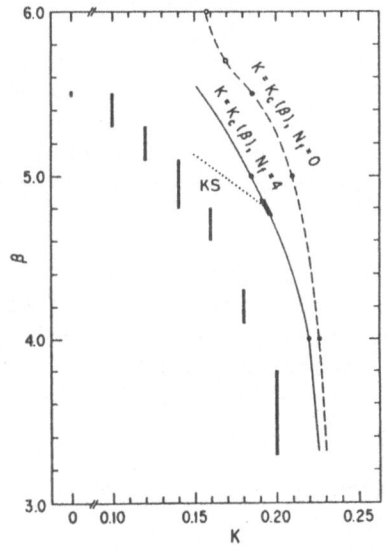

Fig.10 Phase diagram of deconfining transition in the (β,K) plane. KS stands for the expectation from the analysis with the Kogut-Susskind fermion.

$K=K_c(\beta)$ corresponds to $m_q=0$ of the Kogut-Susskind action, then we expect the transition zone crossing with $K=K_c(\beta)$ at $\beta=4.75-4.85$. Of course, it does not make much sense to pursue precise agreement of the values, for two actions are expected to produce the same physics only in the continuum limit and we are obviously far from it. However, we observe here a signature that the physics is very different between the two actions. Fig.10 suggests that the transition zone runs almost parallel with the critical line $K=K_c(\beta)$ running down to the strong coupling limit without crossing.

This means that the critical line $K=K_c$ is always in the high-temperature deconfining phase, irrespective of whether the transition is really a phase transition or not. This has also an important implication on the spectroscopic study. Since the rise of $\langle\Omega\rangle$ towards K_c comes from the quark loop wrapping around the lattice, we also expect a similar phenomenon when the spatial size of the lattice is not large instead of the temporal. Hence, our results for finite temperature may be translated into those on the effect of finite spatial extent in spectroscopic calculations. This implies that, even if the transition is a crossover, spectroscopic quantities may change rapidly in the transition region and those measured close to K_c may reflect the dynamics at high temperature and not that at zero temperature required for spectroscopic studies. This is the reason why we have checked carefully the size effect by generating gauge configurations taking a different boundary condition. As a result we did not find any signature of a significant fake loop contribution, and we can assert that the physics described in the preceding section does not suffer from finite temperature effect.

CONCLUSION

We have shown in this report that the Langevin simulation is practically feasible and works well for the method to simulate QCD with dynamical quark loops. We have shown that the $\Delta\tau$ extrapolation procedure is necessary to remove the finite time step effect for a quantitative analysis. An important point, however, is that systematic errors seem to be controllable.

Our analysis made in this report is still premature as a simulation to give a realistic numbers for hadron physics. I expect, however, that realistic spectroscopic quantities will be calculated in the near future when a gigaflops computer is available.

ACKNOWLEDGEMENT

I would like to thank S. Ohta, Y. Oyanagi and A. Ukawa for a close collaboration which made this work possible.

REFERENCES

1. A.Ukawa and M.Fukugita, Phys.Rev.Lett.55,1854(1985).
2. G.G.Batrouni, G.R.Katz, A.S.Kronfeld, G.P.Lepage, B.Svetitsky and K.G.Wilson, Phys.Rev.D32,2736(1985).
3. M.Fukugita and A.Ukawa, Phys.Rev.Lett.57,503(1986).
4. M.Fukugita, Y.Oyanagi and A.Ukawa, Phys.Rev.Lett.57,953 (1986).
5. M.Fukugita, Y.Oyanagi and A.Ukawa, preprint in preparation.
6. M.Fukugita, S.Ohta and A.Ukawa, Phys.Rev.Lett.57,1974(1986).
7. F.Fucito, E.Marinari, G.Parisi and C.Rebbi, Nucl.Phys.B180, 369(1981).
8. J.Polonyi and H.W.Wyld, Phys.Rev.Lett.51,2257(1983).
9. S.Duane, Nucl.Phys.B257 [FS14],652(1985); S.Duane and J.Kogut, Phys.Rev.Lett.55,2774(1985).
10. M.Fukugita, T.Kaneko and A.Ukawa, Nucl.Phys.B270 [FS16],365 (1986).
11. See e.g., J.Moser, Stable and random motion in dynamical systems (Princeton Univ. Press 1973).
12. S.Duane and J.Kogut, Nucl.Phys.B275 [FS17],398(1986).
13. G.Parisi and Y.-S.Wu, Sci.Sin.14,483(1981).
14. G.G.Batrouni, Cornell University preprint CLNS-85/665(1985); A.S.Kronfeld, Phys.Lett.172,93(1986).
15. M.R.Hestenes and E.Stiefel, J.Res.Nat.Bur.Standards 49,409 (1952).
16. Y.Oyanagi, Computer Phys. Comm.42(1986) (to be published).
17. R.V.Gavai and F.Karsch, Phys.Rev.Lett.57,40(1986).
18. M.Fukugita, et al., preprint in preparation.
19. J.D.Stack, Phys.Rev.D27,412(1983).
20. See ref.3 and references therein.
21. R.D.Pisarski and F.Wilczek, Phys.Rev.D29,338(1984).

ON THE QCD β-FUNCTION

R. V. Gavai

Theory Group
Tata Institute of Fundamental Research
Bombay 400 005, India

ABSTRACT

After presenting evidence to indicate that the pseudo-fermion method is able to incorporate the effects of the dynamical fermions rather well, I describe our results for the β-function of QCD with three flavours of light dynamical quarks, obtained by using the improved ratio method. The β-function appears to have a dip at $6/g^2 \simeq 5.6$ and the deviations from asymptotic scaling appear to be small already for $6/g^2 \geq 5.3$. The possible sources of the dip structure are discussed.

INTRODUCTION

An important step in obtaining physical results from numerical simulations of lattice field theories is that of the continuum limit: only those results which are independent of the limit $a \to 0$, where a is the lattice spacing, are physical. The β-function of a theory, denoted here by $B(g)$, tells us how to approach this limit:

$$B(g) = -a \frac{d}{da} g \tag{1}$$

For small enough coupling g, $B(g)$ can be calculated using the conventional perturbation theory. For QCD, the leading terms in the weak coupling expansion of $B(g)$ are given by

$$B(g) = -b_0 g^3 - b_1 g^5 + O(g^7) \tag{2}$$

where

$$b_0 = (33 - 2N_f) / 48\pi^2, \tag{3}$$

$$b_1 = \left(153 - 19 N_f\right) / 384 \pi^4 \qquad (4)$$

and N_f is the number of massless flavours in the theory. Using (1) and (2) one easily obtains that in this limit

$$a\Lambda = \exp\left(-1/2 b_0 g^2\right) \cdot g^{-b_1/b_0^2} \left[1 + O\left(g^2\right)\right] \qquad (5)$$

where Λ is a constant of integration. Thus the limit $g \to 0$ corresponds to $a \to 0$ for QCD. The coefficients b_0 and b_1 in (2) are known to be universal, while the coefficients of higher order terms, $b_2, b_3 \ldots$ etc., depend on the regularization scheme. For the lattice regularization these coefficients depend on the choice of the lattice action. Depending on the size of a, one can imagine three different regimes for the β-function. If a is very small then clearly $B\left(g\right)$ will be dominated by the two universal terms in (3) and the $O\left(g^2\right)$ terms in (5) will be insignificant. One calls this region of a (or equivalently g) as asymptotic scaling region. For a little larger a, more terms in (2) are needed. One may even have here non-perturbative contributions to the β-function. However, if a unique β-function can still be defined then physical answers can still be obtained from the lattice theory in this regime of a, called the scaling region. For even larger values of the lattice spacing a, no unique β-function can be defined: holding two different physical quantities constant as a is varied yields different functions $g\left(a\right)$. This is the strong coupling regime of the theory.

It is very likely that the present-day numerical simulations of lattice QCD on at least large enough lattices have been performed in the scaling region. If that were so then one would expect the dimensionless ratios of physical quantities to be estimated reliably from these simulations, although (5) is not applicable at those values of g. If, on the other hand, one could ascertain that (5) is valid in this region, many more predictions would follow. It is, therefore, important and perhaps necessary to understand the structure of the β-function of the theory, especially in the regions inaccessible to perturbation theory. One could even envisage that such studies will give us a hint to improve the approach to the continuum limit on finite lattices. At the very least, they would enable us to understand the limit better.

There have been two different methods which have been employed[1] so far, to investigate the β-function of SU(2) and SU(3) gauge theories:

a) the Monte Carlo renormalization group method by blocking, and

b) the ratio method.

While both methods involve comparisons of physical observables which differ by a scale factor, say b, the former needs an explicit integration of appropriately defined groups of the field variables to achieve scale reduction by the factor b. In QCD, the Grassmannian nature of the fermionic variables renders a straightforward application of the first method therefore rather non-trivial. The ratio method can, however, still be used provided that gauge configurations corresponding to the full action of QCD, i.e. including the fermionic determinant, can be generated easily and reliably on the computer. A variety of fermion algorithms now exist which can achieve this. While the field of development and improvement of such algorithms is in a rapidly changing state, I hope to demonstrate below that some of these algorithms already satisfy the necessary criteria for a good fermion algorithm and hence can be used for obtaining physically interesting quantities. In the case of β-function it turns out that agreement (or lack of it) with the asymptotic scaling prediction can be used as a further test of the fermion algorithm, as described below.

THE RATIO METHOD

The early seeds of this method can be found in the work of Creutz.[2] It was revived and refined by Hasenfratz et al.[3] At the basic level, it consists of forming ratios $R_2\left(i_1, i_2, j_1, j_2\right)$, $R_4\left(i_1, i_2, i_3, i_4, j_1, j_2, j_3, j_4\right)\ldots$ which are defined below:

$$R_2\left(i_1, i_2, j_1, j_2\right) = \frac{W\left(i_1, i_2\right)}{W\left(j_1, j_2\right)} \tag{6}$$

such that

$$i_1 + i_2 = j_1 + j_2 \tag{7}$$

and

$$R_4\left(i_1, i_2, i_3, i_4, j_1, j_2, j_3, j_4\right) = \frac{W\left(i_1, i_2\right) W\left(i_3, i_4\right)}{W\left(j_1, j_2\right) W\left(j_3, j_4\right)} \tag{8}$$

such that

$$i_1 + i_2 + i_3 + i_4 = j_1 + j_2 + j_3 + j_4 \tag{9}$$

These ratios are free of perturbative singularities and they satisfy the homogenous renormalization group equations:

$$R_2\left(2i_1, 2i_2, 2j_1, 2j_2; \beta, L\right) = R_2\left(i_1, i_2, j_1, j_2; \beta', L/2\right) \tag{10}$$

$$R_4\left(2i_1, 2i_2, 2i_3, 2i_4, 2j_1, 2j_2, 2j_3, 2j_4; \beta, L\right) =$$
$$R_4\left(i_1, i_2, i_3, i_4, j_1, j_2, j_3, j_4; \beta', L/2\right) \tag{11}$$

Here $\beta = 6/g^2$, $\beta' = 6/g'^2$ and L, $L/2$ are the linear sizes of the four-dimensional symmetric lattices on which these ratios are calculated. Tuning

β' so that (10)-(11) are satisfied, one obtains $\Delta\beta(\beta) = \beta - \beta'$ which is the change in coupling required to compensate the change of scale by a factor of 2. Note that finite size effects are kept to a minimum by having the lattice size on both sides of these equations to be the same in physical units. From the definition of the β-function in (1) one can easily show that the $\Delta\beta$ above is related to it by the following equation

$$\int_{\beta-\Delta\beta}^{\beta} \frac{dx}{x^{3/2}B\left(\sqrt{6/x}\right)} = -\frac{2\ln 2}{\sqrt{6}} \tag{12}$$

In the asymptotic scaling region, one can substitute for B above from equation (2) to obtain

$$\Delta\beta = 12b_0\,\ln 2 + 72b_1\,\ln 2/\beta + O\left(\beta^{-2}\right) \tag{13}$$

which for $N_f = 0$, i.e. quenched QCD or pure SU(3) theory, becomes

$$\Delta\beta = 0.579 + 0.204 \cdot \beta^{-1} + O\left(\beta^{-2}\right), \tag{14}$$

whereas for full QCD with $N_f = 3$:

$$\Delta\beta = 0.474 + 0.128 \cdot \beta^{-1} + O\left(\beta^{-2}\right). \tag{15}$$

It is clear that an agreement with (15), rather than (14), for large enough β would act as a consistency check on the fermion algorithm's capability to incorporate fully the effects of the light dynamical fermions. Using the ratio method described above to extract the β-function, one runs across a problem with small loops: ratios composed of small loops are contaminated by lattice artifacts. One can attempt to correct for them by using perturbation theory. One sees clearly from eqs. (10-11) that if R_i, R_j satisfy the RG equations then so does $R_i + \alpha R_j$ (or any further linear combination such as $R_i + \alpha R_j + \beta R_K$, etc.). Whereas the individual R_i's yield a $\Delta\beta$ in perturbation theory which diverges, one can obtain the right behaviour by tuning the mixing coefficients. Thus an appropriate value of α in $R_i + \alpha R_j$ can ensure that the lattice artifacts are not there in tree level perturbation theory (i.e. $\Delta\beta = 0$). Further systematic improvement of the $\Delta\beta$ in perturbation theory is clearly possible order by order if one mixes more and more ratios.

Hasenfratz et al.[3] used both the basic ratios and the mixed or "improved" ratios to study the β-function of SU(3) lattice gauge theory. Using 8^4 and 16^4 lattices they obtained $\Delta\beta(\beta)$ for $5.8 \leq \beta \leq 6.6$, which has the following general features:

   i) $\Delta\beta(\beta)$ has a dip at $\beta \simeq 6.0$.

ii) The dip covers a range of β from 5.7 to 6.6 which corresponds to a change of correlation length by a factor of four across the dip.

iii) Deviations from asymptotic scaling are negligibly small for $\beta \geq 6.0$, and

iv) approach to asymptotic scaling prediction, i.e. eq. (14), is from below.

$\Delta\beta\,(\beta)$ obtained by other methods, such as the MCRG blocking method, or from the deconfinement temperature or string tension data, was found to agree with these results. While latest calculations[4] at higher values of β seem to cast doubt on physically the most important observation iii) above - one obtains $\Delta\beta$ at higher β values which may not be within one standard deviation of the asymptotic prediction - all the results still do suggest that these methods, particularly the ratio method, may be a very vital tool in exploring the scaling and asymptotic scaling region of QCD.

THE PSEUDO-FERMION METHOD[5]

As may be clear from the discussion above, one can employ the ratio method to investigate the β-function of full QCD, provided one can generate configurations of the full theory. The pseudo-fermion method allows one to do that. For the sake of brevity, I will not discuss the details of the method here. They can be found elsewhere.[5,6] As with the other methods, the pseudo-fermion method involves certain assumptions and can therefore be expected to work well only under certain conditions. The past few years have seen tremendous efforts being invested in determination of these conditions. Most of them involve comparisons with other methods which either have different, perhaps inequivalent, sets of assumptions or approximations, or are exact in principle but are limited by practical constraints of accuracy on a computer. Fig. 1 exhibits an impressive example of the former kind. What is displayed there are the data for $\langle \bar{\psi}\psi \rangle$ and $\langle L \rangle$ as a function of β $(\equiv 6/g^2)$. They were obtained by three different methods: i) the Langevin[7] method, ii) the microcanonical[8] method, and iii) the pseudo-fermion[9] method. All the parameters which govern physics such as the lattice size, the number of flavours, the quark mass and the lattice action were identical in each case. The only difference thus was in the details of each method in incorporating the fermion determinant. All the methods agree with each other very well, as one sees clearly from fig. 1.

The pseudo-fermion method becomes better as:

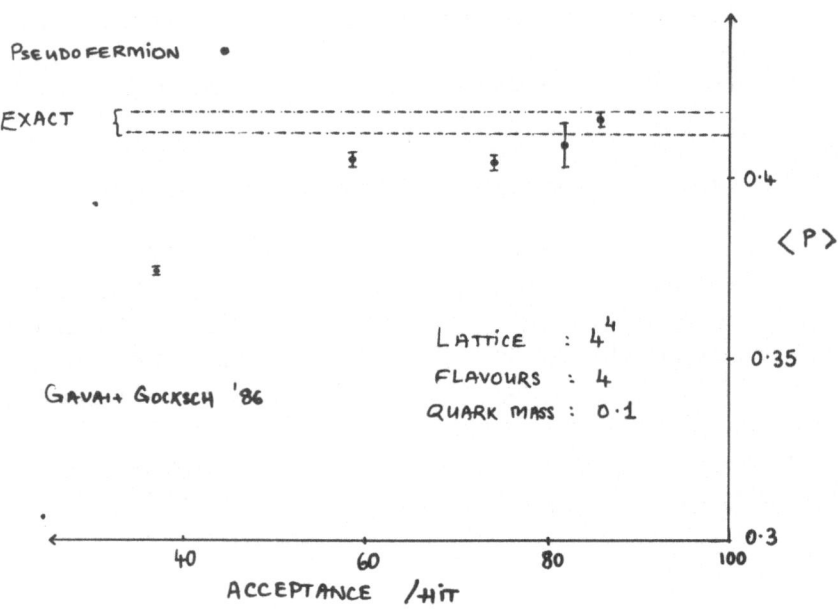

Fig. 1: The chiral condensate $\langle\bar{\psi}\psi\rangle$ and the Polyakov loop $\langle L\rangle$ as functions of $\beta\left(=6/g^2\right)$ on $8^3\times 4$ lattice with 4 flavours of mass 0.1 a^{-1}. The data are taken from refs. 7,8, and 9.

a) the acceptance in the gauge sector $\rho\to 100\%$ (acceptance is defined as the ratio of all the accepted changes in the entire set of gauge variables in a sweep to that of the proposed ones) and

b) the number of pseudo-fermion sweeps $N_{pf}\to\infty$.

In the example above, $\rho\approx 70-75\%$ and $N_{pf}=24$ for the pseudo-fermion case. That with these values, and obviously therefore with both larger ρ and N_{pf}, one obtains a good approximation to the full theory was also confirmed[6] by the comparison of the pseudo-fermion method with the exact algorithm of Scalapino and Sugar.[10] Fig. 2 shows a typical sample of those results. On a small lattice, 4^4, all Wilson loops up to size 3×3 were measured for four (Kogut-Susskind) flavours of mass 0.1 a^{-1}, using the two methods

118

mentioned above. The average plaquette which turned out to be the most sensitive one amongst all those Wilson loops agrees well with the "exact" result for $\rho \geq 80\%$, as one sees in fig. 2. Needless to say that the agreement is equally good, if not better, in the case of other Wilson loops for these values of ρ.

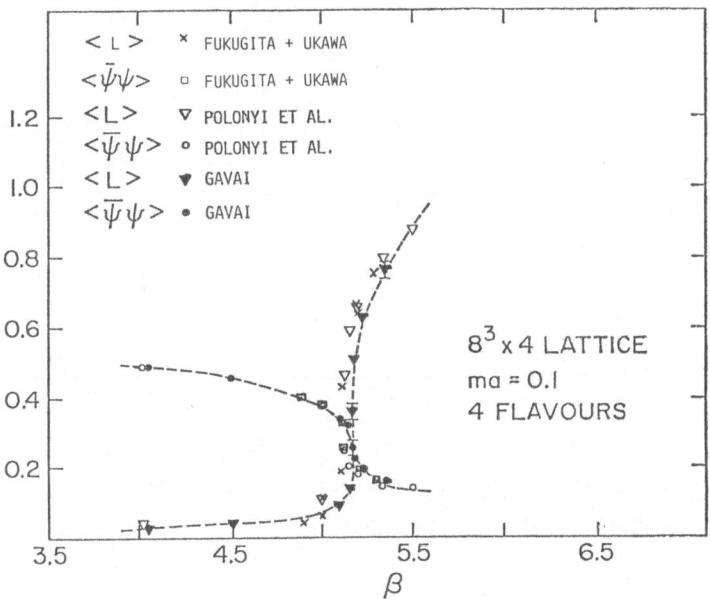

Fig. 2: The average plaquette obtained by the pseudo-fermion method as a function of the acceptance/hit on a 4^4 lattice with 4 flavours of mass 0.1 a^{-1}. The dashed lines indicate the value obtained by an exact algorithm. From ref. 6.

THE β-FUNCTION OF QCD

Encouraged thus by the result that the pseudo-fermion method works quite well in generating the configurations of the full theory, one may use it to obtain the β-function of the full theory. In collaboration with Karsch,[11] I simulated the full theory with 3 light dynamical flavours on 8^4 and 4^4 lattices. We employed the staggered fermions and chose the mass of the fermions to be 0.1 a^{-1} and 0.2 a^{-1} on these lattices respectively. This scaling amounts to neglecting the corrections due to anomalous dimensions. We used the pseudo-fermion method with $\rho \sim 70 - 75\%$ and $N_{pf} = 50$. (The

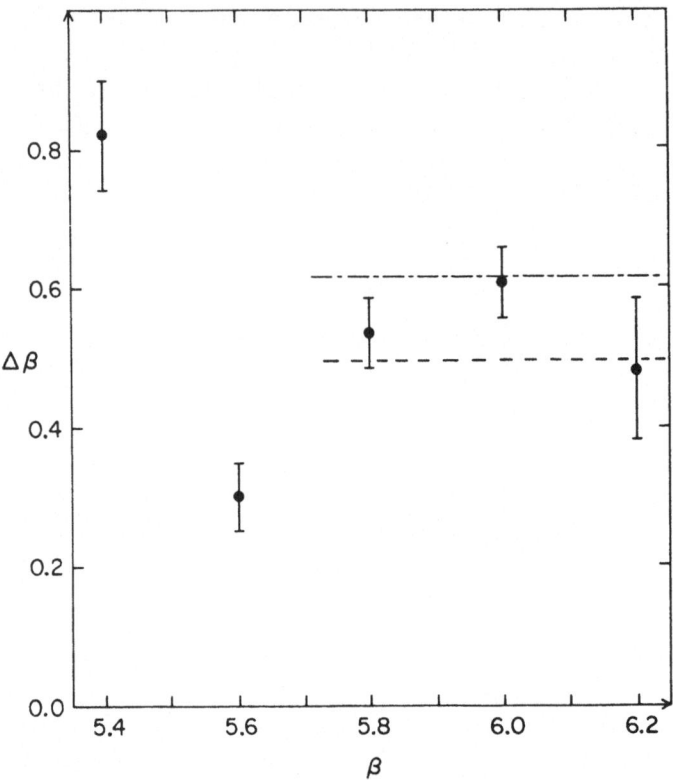

Fig. 3: $\Delta\beta$ vs. β for QCD with 3 light flavours. The horizontal lines are the predictions of eqs. (14) and (15).

"currents" J^μ were obtained by discarding 25 pseudo-fermionic iterations.) These choices were guided by both the practical time constraints and the constraints due to the validity of the method discussed above. At each β we made typically a few thousand iterations, discarding about one thousand to allow for equilibration. On the 8^4 lattice, all Wilson loops up to the size 6×6, and on the 4^4 lattice all loops up to 3×3 were measured every iteration. The errors were then corrected for sweep-to-sweep correlations in Monte Carlo time. 17 basic ratios and about twice as many tree level improved ratios were formed to use in the renormalization group equation. The precise value of β' was obtained by a linear interpolation between the sets of ratios at measured β on the smaller lattice. Figure 3 shows our results for $\Delta\beta$ so obtained. About 20 improved ratios could be selected at each β to have the matching predictions with the error bars displayed, except at $\beta = 5.4$. The smallness of Wilson loops themselves at $\beta = 5.4$, and the ensuing statistical fluctuations compelled us to use basic ratios there which may not be that bad since the improvement is more relevant at higher β values anyway. The dashed line in fig. 3 is the asymptotic scaling prediction of eq. (15) whereas the line above it corresponds to the eq. (14), i.e. to the pure SU(3) or the quenched theory. Note that three light dynamical flavours change the asymptotic scaling prediction by about 20% only.

Remembering the results for $\Delta\beta\,(\beta)$ for pure SU(3) and looking at fig. 3 one can make the following observations. Even in the full theory $\Delta\beta\,(\beta)$ has a pronounced dip but it is shifted to $\beta \simeq 5.6$. Although the quality of our data does not permit any firm statements, it appears that the dip is narrower than the pure gauge case: correlation length seems to be changing across the dip by only a factor of two. Furthermore, the approach to asymptotic scaling appears to be from above. The deviations from asymptotic scaling seem to be negligibly small for $\beta > 5.3$ (remember, $\Delta\beta\,(\beta)$ tells us about the β-function in the range $\beta - \Delta\beta$ to β). The agreement of $\Delta\beta\,(\beta)$ for $\beta > 5.8$ with eq. (15) can also be looked upon as a consistency check on the pseudo-fermion method. While it is true that due to the very small difference between the predictions for the quenched theory and the full theory and also due to our error bars, we cannot distinguish between the two beyond 1σ level, taken together the data does point in the direction of the full theory prediction. Noting that we had typically 5,000 iterations of the full theory on the 8^4 lattice at each β, it may be remarked that a much clearer distinction will be hard to obtain in future studies as well.

In spite of the discussion above, we believe that our data clearly shows the presence of a dip in $\Delta\beta$ for the full QCD. Of course, the details of the dip or even its existence has no direct relevance to calculations aimed at extracting properties or predictions of continuum QCD. For that, the important issue to settle would be whether the onset of asymptotic scaling indeed takes place at $\beta \simeq 5.3$ or whether it is delayed until $\beta \simeq 5.7$ as the overshooting of $\Delta\beta\,(6.0)$ would indicate. Larger lattices and better statistics will surely answer these questions in the near future. However, one would also like to understand how the theory goes over from the strong coupling regime to the scaling regime. One can eventually then look for ways to make this cross-over smoother, say, by modifying the action. In the case of SU(3) theory, the presence of dip had been related[1] to the second order phase transition in the fundamental-adjoint coupling plane. Naively, therefore, one did not expect the dip in the full theory. In the full theory the space of relevant couplings is, of course, enlarged. N_f - the number of light flavours, and m - the quark mass, are two obvious additions. One should, therefore, trace the phase diagram in this enlarged space to understand more about the dip in fig. 3. On the other hand, one can already suspect that a phase transition may be lurking nearby. Indeed, it has been argued that a nontrivial

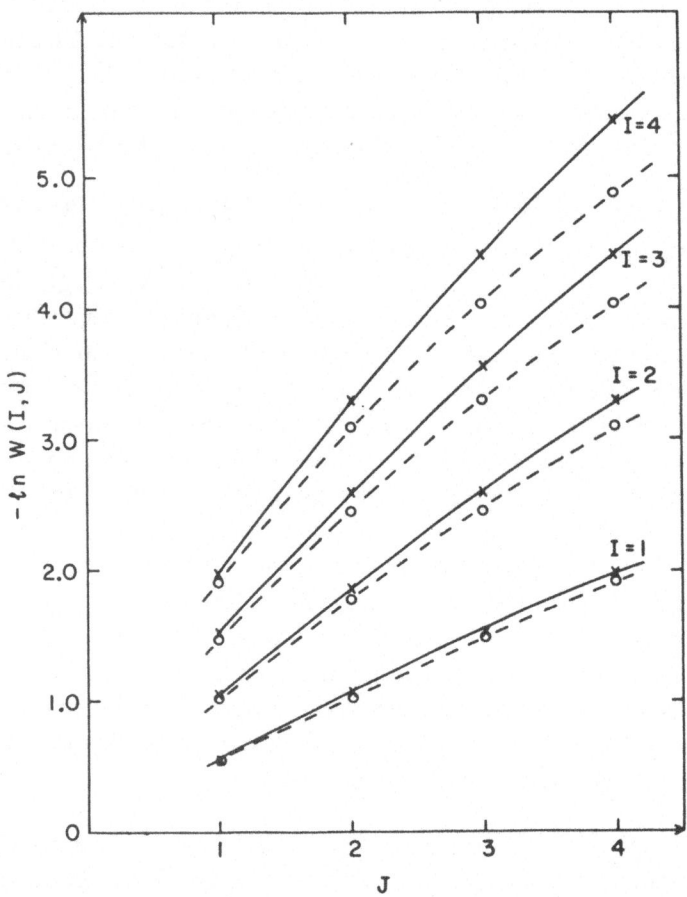

Fig. 4: $-\ln W\,(I,\ J)$ as a function of J for various I. The crosses show the pure SU(3) data of ref. 13 at $\beta = 5.8$ while the circles are our data at $\beta = 5.6$ for the full theory with 3 light flavours.

phase structure appears in the $N_f - \beta$ plane at $(N_f, \beta) \simeq (8,\ 4.67 \pm 0.10)$. $\Delta\beta\,(\beta)$ should be zero at this point and, for $N_f < 8$, it could lead to a dip structure similar to that in fig. 3. In that case, it would be interesting to study $\Delta\beta\,(\beta)$ as $N_f \to 8$. For the dip should become deeper as N_f is increased. It would also be interesting to find out if the dips in the two theories, quenched and full, are related in any way.

The similarities of fig. 3 with the corresponding quenched results, and our failure to distinguish clearly between the asymptotic scaling predictions in eqs. (14) and (15) beyond 1σ level is likely to give rise to suspicions that the pseudo-fermion method has not been able to include the fermionic effects very well. Of course, the shifting of the dip is a clear evidence against it. Unfortunately, one cannot calculate the shift in any approximation to check whether it is of right magnitude. One qualitatively different aspect of the presence of fermions is screening. It is well known[12] that the potential in the SU(3) theory is a linear term in r and a Coulomb term: $\sigma r + \alpha/r$. In full QCD, one expects $\sigma r \to \sigma \left(1 - \exp\left(-\mu r\right)\right)/\mu$ with μ^{-1} as the screening length. It would be nice if this screening effect can be seen in full QCD. Again determining it quantitatively would require a totally new calculation on larger lattices. It turns out, however, that qualitatively it can already be checked from our Wilson loop data, giving one more reason to lay aside the above mentioned suspicions.

Fig. 4 displays our data on Wilson loops at $\beta = 5.6$ on the 8^4 lattice. Also shown is the corresponding data at $\beta = 5.8$, but for the SU(3) theory[13] alone. The lattice size in the two cases is the same, minimizing the finite size effects. The couplings have been so chosen that the plaquette value is the same in the two cases. Since $-\ln W$ is a direct measure of the potential, screening should show up in a systematic downward shift of the full theory data as the loop size increases. Noting that the error bars on all the points are of the size of the points themselves and that they take into account the correlations in MC time already, one clearly sees that screening is present in the configurations which we generated by the pseudo-fermion method. Alternatively, one can state in the following manner the same effect: while a shift of 0.2 in β sufficed to match the plaquettes in the two theories, a shift of 0.3 in β would be necessary to get the 4×4 Wilson loops to agree within the respective errors. It would be nice to confirm these effects more quantitatively on larger lattices.

CONCLUSIONS

Using the staggered fermion formalism and the pseudo-fermion method, QCD with 3 light dynamical flavours was simulated on 8^4 and 4^4 lattices. From the Wilson loops measured on these configurations $\Delta\beta\left(\beta\right)$, the β-function for the full theory was obtained using the improved ratio method. It is found to be very similar to that of the quenched theory, but has a dip at $\beta \simeq 5.6$ instead of $\beta \simeq 6.0$. The approach to asymptotic scaling appears to be from above and its onset may be already at $\beta \simeq 5.3$. The pseudo-fermion method seems to be able to incorporate the effects of the fermion determinant well, both qualitatively and quantitatively, as one can see from figs. 4 and 3 respectively. On the theoretical side, the apparent "universality" of the dip is rather puzzling. Investigations of β-function with different flavours may shed some light on it.

ACKNOWLEDGMENTS

It has been a pleasant experience to collaborate with A. Gocksch, U. Heller and F. Karsch. Most of the original work presented here was done with them. For the computer time on the Cray XMP at NMFECC, Livermore, I would like to thank S. Kahana sincerely. This manuscript was completed during my stay in Bielefeld; I would like to thank the members of the Physics faculty there, especially H. Satz, for the warm hospitality extended to me. This work was largely supported by the U.S. Department of Energy under contract no. DE-AC02-76CH00016.

REFERENCES

1. P. Hasenfratz, CERN preprint TH-3999/84 (1984), to be published in the proceedings of Erice Summer School 1984;
   R. Gupta, in the proceedings of Lattice Gauge Theory Symposium, Wuppertal, (1985).
2. M. Creutz, Phys. Rev. D23 (1981) 1815;
   R.W.B. Ardill, M. Creutz and K.J.M. Moriarty, Phys. Rev. D27 (1983) 1956.
3. A. Hasenfratz, P. Hasenfratz, U. Heller and F. Karsch, Phys. Lett. 143B (1984) 193.
4. K.C. Bowler, in this Procedings.
5. F. Fucito, E. Marinari, G. Parisi and C. Rebbi, Nucl. Phys. B180 [FS2](1981) 369.
6. R.V. Gavai, A. Gocksch and U. Heller, Brookhaven preprint BNL-38449, Santa Barbara preprint NSF-ITP-86-89, Nucl. Phys. B (in press).
7. M. Fukugita and A. Ukawa, Phys. Rev. Lett 57 (1986) 503.
8. J. Poloyni, H.W. Wyld, J. Kogut, J. Shigemitsu, and D.K. Sinclair, Phys. Rev. Lett 53 (1984)644.
9. R.V. Gavai, Nucl. Phys. B269 (1986) 530.
10. D.J. Scalapino and R.L. Sugar, Phys. Rev. Lett. 46 (1981) 519.
11. R.V. Gavai and F. Karsch, Phys. Rev. Lett 57 (1986) 40.
12. S.W. Otto and J.D. Stack, Phys. Rev. Lett. 52 (1984) 2328.
13. K.C. Bowler et al., Nucl. Phys. B257 (1985) 155.

THE PROPAGATOR MATRIX AND THE QUENCHED APPROXIMATION AT FINITE

BARYONIC DENSITY

P.E. Gibbs

Physics Department
University of Edinburgh
Edinburgh EH9 3JZ

INTRODUCTION

Simulations of QCD at finite baryonic density on the lattice have given results which are in complete contradiction to theoretical ideas. It is expected that there will be a chiral symmetry restoration phase transition at nuclear density. I.e. at a chemical potential μ equal to about a third of the nucleon mass $(1/3)m_N$. Monte carlo simulations have however shown a transition at a lower density consistent with a critical μ equal to half the pion mass $(1/2)m_\pi$ [1].

THE PROPAGATOR MATRIX

First we shall consider a method of calculating the chiral condensate at finite density in quenched QCD by means of the propagator matrix[2]. On the lattice the chemical potential μ is introduced by multiplying timelike gauge links in the fermion matrix by $\exp(\mu)$ in the forward direction and by $\exp(-\mu)$ in the reverse direction[3]. Then we can write the fermion matrix in the form,

$$H = G + z^{-1}V + zV^\dagger \quad , \quad z = e^\mu$$

where V is the part of the matrix containing backward going timelike links and G contains only spacelike links and the mass term. We have multiplied by i so that the matrix is hermitian at zero mass and chemical potential.

The chiral condensate is obtained by differentiating the determinant with respect to the quark mass and,

$$\det(H) = \det(zGV + V^2 + z^2)z^{-N}$$

where N is the size of the matrix. We can write this in terms of the determinant of another matrix twice the size,

$$\det(H) = z^{-N}\det(P - z)$$

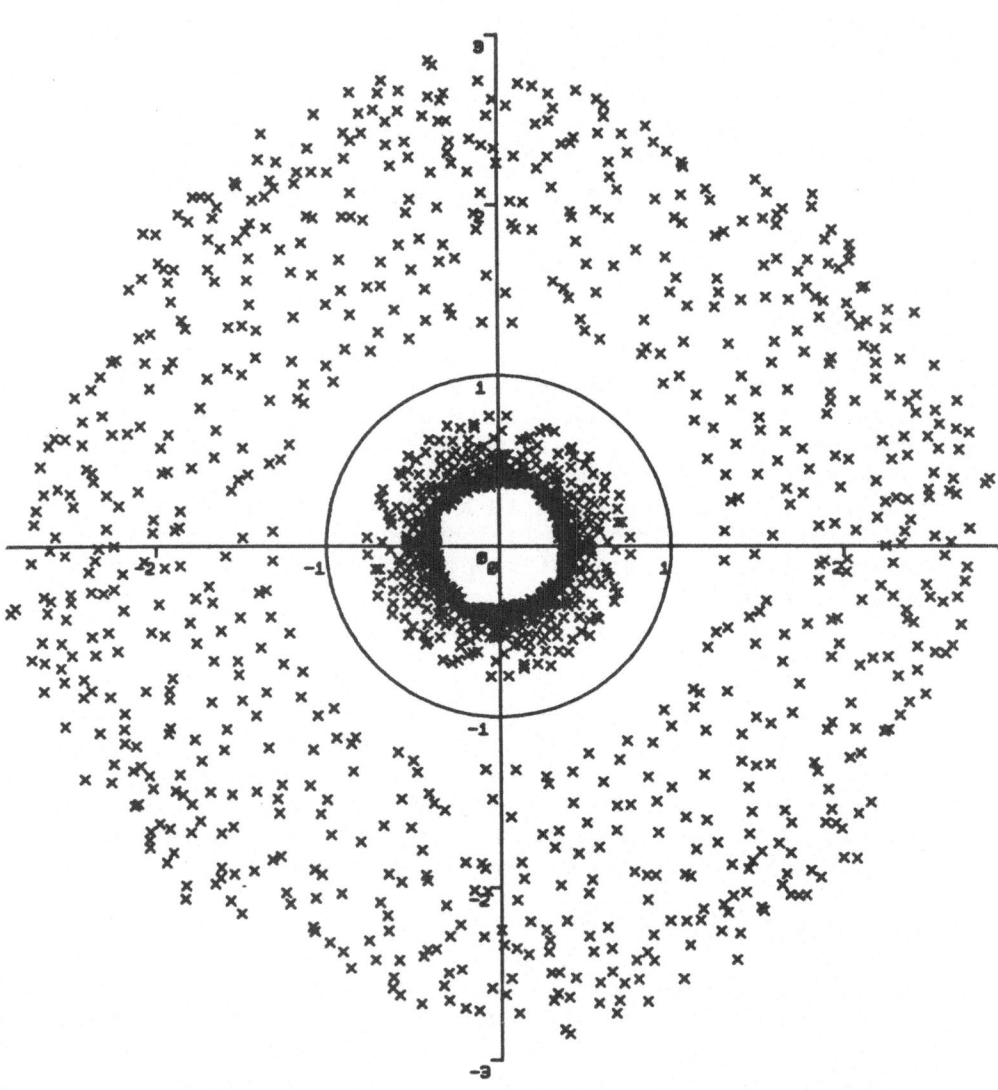

Fig.(1) The eigenvalues of the propagator matrix for a single 4^4 configuration at β=5.3 and m = 0.05.

where

$$P = \begin{pmatrix} -GV & 1 \\ -V^2 & 0 \end{pmatrix}$$

P is called the propagator matrix. If z_a are the eigenvalues of P then

$$\text{Chiral Condensate} = (1/N) \sum_a (dz_a/dm)/(z_a - z) \quad , \quad z = e^{\mu}$$

This formula is useful since once we have the eigenvalues z_a at a given mass on one configuration then we can quickly plot the chiral condensate as a function of μ. We can also derive analytic results. By summing over a symmetry of the eigenvalues

$$\text{Chir. Cond.} = (1/N) \sum_a (dz_a/dm) (1/z_a) (1/\{1 - [z/z_a]^n\})$$

where n = number of sites in the time direction. In the zero temperature limit $n \to \infty$,

$$\text{Chir. Cond.} = (1/N) \sum_{|z_a|>|z|} (d/dm)\ln(z_a)$$

Fig.(1) shows a plot of the eigenvalues of the propagator matrix for one SU(3) 4^4 configuration at $\beta = 5.3$ and a mass of 0.05. Only eigenvalues within a circle of radius $\exp(\mu)$ contribute to the chiral condensate and it can be seen that there is an area about the unit circle which contains no eigenvalues. Clearly there will be a phase transition when the μ reaches the edge of this area. Fig.(2) shows a plot of the chiral condensate plotted from these eigenvalues. The eigenvalues of the propagator matrix can be related to the hadron spectrum[4]. In particular the eigenvalues nearest the unit circle represent states of the pion and the critical chemical potential is therefore equal to a half of the mass of the pion.

PROBLEMS WITH THE QUENCHED APPROXIMATION

We can identify three aspects of quenched simulations which are at odds with theoretical expectations.

(1) The chiral transition is at a chemical potential $(1/2)m_{\pi}$ instead of $(1/3)m_N$

(2) There is no deconfinement transition at finite density whereas in theory there should not be a chiral transition if there is not a deconfinement transition.

(3) The signal for the transition in terms of the distribution
     of eigenvalues of the fermion matrix is not the same as at
     finite temperature.

   We have already discussed point (1) but note that for SU(2)
this is no problem since mesons and baryons are indistinct. Point
(2) is clearly due to the quenched approximation since the chemical
potential cannot affect the Polyakov loops which are used to measure
deconfinement. This is the case for SU(2) and SU(3). We shall
return to point (3) later.

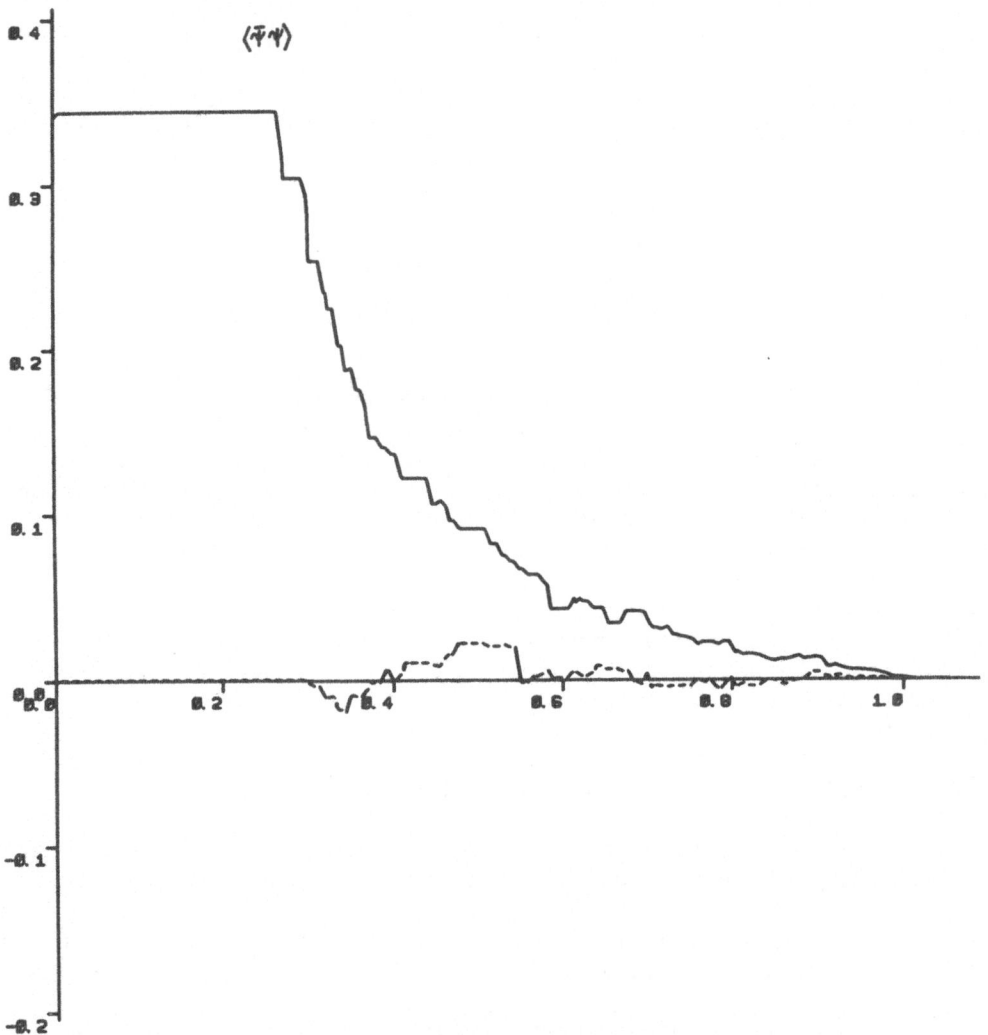

Fig.(2) The chiral condensate plotted as a function of the chemical
potential on the same β = 5.3 configuration.

We can also identify 3 possible reasons for the discrepancies.

(A)   The lattice simulations actually represent real physics and other theoretical ideas are wrong.

(B)   Our formulation of finite density physics on the lattice is wrong.

(C)   The quenched approximation is wrong.

Calculations done in the strong coupling region[5] show that (A) and (B) cannot be the full story and this leaves us with possibility (C).

A U(1) MODEL

By investigating a very simple one-dimensional U(1) gauge model it is possible to see how and why the quenched approximation fails. The NxN fermion matrix is,

$$
H = \begin{pmatrix}
m & z & & & & & -z^{-1} \\
& & & & & & \\
-z^{-1} & m & z & & & & \\
& & & & & & \\
& -z^{-1} & m & & & & \\
& & & \cdot & & & \\
& & & & \cdot & & \\
& & & & & \cdot & z \\
z & & & & & -z^{-1} & m
\end{pmatrix}
$$

where $z = \exp(i\theta + \mu)$. The number of field variables has been reduced to just one θ by a gauge transformation and there is no gauge action. The determinant can be found analytically,

$$
\det(H) = z^{-N} - 2\cosh(Nm'/2) + z^{N}
$$

where $m'/2 = \sinh^{-1}(m/2)$

This enables the chiral condensate to be worked out exactly. In the unquenched theory,

$$
\text{Chir. Cond.} = (1/N)\,(d/dm)\,\ln\!\int (d\theta/2\pi)\det(H)
$$

Taking $N \to \infty$,

$$
= 1/\sqrt{(m^2 + 4)}
$$

This is independent of μ which is what we should expect since there are no baryons in a U(1) gauge theory. However in the quenched theory we find,

$$\text{Chir. Cond.} = (1/n) \int (d\theta/2\pi) \, (d/dm) \ln \det(H)$$

which gives an integral round poles. For small μ the same result as in the unquenched theory is obtained but at $\mu = m'/2$ there is a transition above which the chiral condensate is zero. Thus the quenched theory has a spurious transition.

By examining the form of the determinant we can uncover the source of this failure. The determinant is not a real quantity but if N is large then for $\mu < m'/2$ it is approximately real. However for $\mu > m'/2$ it has a very variable complex phase. Clearly the quenched theory fails because it is wrong to try an approximate a variable phase by a real quantity such as one. In SU(2) the determinant is real because of an extra symmetry. This is consistent with the observation that the quenched approximation works better for SU(2) in the sense that the phase transition is in the correct place. In SU(3) theories the determinant is again not real at nonzero chemical potential and indeed we might expect it to have a very variable phase for $\mu > (1/2)m_\pi$ since the determinant is then sensitive to small fluctuations of the eigenvalues close to z. This gives us every reason to believe that the transition seen in the quenched SU(3) theory is spurious and would not be present in a correct treatment of the full theory.

DYNAMICAL FERMION SIMULATIONS

To find the physical phase transition we must investigate the full theory with dynamical fermions. The large amount of computation required restricts us at present to small lattices. For SU(2) there are hopeful signs that such calculations will be successful but for SU(3) a further complication is encountered[6]. Configurations can only be generated with a real positive probability distribution but the fermion determinant is complex. We must therefore use something like the modulus of the determinant as the weight but it would be wrong to ignore the phase of the determinant since that would only lead to the same problems encountered with the quenched theory. It must therefore be included with the observables.

$$\langle O \rangle = \langle O e^{i\theta} \rangle_c / \langle e^{i\theta} \rangle_c$$

The expectation values of The RHS are averages over the generated configurations. A more detailed description of this method and some results were given by Ian Barbour in his talk.

Unfortunately this method has a serious problem. We have already seen how the phase varies above the spurious transition. This means that the numerator will be small

$$\langle e^{i\theta} \rangle_c \rightarrow 0$$

and its value which must be calculated via cancellations of values of order unity will be subject to large statistical errors. The problem will get worse as the lattice size increases indicating that in fact the principle of importance sampling on which the monte carlo method relies has effectively broken down. There is no problem below the transition where the determinant is nearly real, but this is not an interesting region. Although the physical theory does not pass through a phase transition at $\mu = (1/2)m_\pi$ the monte carlo procedure does and it would be an easy mistake to misinterpret this transition as real.

AN OPTIMISTIC VIEW

So far this talk has been very negative and seems to imply that there may be no way of measuring the phase transition in SU(3). I would like to finish with a more optimistic outlook. Recall point (3): The signal for the chiral phase transition in terms of the eigenvalues of the fermion matrix is different at finite density to the one seen at finite temperature in the quenched theory. At finite temperature the transition is indicated by a movement of the eigenvalues away from the real axis[7]. This does not happen at finite density in the quenched theory yet the transitions are thought to be parts of the same line of transitions in the μ,T phase diagram. Let us suppose that at finite density with dynamic

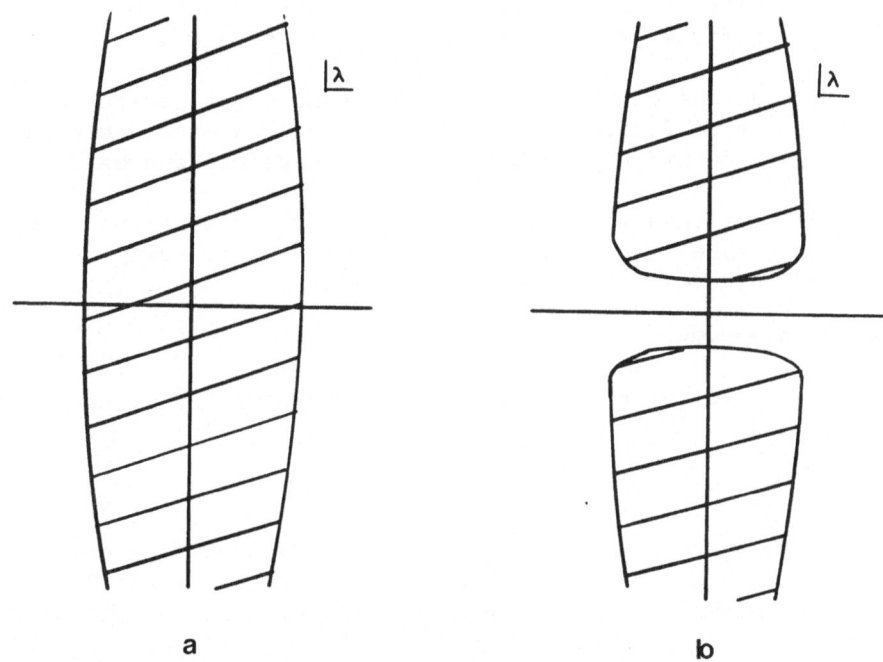

a   b

Fig.(3) Distributions of the eigenvalues of the fermion matrix at finite baryonic density, a; in the chiral symmetry broken phase, b; in the chiral symmetric phase.

fermions there will be a similar movement of eigenvalues away from the real axis above the true transition Fig.(3). It may then be that for larger lattices the phase of the determinant in this region does not fluctuate so much since the eigenvalues have moved clear of the mass pole. Monte carlo simulations will become possible again although reasonably large lattices and statistics will be necessary to measure observables. There still remains a region,

$$(1/2)m_\pi < \mu < (1/3)m_N$$

where calculations are impossible but this is a less interesting region and at least the position of the transition may be measurable.

CONCLUSIONS

By an analysis of the eigenvalues of the propagator matrix we can show that in the quenched approximation there is a transition at "the density of mesonic matter" rather than at "the density of nuclear matter" where it was expected.

The quenched approximation can exhibit spurious transitions at finite density which are not present in the full theory. Thus in SU(3) the quenched approximation is wrong above $\mu > (1/2)m_\pi$.

Simulations of QCD with dynamic fermions at finite density are likely to be difficult and it is essential that the distribution of eigenvalues be understood before any conclusions can be drawn.

The real transition at finite density may be signaled by a movement of the eigenvalues of the fermion matrix away from the real axis and this may make simulations easier.

ACKNOWLEDGEMENTS

I am particularly grateful to Ian Barbour for discussions on this work.

REFERENCES

1.   I.Barbour, N.Behilil, E.Dagotto, F.Karsch, A.Moreo, M.Stone
          and H.W.Wyld, Nucl. Phys. B275 (FS17) (1986) 296.
2.   P.E.Gibbs, 'Lattice Monte Carlo Simulations of QCD at Finite
          Baryonic Density', (To be published in Phys. Lett. B).
3.   J.Kogut, H.Matsuoka, M.Stone, H.W.Wyld, S.Shenker,
          J.Shigemitsu and D.K.Sinclair, Nucl. Phys. B255 (FS9)
          (1983) 93.
      P.Hasenfratz and F.Karsch, Phys. Lett. 125B (1983) 308.
4.   P.E.Gibbs, Phys. Lett. 172B (1986) 53.
5.   E.Dagotto, A.Moreo and U.Wolff, 'Study of lattice SU(N) QCD
          at Finite Baryon Density', Illinois preprint,

ILL-(337)-86-12 (1986).

6.   B.Berg, J.Engels, E.Kehl, B Waltl, H.Satz, Bielefeld preprint, BI-TP 86/05.

7.   I.M.Barbour, P.Gibbs, J.Gilchrist, H.Schnieder, G.Schierholz and M.Teper, Phys. Lett. 136B (1984) 80.

THE QCD GLUEBALL MASS IN THE PRESENCE OF DYNAMICAL

FERMIONS

A. Gocksch

University of California at San Diego
La Jolla, CA 92093

ABSTRACT

First results for the masses of the 0^{++} and 2^{++} glueballs obtained from a simulation of lattice QCD on a $4^3 \cdot 24$ lattice with the effects of dynamical fermions included using the pseudo-fermion method are reported. The masses are determined from the exponential falloff of adjoint line correlation functions. We find that $M_{2^{++}} < M_{0^{++}}$, extending an earlier result by Berg, Billoire and Vohwinkel for the pure gauge case to the full theory.

INTRODUCTION

In this talk I would like to report on some results I have obtained in collaboration with R.V. Gavai and U. Heller on the masses of the lightest glueballs in QCD with fermions. This work was motivated by a recent paper [1] by Berg, Billoire and Vohwinkel who presented evidence that at least in quenched QCD the 0^{++} and 2^{++} glueballs are nearly degenerate. Actually, for all values of the parameters they considered, the $J^{PC} = 2^{++}$ state came out *lighter* by a factor of about 1.2. In the light of earlier estimates of this mass ratio using the Monte Carlo Variational Method [2] this result came as a surprise. However these early calculations were plagued by large statistical errors and the results can only be considered to give upper bounds on the masses. The best estimate of the scalar glueball mass using *local* operators comes from the so called source method [3]. Measuring the correlation of various Wilson loops against a source in the lattice one is able to follow the signal for many time slices and extract accurate results. For the tensor glueball the situation is worse. There the signal dies very quickly in the noise and one is unable to reach any conclusion. A method that gives a good signal in both the scalar and the tensor channel is clearly called for.

In analogy to measurements of the string tension using Wilson lines [4] it was proposed in ref. [5] to use correlation functions of spatial *adjoint lines* to extract the glueball mass, the idea being that an extended object might have a better overlap with the glueball state. An adjoint line is a straight line Wilson loop wrapping around the lattice in one particular direction. The line is closed by the periodic boundary conditions and a trace is taken in the adjoint representation of the gauge group. The authors of ref. [1] appropriately combined adjoint lines in different spatial directions to project onto states with definite quantum numbers. The signal obtained from correlation functions of such operators was good in both the 0^{++} and the 2^{++} channels leading to the result in the mass ratio quoted above. We have essentially repeated their calculation however with the effect of dynamical fermions included. To this end we employed the pseudo-fermion method, the usefulness of which I will comment on later on in this talk. But first of all I would like to discuss some of the theory that is behind the adjoint line correlation function method. The relevant theoretical framework is the formulation of QCD on a torus as originally conceived by t'Hooft [6] and developed into a calculational tool by Lüscher [7]. Both Berg and van Baal talked about this topic in some detail at this conference, so I will be brief in my presentation. The interested reader may consult their contributions to these procedings for more information.

Adjoint line correlation functions are measured on lattices of dimensions $L_s{}^3 \cdot L_t$ with $L_t \gg L_s$. Correlations are measured along the long direction which we call the "time" direction. For the particular kind of geometry described here we can consider correlation functions in the time direction as probing the spectrum of the theory defined in $L_s{}^3$ periodic box. If L_t is large, expectation values, when using the timelike transfer matrix, will actually be given by *vacuum expectation values* with respect to the vacuum in the box. Taking the continuum limit, we expect the spectrum to go over into the spectrum of a *continuum* box, which is perturbatively calculable when the physical size of the box is small [7]. In a box with periodic boundary conditions, the spectrum of QCD (without fermions) factorizes into distinct sectors labeled by *electric flux,* originally introduced by t'Hooft [6]. To all orders in perturbation theory the different electric flux sectors are degenerate but tunneling will split them leading to a unique ground state- the lowest state in the zero electric flux sector [8]. The glueball masses are the lowest energies above the vacuum of excited states in the zero electric flux sector with appropriate quantum numbers. The effect of fermions has not yet been included analytically. Qualitatively we can say the following. When fermions are added to the theory electric flux ceases to be a good quantum number. The fermions will induce mixing between the different electric flux sectors and lift their degeneracy.

Measurements of correlations of Wilson lines, i.e. spatial lines with traces in the fundamental representation will give the *energy of electric flux* which is the energy of the lowest state in the electric flux $|\vec{e}\,|=1$ sector above the vacuum. For large box sizes this energy is related to the string tension by $\Delta E = \sigma \cdot L_s$. Adjoint line correlation functions on the other hand determine the lowest excited state of given quantum numbers in the zero electric flux sector, i.e. the corresponding glueball. In the perturbative calculation [7] the lightest glueball has $J^{PC} = 2^{++}$.

Since the expectation value of an adjoint line decreases exponentially with it's length the method is useful only for rather moderate values of the spatial size L_s. The largest lattice considered in ref. [1] had $L_s = 8$. For this reason the glueball interpretation of the lowest state in the zero electric flux sector has recently been called into question [9]. There it is argued that in the spectrum of the Hamiltonian there should be states made up of 2 flux lines, running in opposite directions. The author of ref.[9] claims that such states, which he calls "torelon-antitorelon pairs" [†], can easily dominate the signal of adjoint line correlation functions if their mass is smaller than that of the glueball. Since their mass is given by $M_{T\bar{T}} = 2\sigma L_s$, *neglecting* interactions, this scenario could actually apply to the calculations of ref. [1].

We do not believe, that what is measured in the correlation functions of adjoint lines are "torelon-antitorelon pairs" for the following two

---

[†]In the earlier version of the paper these states were called "toron-antitoron pairs". The word "toron" however already has a different meaning. A toron is a zero action solution of the classical field equations on the torus.

reasons. The data of ref.[5] do not show the L_s dependence of the masses one would expect for such objects. Furthermore, the mass obtained by Gupta *et. al.* ref.[3] for the 0^{++} state using the the source method *agrees* with the one obtained using the adjoint line method. The source method can be used for lattices of larger spatial size L_s and hence can avoid the problem with a possible contamination from "torelon-antitorelon pairs". Moreover in the source method one uses *local* operators which presumably have little overlap with these objects. Although we are confident that one is actually measuring "glueballs" using adjoint line correlation functions, one must nevertheless be careful in extrapolating the results obtained on small lattices to the infinite volume limit. Only when the size of the box becomes large in *physical* units, can we hope to extract realistic numbers. A more detailed discussion of the problems involved in interpreting results obtained on small lattices of the type discussed above can be found in ref.[10]. There the issue of T_C which was discussed in Berg's lecture will also be addressed.

The results that I am presenting here constitute only a small first step torwards a real understanding of the glueball in full QCD. The problems to be addressed in the future are both difficult and fascinating. These include the amount of mixing with flavour singlet mesons of the same quantum numbers and decays of these particles. We are presently studying possible ways to tackle these questions. Theoretically, not much is known about the effect of fermions on the glueball spectrum. Naively, one might think that their effect will be small, for example on the basis of large N arguments.

THE CALCULATION

We have simulated lattice QCD defined by the following action:

$$S = \frac{\beta}{2N} \sum_P (tr U_P + h.c.) + S_F \qquad (1)$$

where S_F is given by

$$S_F = \frac{1}{2} \sum_{x,y,\mu} \eta_\mu(x) \overline{\chi}(x) \left[U_\mu(x) \delta_{x,y-\mu} - U^\dagger_\mu(y) \delta_{x,y+\mu} \right] \chi(y) + \sum_x m \, \overline{\chi}(x) \chi(x).$$

In (1) β is related to the gauge coupling constant by $\beta = \frac{6}{g^2}$ and U_P is the product of links $U_\mu(x)$ around an elementary plaquette in the lattice. S_F in (1) and (2) is the fermionic action depending on the anti-commuting

Grassmann fields $\chi(x)$ and $\bar{\chi}(x)$. The Kogut-Susskind phase factors are given by $\eta_\mu(x)=(-1)^{x_1+x_2+\cdots+x_{\mu-1}}$ and the fermion mass is given by m. Integrating out the fermion fields in (2) leads to the determinant of the lattice Dirac operator which we have in our simulation raised to an appropriate power so that the number of flavours is effectively three. The pseudo-fermion method that was used here is discribed in detail in ref.[11]. From our own investigations [12] we know that the method can be used reliably when the quark mass is not too small. Here the quark mass was chosen to be $m=0.1$ and the acceptance rate in the gauge field update was 80 per cent. The number of pseudo-fermionic hits was 24 with 4 hits discarded for thermalization. Our lattice size was $4^3 \cdot 24$ with boundary conditions periodic in space and antiperiodic in time for fermions and periodic in all directions for the gauge fields. We ran at three different values of β, $\beta=5.8$, 6.0 and 6.2. To check our program we also ran at $\beta=6.0$ in the pure gauge theory. Our results in that case for the $M_{2^{++}}$ to $M_{0^{++}}$ ratio as well as the ratio of the "string tension" to the square of the glueball mass are consistent with the numbers quoted in ref.[1] obtained on the same size lattice. Denoting the adjoint line in direction i at time slice t by $A_i(t)$ we measured the following correlation functions:

$$C_0(t)=<\left[\sum_{i=1,3} A_i(t)\right]\cdot\left[\sum_{j=1,3} A_j(0)\right]>-<\sum_{i=1,3} A_i>^2 \qquad (2)$$

$$C_2(t)=<\left[A_1(t)-A_2(t)\right]\cdot\left[A_1(0)-A_2(0)\right]> \qquad (3)$$

From C_0 we determine the 0^{++} mass while C_2 is used to get the mass of the 2^{++} glueball. We also measured the connected correlation function C_f of Wilson lines whose exponential falloff determines the energy of electric flux. While in the pure gauge system we can use $\Delta E=\sigma(L_s)\cdot L_s$ to determine the "string tension", we here expect this formula to be modified due to the screening effect of the fermions. The results of our measurements of C_f can be found in our paper [13].

Our statistics for the pseudo-fermionic runs is summarized in table 1. Note that it was necessary to discard a large number of sweeps. The correlation functions were measured every fifth sweep.

Note that in the presence of dynamical fermions it is *not* possible to use DLR improved measurements [4] of the correlation functions due to the non-local nature of the interaction. We also measured in addition to the above mentioned functions correlations of plaquettes and 2×2 Wilson loops. We were however unable to extract a signal from them. Our results

Table 1. Summary of our Statistics

β	*sweeps*	*sweeps discarded*
5.8	50000	10000
6.0	40000	10000
6.2	30000	15000

for the masses are shown in table 2. All masses quoted are obtained from least square fits to a cosh function. All fits are generally very good. We always included the value of the correlation function at time slice $t = 1$ in our fits, excluding it did not significantly alter our results. This means that the contribution of higher states is small. We determined our errors on the masses in the following way. Partitioning our data into bins of 1000 measurements each, we fitted a mass in each bin and used the standard error of the mean as an error estimate for the masses. All errors in table 2 were obtained that way. One should note that an error estimate coming from a single fit through *all* data will be unrealistically small since the data points are correlated. The fluctuations of the masses from bin to bin were significant leading to a typically 3–4 fold increase of the error over the naive estimate coming from a single fit. The autocorrelation length in the correlation functions was typically of $O(50)$ sweeps.

The statistics of the data that I have presented here is not good enough to warrant a plot of our data versus the scaling variable $z = M \cdot L_s$. Even if there was the kind of "z–scaling" advocated in ref.[1] for ratios of physical quantities, we would not be able to see it with our poorer statistics. The value of $M_{2^{++}}$ at $\beta = 6.2$ is larger than at $\beta = 5.8$ contrary to what one might expect in an infinite volume from the renormalization group flow. But with increasing β the physical volume becomes smaller and therefore finite size effects larger. So our mass estimates are clearly not yet asymptotic. However, the quality of our signal clearly indicates that there exist stable states in QCD that couple to purely gluonic operators. In addition, the 2^{++} state is significantly lighter than the 0^{++} state.

Whether this result holds up on larger lattices and weaker couplings remains to be seen. We have no reason to doubt it based on the experience with the pure gauge theory. But it *is* possible that finite volume distortions are more significant in the full theory. Our masses at $\beta = 5.8$ are actually larger than the masses extracted in the pure gauge theory at $\beta = 6.0$. If a simple shift in β of that order sufficed to take into account most of the effect of dynamical fermions, we would expect the masses to

Table 2. Glueball masses

β	$M_{0^{++}}$	$M_{2^{++}}$	$\dfrac{M_{0^{++}}}{M_{2^{++}}}$
5.8	0.88(0.15)	0.52(0.1)	1.69(0.4)
6.0	0.81(0.1)	0.50(0.07)	1.62(0.3)
6.2	0.77(0.06)	0.59(0.05)	1.31(0.17)

go down on the basis of mixing with mesonic states. Finite volume effects however will tend to bring the masses up with respect to the pure gauge theory if they are more severe when the effect of dynamical fermions is included.

Let me summarize. We have studied the masses of the low lying glueballs in QCD using the adjoint line correlation function method. In terms of the scaling variable z we are working relatively far from infinite volume, continuum physics. Here z is about 2.5. To see the physical masses, one should probably go past $z = 5$. We do believe however that we have demonstrated the feasability of a more detailed study. The signal in adjoint line correlation functions is very good even when the fermions are added. We see no problem in going to larger lattices and hence larger values of z.

ACKNOWLEDGMENTS

I would like to thank the organizing comittee for making this a very stimulating and enjoyable meeting. The computations were carried out on the Cray XMP-48 of the San Diego Super Computer Center. The computations reported here took about 100 hours of CPU time. This work was supported by the U.S. Department of Energy under contract number DE-AT03-81-ER40029.

REFERENCES

1. B.Berg, A. Billoire and C. Vohwinkel, Phys. Rev. Lett. 57, 400 (1986).
2. K. Ishikawa, G. Schierholz and M. Teper, Z. Phys. C19, 327 (1983);
   B. Berg, A. Billoire, S. Meyer and C. Panagiotakopoulos, Comm. Math. Phys. 97, 31 (1985).
3. Ph. deForcrand, G. Schierholz, H. Schneider and M. Teper, Phys. Lett 152B, 107 (1985);
   R. Gupta, G. Guralnik, G.W. Kilcup, A. Patel and S.R. Sharpe, preprint UCSD-10P10-260, 1986);
   T.A. DeGrand and C. Peterson, preprint COLO-HEP-117, (1986).
4. G. Parisi, R. Petronzio and F. Rapuano, Phys. Lett. 128B, 418 (1983).
5. B. Berg and A. Billoire, Phys. Lett. 166B, 203 (1986).
6. G. 't Hooft, Nucl. Phys. B153, 141 (1979).
7. M. Lüscher, Nucl. Phys. B219, 233 (1983);
   M. Lüscher and G. Münster, Nucl. Phys. B232, 445 (1984).
8. P. van Baal and J. Koller, preprints ITP-SB-86-31 (1986); ITP-SB-85-56 (1986), final version.
9. C. Michael, Urbana preprint 86-0413 (1986), Liverpool preprint LTH 147 (1986), revised version..
10. A. Gocksch, J. Kuti and F. Niedermayer (in preparation).
11. F. Fucito, E. Marinari, G. Parisi and C. Rebbi, Nucl. Phys. B180 [FS2], 369 (1981).
12. R.V. Gavai, A. Gocksch and U.M. Heller, preprint NSF-ITP-86-89 (1986); to be published in Nucl. Phys. B.
13. R.V. Gavai, A. Gocksch and U.M. Heller, preprint UCSD-10P10-268, (1986).

MORE ON THE FIRST ORDER CHIRAL SYMMETRY

TRANSITION IN QCD

Rajan Gupta

J. Robert Oppenheimer Fellow
Theoretical Division, MS-B285
Los Alamos National Laboratory,Los Alamos, NM 87545

ABSTRACT

We present evidence from numerical simulations for a first order chiral symmetry restoration transition in QCD at finite temperature. Also, for quarks in the fundamental representation, there is a simultaneous deconfinement transition. These transitions are present only for small quark masses. We use an exact algorithm to incorporate the dynamical quarks.

In [1] we presented evidence for the chiral symmetry restoration and deconfinement transition at $6/g^2 = 4.9$ and $m_q = 0.025$. In this talk I shall present results at $6/g^2 = 4.95$. The work was done in collaboration with G. Guralnik, G. Kilcup, A. Patel and S. Sharpe at Los Alamos.

We are pursuing a program aimed at developing an algorithm efficient at small g and m_q. In particular we want to investigate how and when systematic biases become dominant in various approximate algorithms. To do this we have made extensive runs using the exact algorithm of Scalapino and Sugar [2] to act as a standard for comparison. The largest lattice on which this can be done with reasonable statistics is $4 \times 4 \times 4 \times 4$. In exploring the small quark mass limit we find evidence for the chiral transition. In this talk I will concentrate on this physics result. Due to the small lattice size, 4^4 , and the fact that $N_{spatial} = N_{time}$, we do not claim any quantitative results. Clearly, in a realistic finite temperature study, the transition temperature will be shifted. However, we think that the clarity of the signal of the transition in our data, together with our use of an exact algorithm, make our qualitative result interesting.

Simulations of pure gauge SU(3) show a strong first order transition at a temperature $T_c \approx \Lambda_{\overline{MS}}$ [3]. At this transition the global $Z(3)$ symmetry of the theory is spontaneously broken. A non-zero expectation value of the Polyakov line $\langle L \rangle$ in the high temperature deconfined phase implies a finite free energy for the quarks. A second order parameter, the chiral condensate $\overline{\chi}\chi$ measured in the quenched approximation, is also discontinuous at the transition. $\overline{\chi}\chi$, when extrapolated to $m_q = 0$, changes from a non-zero value at low T to zero in the high T phase.

Dynamical quarks act as external fields and explicitly break the $Z(3)$ symmetry. $\langle L \rangle$ is still a measure of the quark free energy but it is non-zero for all temperatures due to vacuum polarization. $\overline{\chi}\chi$ remains a good order parameter to study chiral symmetry. The only theoretical understanding of the realization of chiral symmetry comes from a renormalization group analysis of an effective spin model in $4 - \epsilon$ dimensions [4]. The conclusion is that QCD has a first order chiral symmetry transition for $N_f \geq 3$ and at $m_q = 0$. For $T < T_c$, one expects $\overline{\chi}\chi \neq 0$ when extrapolated to $m_q = 0$. For $T > T_c$ the chiral symmetry is restored, consequently $\overline{\chi}\chi \propto m_q$ for small m_q. This needs to be verified. Also, if, as in the pure gauge theory, there is a discontinuity in $\langle L \rangle$, then we expect to see interesting thermodynamical properties of the quark-gluon plasma [5] created in heavy ion collisions.

The expected phase diagram for QCD is as follows: The confinement transition at $m_q = \infty$ extends to some finite m_q in the $m_q - T$ phase plane, and similarly the chiral transition at $m_q = 0$ extends to some non-zero m_q. The questions to settle are whether these two transitions are connected and whether the chiral transition with the three physical light flavors is first order. The status of the chiral transition is not clear. We summarize the results for 4 flavors of staggered fermions obtained using approximate algorithms. The most detailed calculations are by Kogut *et al.* [6] who find a rapid crossover for $m_q = 0.1$ and 0.05. They extrapole the center of the cross-over to $m_q = 0$, and thereby estimate the transition coupling $6/g^2$ for $N_t = 6$ (4) to be 5.01 ± 0.025 (≈ 4.9). Further, assuming that asymptotic scaling is valid, they estimate the transition temperature to be $T_c = (2.14 \pm 0.1) \, \Lambda_{\overline{MS}}$. These results are obtained with both the micro-canonical and the hybrid algorithm. Similarly, Gavai, [7] does not find evidence for a first order transition using the pseudo-fermion algorithm. On the other hand, Fucito and Solomon [8] claim to see a first order transition at $m_q = 0.1$. They used perturbation theory to reduce the number of flavors to 3. Fukugita and Ukawa, [9] use the Langevin algorithm and also claim that at $m_q = 0.1$ the transition is already first order. The chief criticism against the last two calculations which found a hysteresis, *i.e.* two metastable states, has been that the runs were not long enough for complete thermalization. Also, the algorithm used to simulate fermions is different in each study so the reason for conflicting results may lie in the nature of the bias introduced due to the approximation.

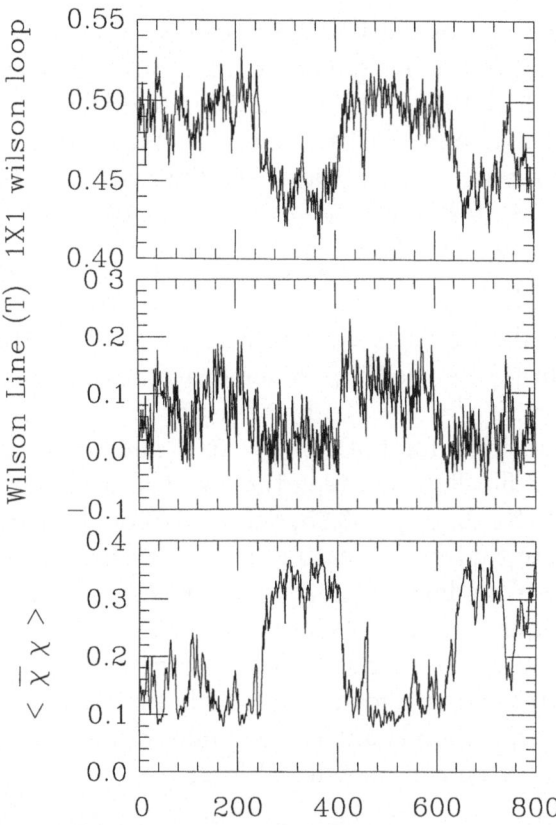

Fig.1   Plot of a) 1×1 Wilson loop, b) $\langle L \rangle$ and c) $\overline{\chi}\chi$ with $\beta = 4.95$, $m_q = 0.05$ and $N_{cg} = 60$ versus Monte Carlo Sweeps.

In the exact algorithm the ratio of determinants $R \equiv det(1 + M^{-1}\delta M)$ is calculated at each link update. Since we use staggered fermions (4 flavors), the algorithm requires a calculation of a 6×6 block of M^{-1}. Because M^{-1} is calculated with the conjugate gradient (CG) iterative algorithm to some approximation, there can be a systematic bias. We discuss this later. In a Metropolis update, a link can be changed many times without having to recalculate M^{-1}. The multi-hit algorithm we use is that described in detail by Gavai and Gocksch [10]. We use antiperiodic boundary conditions in all directions. We update each link with 50 hits and the acceptance is adjusted to $\approx 30\%$. In solving $A x_{even} = M^{\dagger} M x_{even} = b$, we define the convergence by $C = \frac{\langle b - A x | b - A x \rangle}{\langle x | x \rangle}$, which depends on the number of CG iterations (N_{cg}).

In our data all the observables, $\overline{\chi}\chi$, $\langle L \rangle$, Wilson loops and R are correlated. In [1] we used $\overline{\chi}\chi$ to demonstrate that the transition for $m_q = 0.025$ exists at $6/g^2 = 4.9$. In Figs. 1 and 2 we show that with $N_{cg} = 60$ there is a transition for both $m_q = 0.05$ and $m_q = 0.025$ at $6/g^2 = 4.95$. The correlated flip-flops between the 2 states can be seen in $\overline{\chi}\chi$, $\langle L \rangle$ and in Wilson loops. So for $m_q = 0.05$, the transition appears between $6/g^2 = 4.9$ and 4.95. Note that on increasing $6/g^2$, the physical temperature increases for fixed N_t and so does the quark mass when the bare lattice quark mass is held fixed. Thus, for any given constant m_q, and N_t we expect to go from the confined to the deconfined phase on increasing $6/g^2$. This is what we see with the transition remaining strongly first order. We also find that the nature of the transition is unchanged under changing the 3 spatial boundary conditions to periodic as shown in Fig 2c.

To further analyze the transition we study $\overline{\chi}\chi$ as a function of m_q. The estimates for $\overline{\chi}\chi$ in the two phases are shown in Fig. 3. In the confined phase we estimate $(\overline{\chi}\chi - 0.3) \propto m_q$ from the data at $6/g^2 = 4.9$. To study $\overline{\chi}\chi$ in the high temperature phase, we made runs with $N_{cg} = 60$ at $6/g^2 = 4.95$ and $m_q = 0.05$, 0.025, 0.02 and 0.015. We find a flip-flop at the two heavier m_q, while the system is predominately in the deconfined phase at the two smaller m_q. The agreement with the expected behavior $\overline{\chi}\chi \propto m_q$ in the deconfined phase is good. Unfortunately, this behavior though significant is not sufficient proof of the order of the transition. Thus a corraboration on lattices with larger N_t and with $N_s \gg N_t$ is necessary. Further shortcomings of using Staggered fermions are that the flavor symmetry at finite lattice spacing is not $SU(n_f)$ and second the phenomenologically interesting case of 3 flavors is hard to achieve.

In [1] we presented an analysis of the systematic biases in our simulation. Our implementation of the CG algorithm tends to underestimate the effects of the fermions, i.e. it tends to give too small a value for $S \equiv |ln(\mathrm{R})|$. We have studied this by changing a single link and comparing the exact R with that calculated with a variety of CG sweeps. The exact R is obtained by calculating the determinants, before and after changing the link, using gaussian elimination. To study if there is an accumulation of the bias, we compare the product of the accepted determinant ratios $(A \equiv ln R_{acc})$ with the exact answer (T).

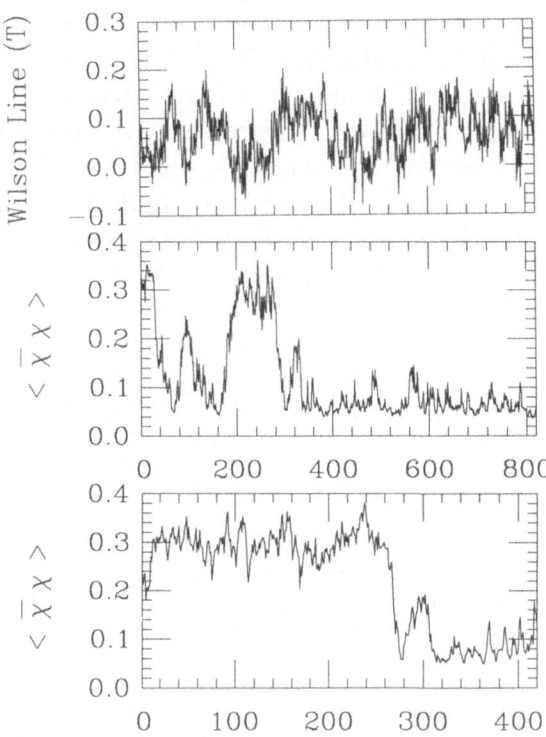

Fig.2   Plot of a) $\langle L \rangle$, b) $\overline{\chi}\chi$ with anti-periodic b.c. in all directions and c) $\overline{\chi}\chi$ with periodic b.c. in space –– at $\beta = 4.95$, $m_q = 0.025$ and $N_{cg} = 60$ versus Monte Carlo Sweeps.

The bias at $6/g^2 = 4.95$ is similar to that at 4.9. The disagreement between A and T gets progressively worse with decreasing N_{CG}, but C is consistently a factor of ≈ 20 smaller in the high temperature phase. Conversely, the bias decreases as m_q increases. We again find that $C \leq 10^{-7}$ is necessary to avoid a bias at $m_q = 0.025$ in both phases.

In Fig. 4 we show a distribution of the generated values of R and those accepted in the Metropolis algorithm. We expect this distribution of R to be a gaussian for a single link change in a fixed environment. That this is still approximately true when averaged over all links, as shown by the data, suggests that in a local update algorithm, the environment of any link is well described by a mean field approximation. Some properties of the distribution of the generated values of R are; 1) the mean which is always less than one becomes smaller and 2) the width increases with decreasing quark mass and increasing $6/g^2$. However, the change is small on going from $m_q = 0.05$ to $m_q = 0.015$. The relevance of this data is that for a method that gives an unbiased estimator for R to be efficient, the spread of the estimate must be smaller than the narrow width shown.

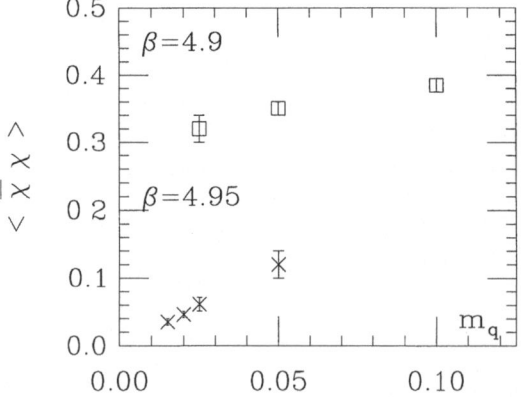

Fig.3   Plot of $\bar{\chi}\chi$ versus the bare quark mass in lattice units. The expected chiral behavior is seen in both phases.

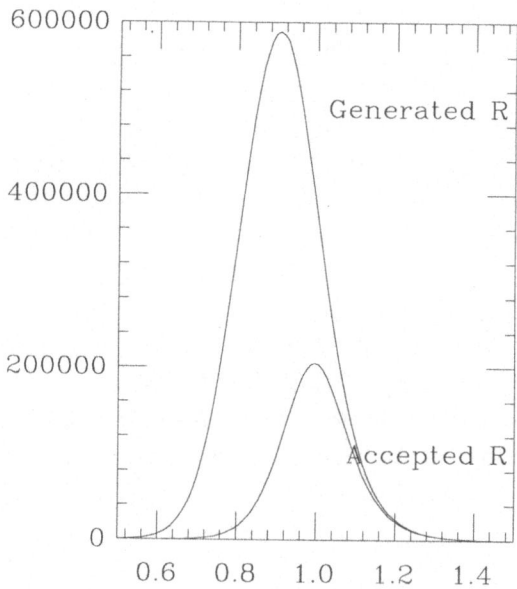

Fig.4   Distribution of the value of R generated at each hit per link update. Also shown is the distribution of R for the hits accepted in the Metropolis algorithm. The data is at $\beta = 4.95$, $m_q = 0.025$ and $N_{cg} = 60$.

To conclude, we present evidence that there does exist a first order chiral/thermal transition at small m_q. Since hadronic matter at high temperature and density is expected to undergo a transition to quark-gluon plasma, which can be checked by the planned heavy ion experiments, it is therefore very important to understand the nature of the transition and determine the transition temperature. Further, a first order transition in standard cosmology (thermal history) would dump entropy and affect the cosmological constant unless it occurs without appreciable supercooling. Thus one needs to know the transition rate. I hope these questions can be addressed by future lattice calculations.

References

[1] R. Gupta *et. al. Phys. Rev. Lett.* **57** (1986) 2623.
[2] D. J. Scalapino and R. L. Sugar, *Phys. Rev. Lett.* **46** (1981) 519.
[3] J. B. Kogut, H. Matsuoka, S. H. Shenker, J. Shigemitsu, D. K. Sinclair, M. Stone and H. W. Wyld, *Phys. Rev. Lett.* **51** (1983) 869.
[4] R. D. Pisarsky and F. Wilczek, *Phys. Rev.* **D29** (1984) 338.
[5] B. Svetitsky, Proceedings of the Fifth International Conference on Ultra-Relativistic Nucleus-Nucleus Collisions, *Quark Matter, 86* Asilomar; and M. Matsua, *ibid.*
[6] J. B. Kogut and D. K. Sinclair, Urbana Preprint ILL-(TH)-86-86-46.
[7] R. Gavai, *Nucl. Phys.* **B269** (1986) 530.
[8] F. Fucito and S. Solomon, *Phys. Rev. Lett.* **55** (1985) 2641.
[9] M. Fukugita and A. Ukawa, *Phys. Rev. Lett.* **57** (1986) 503.
[10] R. Gavai and A. Gocksch, *Phys. Rev. Lett.* **56** (1986) 2659.

THE CONTINUUM LIMIT OF THE STANDARD SU(2) GAUGE–HIGGS MODEL

Anna Hasenfratz

Supercomputer Computations Research Institute
Florida State University
Tallahassee, FL 32306

ABSTRACT

It is shown that the continuum limit of the standard model defined around the Gaussian fixed point is a free field theory.

INTRODUCTION AND SUMMARY

In recent years many Monte-Carlo (MC) studies dealt with the phase structure of gauge-Higgs models [1] . One of the most interesting one is the SU(2) gauge system coupled to fundamental Higgs field as it describes the scalar sector of the Weinberg-Salam model. The MC calculations indicate first order phase transition between the Higgs and symmetric phase at every finite value of the gauge coupling (g^2). This suggest that a continuum limit can be defined only on the $g^2 = 0$ hyperplane. However, at $g^2 = 0$ one is left with a scalar field theory which triviality is almost rigorously proven. It means that the only fixed point around which a continuum limit can be defined on the $g^2 = 0$ hyperplane is the Gaussian fixed point (at zero scalar coupling). The basic question then is the following: Although the bare gauge coupling is tuned towards zero, physical gauge coupling should be finite. Can this non-zero renormalized coupling create a non-trivial effective potential? If so, the Higgs-boson mass can be determined from the W-boson mass and the value of the renormalized gauge coupling [2].

One should emphasize that the behaviour of a field theory around a Gaussian fixed point (FP) is perturbative problem. The answer to the questions above can be given with the help of perturbation theory and renormalization group considerations.

In Ref.[3] we showed that in the Higgs phase the renormalized gauge coupling will be zero as the cut-off is removed to infinity, therefore the scalar potential will be trivial. The continuum theory defined on the Gaussian FP is a free field theory with an arbitrary ratio $R = \frac{m_H}{m_W}$. In this paper I will summarize these results.

NOTATIONS

The Lagrangean has the form ($d = 4$, Euclidean space):

$$\mathcal{L} = \frac{1}{4} \sum_{a=1}^{3} F_{\mu\nu}^a F_{\mu\nu}^a + \sum_{i=1}^{2} (D_\nu \Phi)_i^* (D_\nu \Phi)_i + \frac{1}{2} r_o \sum_{\alpha=1}^{4} \varphi_\alpha^2 + \frac{u_o}{4!} \left(\sum_{\alpha=1}^{4} \varphi_\alpha^2 \right)^2 , \quad (1)$$

where

$$\Phi = \frac{1}{\sqrt{2}} \begin{pmatrix} \varphi_1 + i\varphi_2 \\ \varphi_3 + i\varphi_4 \end{pmatrix}$$

is the scalar doublet, while the covariant derivative is defined as

$$D_\nu = \partial_\nu \mathbf{1} - i g_o A_\nu^a \frac{\tau^a}{2} .$$

We shall use dimensional regularization, Landau gauge and always work in the Higgs phase. The renormalization mass is denoted by Λ, and all dimensions are carried by this arbitrary scale. When $\Lambda \longrightarrow \infty$, it plays the role of the cut-off and the parameters r, u, g^2 can be considered as bare parameters. We shall call the parameters on the scale of the W-mass "physical", and denote them by $r_p \, (= r \, (m_W))$, u_p and g_p^2.

EFFECTIVE POTENTIAL AND THE RENORMALIZATION GROUP EQUATIONS

We want to know under what condition the system will be in the Higgs phase at the physical scale and what are the renormalized couplings there.

The renormalization group equations and their solutions, which tell us how the couplings change under the scale change $\Lambda \longrightarrow e^{-t}\Lambda$ ($t > 0$) at 1-loop level are well known [4]. The equations

$$\frac{d\bar{g}^2}{dt} = \beta_0 \bar{g}^4 \quad , \qquad \beta_0 = \frac{1}{16\pi^2}\frac{86}{6} \quad , \tag{2}$$

$$\frac{d\bar{u}}{dt} = -\frac{1}{16\pi^2}\left[4\bar{u}^2 - 9\bar{u}\,\bar{g}^2 + \frac{27}{4}\bar{g}^4\right] \quad , \tag{3}$$

$$\frac{d\bar{r}}{dt} = \left(2 + O(u, g^2)\right)\bar{r} \quad , \tag{4}$$

$$\frac{d\bar{x}^2}{dt} = \left(2 - O(g^2)\right)\bar{x}^2 \quad , \tag{5}$$

and their solutions

$$\bar{g}^2 = \frac{g^2}{1 - \beta_0 g^2 t} \quad , \tag{6}$$

$$\bar{u} = \bar{g}^2\left\{\frac{a}{2}\tan\left[\frac{a}{b}\ln(1 - \beta_0 g^2 t) + \delta\right] - \frac{2}{3}\right\} \quad , \tag{7}$$

$$\bar{r} = re^{2t}, \tag{8}$$

$$\overline{M}^2 = M^2 e^{2t} \quad , \tag{9}$$

where

$$\delta = \delta(u, g^2) = \arctan\left(\frac{2u/g^2 + \frac{4}{3}}{a}\right) \quad , \tag{10}$$

and the constants a and b are given by

$$a = \sqrt{\frac{27}{4} - \frac{16}{3}} \quad , \tag{11}$$

$$b = 8\pi^2\beta_0$$

To study the phase structure we will investigate the scalar effective potential [5]. It has the form

$$V_{eff}(x) = \frac{1}{2}rx^2 + \frac{1}{4!}ux^4 + \text{1-loop corrections } O(u^2, ur, r^2, ug^2, rg^2, g^4)$$

$$+ \text{ higher order terms} \tag{12}$$

Although a simple calculation gives all the terms at 1-loop level, we do not write it out explicitly as we will not need it in the following.

Starting at the scale of the cut-off, Λ, we scale down to physical energies. The renormalization group equation for the effective potential is

$$V_{eff}(x; r, u, g^2) = e^{-4t} \quad V_{eff}\left(\bar{x}(t); \bar{r}(t), \bar{u}(t), \bar{g}^2(t)\right) , \tag{13}$$

While changing the scale we have to make sure that a) the running coupling in Eq's (6-9) are small so we can use perturbation theory and b) the logarithmic terms in the effective potential (which typically have the form $ln\frac{m_H}{\Lambda_{cut}}$, $ln\frac{m_W}{\Lambda_{cut}}$) are under control [6]. This second constraint suggest running the renormalization group equations until the (dimensionless) W-mass is in the order of 1. At this energy scale we can use the tree-level relation $\bar{m}_W^2 = \frac{1}{4}\bar{g}^2\bar{M}^2 = 1$, where $\bar{M}^2 = <\phi^2>$.

To be specific, we shall take $t = t^*$ where

$$\ln\left(\frac{1}{4}\bar{g}^2(t^*)\bar{M}^2(t^*)\right) = 0 \quad . \tag{14}$$

Eq.'s (6-9) tell us the couplings at this energy scale. They will be indexed by "p" in the following $\left(\bar{g}^2(t^*) \equiv g_p^2, \ldots\right)$.

The next question is under what condition the system will be in the Higgs phase. If there exist a non-trivial minimum at $\bar{M}$, we have

$$\frac{dV_{eff}}{d\bar{x}}\bigg|_{\bar{x} = \bar{M}} = 0$$

Using the tree level form of the effective potential we obtain

$$r_p = -\frac{2}{3}\frac{u_p}{g_p^2} + O(u) \tag{15}$$

Although this equation looks like a constraint between the couplings, it is not. It just expresses the fact that we have selected a specific energy scale, it is the equation of the surface in the parameter space where $m_W = 1$.

The system is in the Higgs phase if

$$V_{eff}(\overline{M}) \leq V_{eff}(0) \quad , \tag{16}$$

From Eq.'s (12) and (14) we obtain the constraint

$$\frac{u_p}{g_p^2} \geq O(g_p^2) \tag{17}$$

Using the solution of the renormalization group equations (6-9) we express the left hand side in terms of g_p^2 and g^2. On the critical surface separating the Higgs and symmetric phase

$$\frac{u_p}{g_p^2} = \frac{a}{2} \tan \left[\delta(u, g^2) - \frac{a}{b} \ln \frac{g_p^2}{g^2} \right] - \frac{2}{3} = O(g_p^2), \tag{18}$$

The solution of Eq(18)

$$\frac{g_p^2}{g^2} = e^{\frac{b}{a}\left(\frac{\pi}{2} - \arctan \frac{4}{3a} \right)} \quad . \tag{19}$$

Since δ is bounded by $\pi/2$, we obtain

$$\lim_{g^2 \to 0} \left(g_p^2/g^2 \right) \leq e^{\frac{b}{a}\left(\frac{\pi}{2} - \arctan \frac{4}{3a} \right)} \approx 27.6 \quad , \tag{20}$$

which implies that $g_p^2 \to 0$ as g^2 is tuned towards the Gaussian point along the singular surface.

Inside the Higgs phase, away from the singular surface, the left hand side of Eq(20) gets larger , therefore the $\frac{g_p^2}{g^2}$ becomes smaller. It follows that in the $g^2 \to 0$ limit the physical gauge couplings g_p^2 becomes zero, no gauge interaction remains. Without gauge interaction the scalar field cannot sustain non-zero self interaction u_p either.The continuum limit will be a trivial free field theory.

Until now we have not used the explicit form of the 1-loop effective potential. However,if we keep the $O(g^4)$ term, we easily get the Linde-Weinberg lower bound for R [7].

The $O(g^4)$ contribution to the effective potential is

$$\frac{1}{(8\pi)^2} \frac{9}{16} g^4 x^4 \left[\ln \left(\frac{1}{4} g^2 x^2 \right) - \alpha \right] \tag{21}$$

where α depends on the subtraction procedure used. Eq.(17) is modified

$$\frac{u_p}{g_p^2} \geq -\frac{27}{256\pi^2}g_p^2 \tag{21}$$

We can calculate the Higgs mass from the second derivative of the effective potential,while $m_W = 1$ at this scale. For the ratio we get

$$R^2 \equiv \frac{m_H^2}{m_W^2} = \frac{4}{3}\frac{u_p}{g_p^2} + \frac{9}{32\pi^2}g_p^2 \quad , \tag{22}$$

Using Eq.(21) we obtain

$$R_{\min}^2 = \frac{9}{64\pi^2}g_p^2 . \tag{23}$$

The ratio R takes its minimum value on the singular surface. The bound Eq.(23) is the well known Linde-Weinberg bound.

Although the cut-off independent theory in the $g^2 \to 0$ limit is a massive free field theory, one can define an interacting effective theory with a finite cut-off [8]. Such an effective theory could predict an upper bound for the ratio R [9], but this is clearly outside the range of perturbation theory.

ACKNOWLEDGEMENTS

This work was done in collaboration with Peter Hasenfratz [3]. Work supported in part by a grant from the U.S. Department of Energy (DE-FC05-85ER250000).

REFERENCES

(1) C.B. Lang, C. Rebbi and M. Virasoro, *Phys. Lett.* **104B** (1981) 294.

M. Tomiya and T. Hattori, *Phys. Lett.* **104B** (1984) 370.

J. Jersak, C.B. Lang, T. Neuhaus, and G. Vones, *Phys. Rev.* **D32** (1985) 2761.

V.P. Gerdt, A.S., Ilchev, V.K. Mitrjushkin, I.K. Sobolev and A.M. Zadorozhny, *Nucl. Phys.*, **B265**, [FS15] (1986) 145.

V.P. Gerdt, A.S., Ilchev, V.K. Mitrjushkin, I.K. Sobolev and A.M. Zadorozhny,*Phys. Lett.*, **172B** (1986) 65.; Dubna Preprint **E2-85-104**.

H.G. Evertz, J. Jersák, C.B. Lang and T. Neuhaus, Aachen Preprint **PITHA 85/23**.

H.G. Evertz, V. Grösch, J. Jersák, H.A. Kastrup, D.P. Landau, T. Neuhaus and J.-L. Xu, *Phys. Lett.*, **175B** (1986) 335.

K. Decker, I. Montvay and P. Weisz, Preprint **DESY 85-123**.

For a recent summary, see J. Jersák, Preprint **PITHA 85/25**.

D.J.E. Callaway, R. Petronzio, *Nucl. Phys.*, **B267** (1986) 253.

(2) I. Montvay, Preprint *Nucl. Phys.*, **B269** (1986) 170.

W. Langguth and I. Montvay, *Phys. Lett.*, **165B** (1985) 135.

I. Montvay, Talk at the Conference "Advances in the Lattice Guage Theory, April 1985, Tallahassee, Fl., Preprint **DESY 85-050**.

I. Montvay, W. Langguth, P. Weisz, *Nucl. Phys.*, **B277** (1986) 11.

(3) A. Hasenfratz, P. Hasenfratz, *Phys. Rev.*, **D34** (1986).

(4) D. Gross and F. Wilczek, *Phys. Rev.*, **D8** (1976) 3633.

(5) The effect of radiative corrections on spontaneous symmetry breaking is discussed in:

S. Coleman and E. Weinberg, *Phys. Rev.*, **D7** (1973) 1888.

(6) There exists a broad literature on this point. See ex.

I.D. Lawrie, *Nucl. Phys.*, **B200** [FS4] (1982) 1.

(7) S. Weinberg, *Phys. Rev. Lett.*, **36** (1976) 294.

Linde, *JETP*, **23** (1976) 64.

(8) For a discussion on isolated versus effective theories, see for instance:

A. Hasenfratz, P. Hasenfratz, *Nucl. Phys.*, **B270** [FS16] (1986) 687.

(9) There are many recent attempts to derive an upper bound on R:

R. Dashen and H. Neuberger, *Phys. Rev. Lett.*, **50** (1983) 1897.

M.A. Beg, C. Panagiotakopoulos and A. Sirlin, *Phys. Rev. Lett.*, **52** (1984) 883.

D.J.E. Callaway, *Nucl. Phys.* **B223** (1984) 189.

A. Bovier and D. Wyler, *Phys. Lett.*, **154B** (1985) 43.

THE HIGGS MESON MASS AND THE SCALE OF NEW
PHYSICS IN THE STANDARD MODEL

P. Hasenfratz

Universität Bern
CH-3012 Bern, Switzerland

ABSTRACT

There are indications for the Weinberg-Salam model being an effective theory only. This would imply a relation between the scale $\left(\Lambda^{\mathrm{cut}}\right)$ before which new physics should be observed and the Higgs meson mass m_H. The relation $\Lambda^{\mathrm{cut}} = \Lambda^{\mathrm{cut}}\left(m_H\right)$ is studied in an approximate, but nonperturbative way. For $m_H/m_W \gtrsim 10$, Λ^{cut} is of the order of the Higgs meson mass itself (i.e. this is an upper bound), while for $m_H/m_W \lesssim 6$, Λ^{cut} can be very large and no practically interesting constraint emerges.

This contribution will be published in Acta Physica Hungarica. Preprint BUTP-86/20 (1986, Bern University).

THE FUNDAMENTAL $SU(2)$ HIGGS MODEL AT FINITE TEMPERATURE

Urs M. Heller

Institute for Theoretical Physics

University of California

Santa Barbara, California 93106

ABSTRACT

The fundamental $SU(2)$ Higgs model is studied at finite temperature by means of Monte Carlo simulations. We find when raising the temperature in the confined phase, a transition to a deconfined phase. But the Higgs phase transition remains roughly unchanged at finite temperature, and we find no evidence for symmetry restoration from the Higgs phase at finite temperature.

I. INTRODUCTION

The study of the fundamental $SU(2)$ Higgs model, particularly via its non perturbative lattice formulation, has recently attracted a lot of attention.[1] Such investigations are the first steps towards a non perturbative understanding of the phenomenologically successful weak-coupling Weinberg-Salam model. By now the phase diagram of the lattice model is known with rather good accuracy, though the order of the Higgs phase transition is still not known in the whole parameter range. The question of the existence of a non-trivial continuum limit of the lattice model is still open.

In this talk I shall present and discuss some results from a Monte Carlo simulation of the model at finite temperature.[2] More than a decade has now passed since the prediction, based on weak coupling perturbation theory, of a finite temperature phase transition from the Higgs phase to a "symmetry restored" phase.[3] However, there are problems with the perturbative prediction. Some are of technical nature within the perturbation expansion itself. But more fundamentally one should expect that non perturbative effects would play an important role, especially in the transition region, since the transition is supposed to go into a "symmetry restored" phase of a confining gauge matter theory. Furthermore, a confining gauge matter theory deconfines at a certain temperature when it is heated up. This raises the question whether the symmetry restoration temperature is higher or lower than the deconfinement temperature. In the latter case one would have a rather peculiar scenario for the Weinberg-Salam model at finite temperature, in which the W's, the Z and all the leptons would be first in their ordinary "Higgs" realization, then would be confined in bound states and finally liberated above the deconfinement temperature.

These questions can be answered only with non perturbative methods. We have opted for a search of possible phase transitions or rapid crossovers, such as presumably the deconfining transition in a gauge theory with matter fields, via Monte Carlo simulations. Note that working out of necessity with a finite system, we can't make rigorous statements about true phase transitions in the mathematical sense (with a singularity in the infinite volume limit). We will use the word phase transition throughout in a rather loose sense.

II. THE MODEL AT FINITE TEMPERATURE

To simulate the finite temperature theory we work on an asymmetric lattice of size $N_\sigma^3 \times N_\tau$ with N_σ sites in each of the spatial directions and N_τ sites in the temporal or time direction. We use a uniform lattice spacing a set equal to 1. The physical temperature is then given by $T = (aN_\tau)^{-1}$. We use the customary lattice parametrization of the model, with action $S = S_G + S_H$:

$$S_G = \frac{\beta}{4} \sum_P \text{tr}(2 - U_P - U_P^\dagger), \qquad \beta = 4/g^2$$

$$S_H = \sum_x \left\{ \phi^\dagger(x)\phi(x) - \kappa \sum_\mu (\phi^\dagger(x)U_\mu(x)\phi(x+\mu) + \text{h.c.}) \right. \tag{1}$$

$$\left. + \lambda(\phi^\dagger(x)\phi(x) - 1)^2 \right\}$$

with

$$\kappa = \frac{1 - 2\lambda}{(ma)^2 + 8}, \qquad \lambda = \kappa^2 \lambda_c \tag{2}$$

where m and λ_c are the corresponding continuum bare mass and Higgs self-coupling.

The unrenormalized energy density ϵ can be written as $\epsilon = \epsilon_G + \epsilon_H$, with ϵ_G and ϵ_H given by the expectation values[2]

$$a^4 \epsilon_G = \frac{6}{g^2} \left\langle \text{tr}(1 - U_{P_\sigma}) - \text{tr}(1 - U_{P_\tau}) \right\rangle + \cdots \tag{3}$$

162

$$a^4 \epsilon_H = \Big\langle \kappa[\phi^\dagger(x) U_0(x) \phi(x+0) + \text{h.c.}] - 3\kappa[\phi^\dagger(x) U_j(x) \phi(x+j) + \text{h.c.}]$$

$$+ (1 - 2\lambda - 4\kappa)\phi^\dagger(x)\phi(x) + \lambda(\phi^\dagger(x)\phi(x))^2 \Big\rangle \tag{4}$$

In (3) U_{P_σ} and U_{P_τ} are the plaquettes in space-space and space-time directions, respectively. To obtain the physical energy density the zero temperature contribution has to be subtracted. This is usually simulated on a symmetric lattice and hence vanishes for ϵ_G.

III. THE DECONFINEMENT TRANSITION

At small values of κ one is at zero temperature in the confinement phase of a gauge matter theory. Raising the temperature we expect, as in a gauge fermion system, a rather sharp transition to a deconfined phase consisting of a plasma of weakly interacting gauge bosons and massive Higgs scalars.

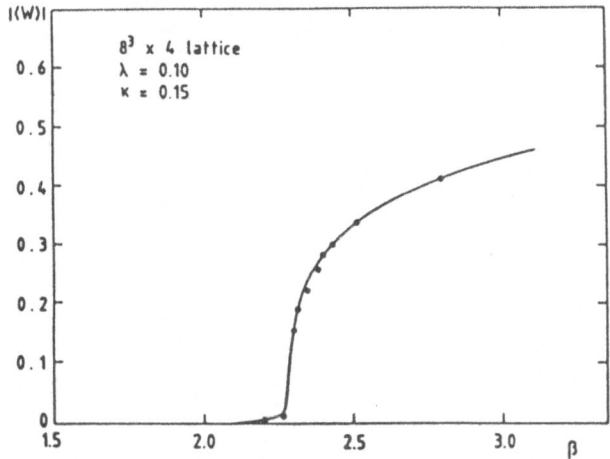

Figure 1: The deconfinement transition as seen in $\langle W \rangle$.

The transition is clearly visible in the behavior of the expectation value of the Polyakov line, $\langle W \rangle$, fig. 1. In fact $\langle W \rangle$ satisfies rather well

$$\langle W \rangle \simeq c(\beta - \beta_c)^{1/3}, \qquad \beta > \beta_c \tag{5}$$

near the transition region (fig. 2), as suggested from $Z(2)$ universality arguments. Both ϵ_G, fig. 3, and ϵ_H, fig. 4, change rapidly in the transition region, with ϵ_G approaching at high temperature the (lattice) Stefan-Boltzman (SB) limit, $a^4 \epsilon_G = 0.2128$ on an $8^3 \times 2$ and $\epsilon_G = 0.0108$ on an $8^3 \times 4$ lattice.[4] At the κ value considered here ϵ_H on the other hand remains far from the SB limit, an indication that the Higgs scalars at this κ are rather massive.

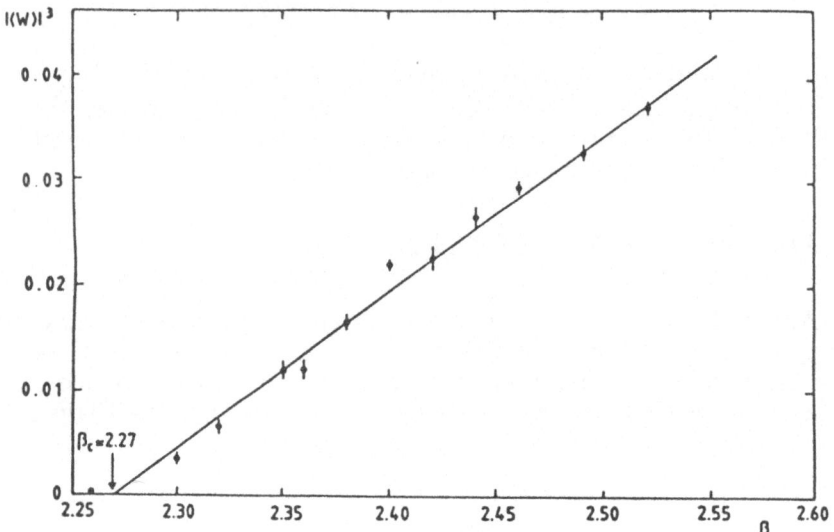

Figure 2: $Z(2)$ universality behavior of $\langle W \rangle$, eq. (5).

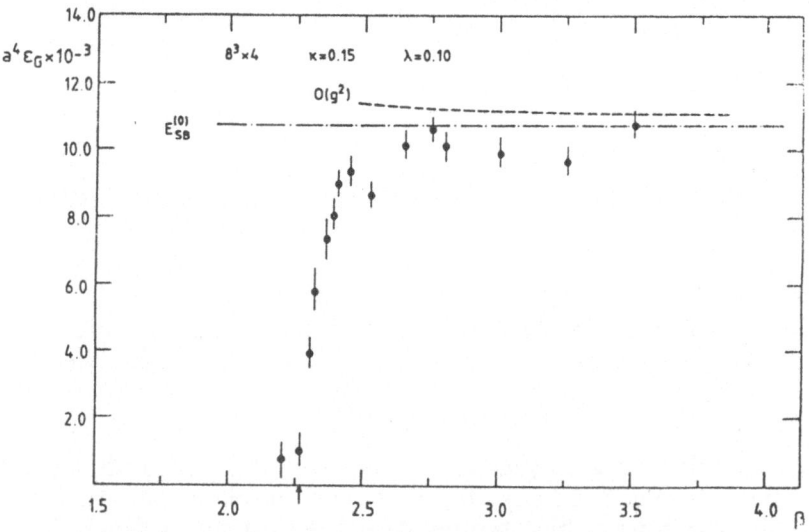

Figure 3: The contribution ϵ_G to the energy density.

164

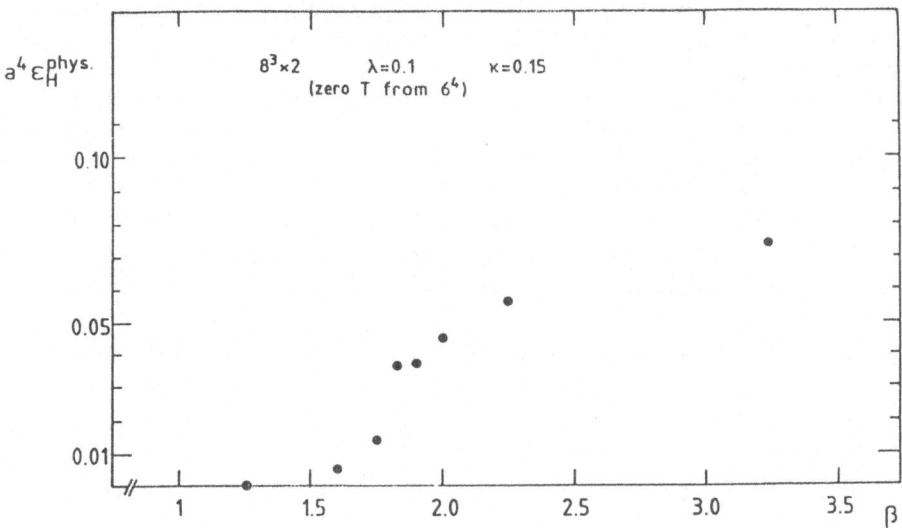

Figure 4: The contribution ϵ_H to the energy density.

IV. THE HIGGS TRANSITION AT FINITE TEMPERATURE AND THE HIGGS PHASE AT HIGH TEMPERATURE

To detect the Higgs transition, we made runs on the asymmetric lattice at fixed values of λ and β, but varying κ. We found the same signals for the Higgs transition from the confined region for $\beta < \beta_c$ (β_c the deconfinement

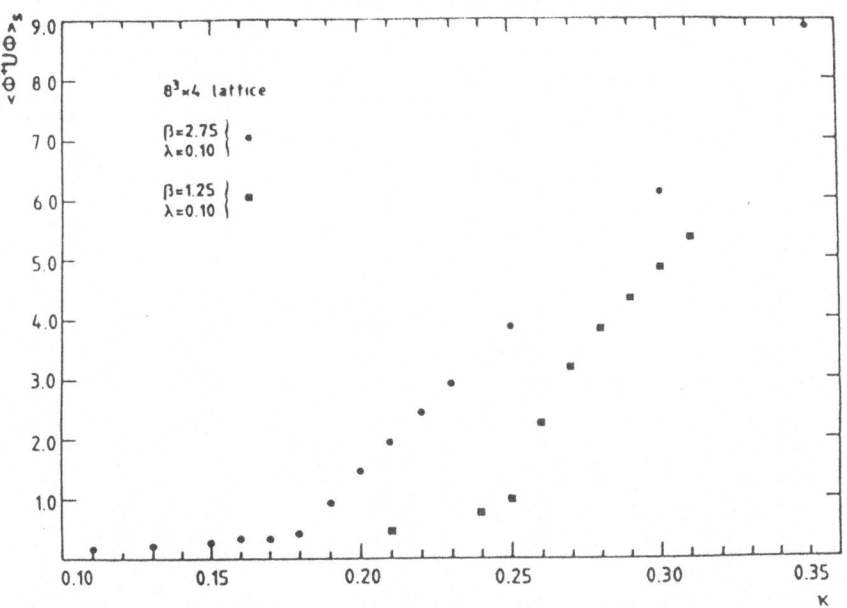

Figure 5: Evidence for the Higgs transition at $\beta = 1.25 < \beta_c$ and $\beta = 2.75 > \beta_c$.

165

transition at low κ), and from the deconfined region for $\beta > \beta_c$, fig. 5. The critical values κ_c agree within our accuracy with the corresponding ones at zero temperature.

To investigate the Higgs phase at high temperature, we made runs at fixed κ, chosen to be in the Higgs phase at zero temperature, and fixed λ, but increasing β up to $\beta = 50$.* As can be seen in fig. 6, no structure reminiscent of a crossover or a transition is found. Even at $\beta = 50$, when varying κ, the Higgs transition from the deconfined phase shows up, fig. 7, just as for $\beta = 2.5$, fig. 8, which is much closer to the deconfinement transition. Again the κ_c's agree roughly with the corresponding values at zero temperature.

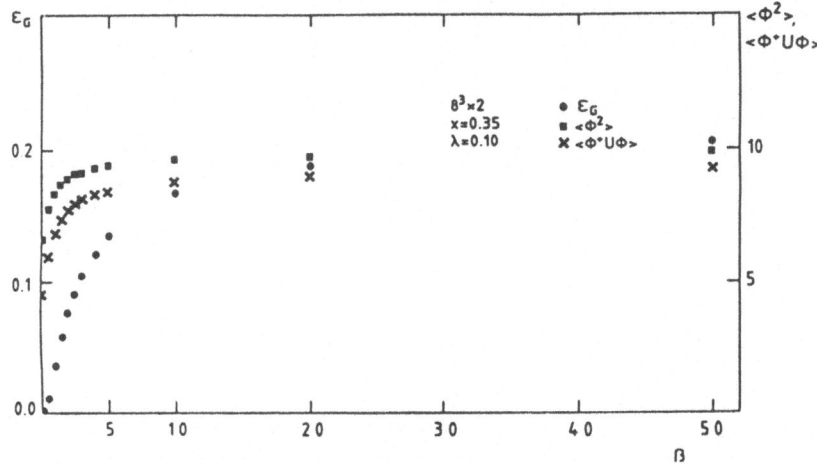

Figure 6: Results of runs at fixed κ and λ.

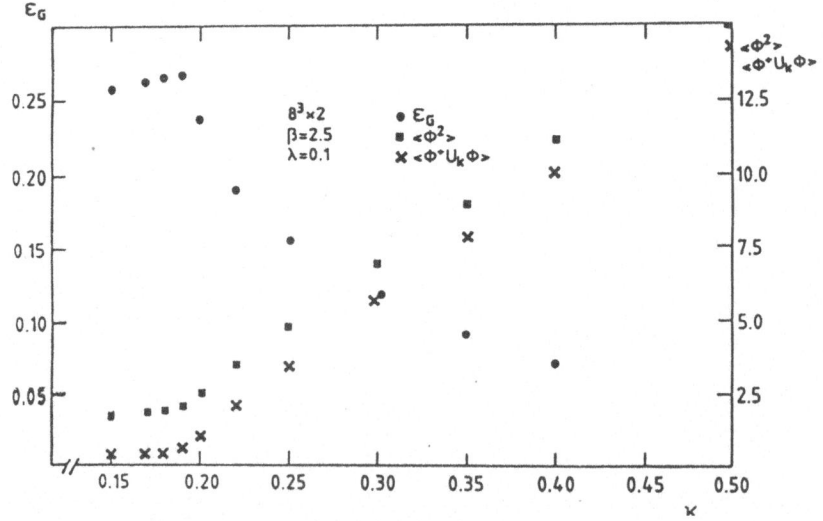

Figure 7: The Higgs transition at $\beta = 50$.

* This was motivated by the fact that "lines of constant physics" seem to be almost parallel to the β axis, at least for large β.[5]

166

Unfortunately the relation between β and the lattice spacing, and hence the temperature, is not very clear in the Higgs phase,[2] and might be much weaker than in the confined phase.[6] But there is a marked difference between the behavior of the contribution ϵ_G to the energy density, see figs. 7 and 8, between $\beta = 2.5$ and 50. In the latter point ϵ_G stays close to the SB limit throughout the range of κ investigated. We take this as evidence that the temperature has increased between the two β values, being larger than the gauge boson mass m_W at $\beta = 50.0$, while at $\beta = 2.5$ the ratio m_W/T decreases rapidly with increasing κ (in the Higgs phase). ϵ_H, needed to complete the physical energy density, is much harder to measure since it requires the subtraction of the zero temperature contribution. We have checked at a few points that ϵ_H behaves very similar to ϵ_G.

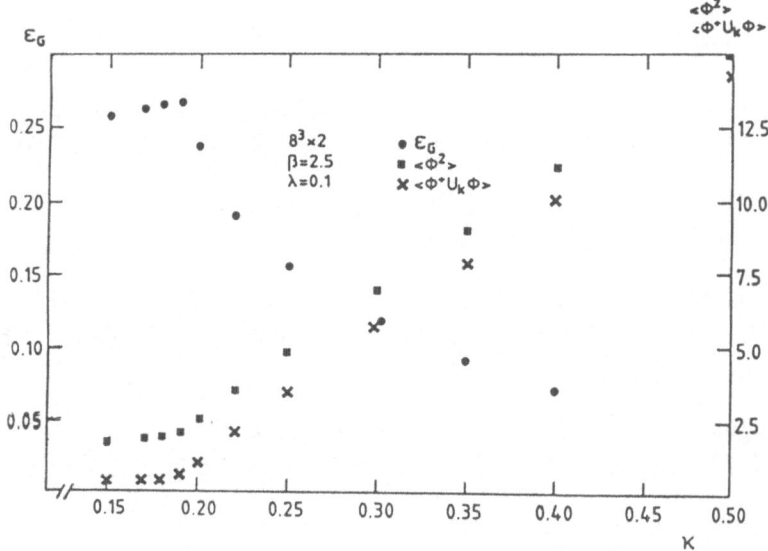

Figure 8: The Higgs transition at $\beta = 2.5$.

V. CONCLUSIONS

The finite temperature effect in the confinement part of the phase diagram nicely confirmed our expectation with the emergence of a deconfinement transition at finite temperature. The Higgs transition remained roughly unchanged at finite temperature, even from the now deconfined regime.

Within our accuracy, we have found no evidence for symmetry restoration. However we can not exclude the possibility of a tiny upward shift in the critical κ for the Higgs transition when going from a symmetric to an asymmetric lattice. The existence of such a shift could open the possibility of symmetry restoration if lines of constant physics (at zero temperature) pierce at finite temperature through the shifted Higgs transition. Work by P.H. Damgaard and myself is in progress to determine the critical κ's more accurately and possibly detect such a shift in κ_c. Some evidence of a possible shift for a larger λ than considered here has been presented at this conference by K. Kanaya.

In view of the importance of the existence of the symmetry restoration transition in models of the early universe more work is clearly needed to clarify this issue.

ACKNOWLEDGEMENT

I would like to thank J. Jersák, K. Kanaya, I. Montvay and especially my collaborator P.H. Damgaard for interesting discussions and comments.

REFERENCES

1. J. Jersák, Lattice Higgs Models, talk given at the Wuppertal Workshop on Lattic Gauge Theory, Aachen preprint PITHA 85/25 (1985).

2. P.H. Damgaard and U.M. Heller, Phys. Lett. 171B:442 (1986), and CERN preprint CERN-TH. 4481/86 (1986).

3. D.A. Kirschnitz and A.D. Linde, Phys. Lett. 42B:471 (1972); L. Dolan and R. Jackiw, Phys. Rev. D9:3320 (1974); S. Weinberg, Phys. Rev. D9:3357 (1974).

4. J. Engels, F. Karsch, I. Montvay and H. Satz, Nucl. Phys. B205[FS5]:545 (1982); U. Heller and F. Karsch, Nucl. Phys. B251[FS13]:254 (1985).

5. I. Montvay, Nucl. Phys. B269:170 (1986).

6. Evertz, J. Jersák, C.B. Lang and T. Neuhaus, Phys. Lett. 171B:271 (1986).

HADRON SPECTRUM ON A 16^3 x 48 LATTICE

Y. Iwasaki

Institute of Physics, University of Tsukuba
Ibaraki 305, Japan

Abstract

Hadron masses are calculated in the quenched approximation to lattice gauge theories with a renormalization group improved lattice SU(3) gauge action and Wilson's quark action on a 16^3 x 48 lattice. The ground state masses of hadrons calculated for several quark masses are free from finite size effects and the contamination of excited states. When the results are extrapolated to the quark mass where the pion and rho masses take the physical values, the proton and delta masses agree with the physical values with at most 10~15 % errors.

Our primary concern is what the correct masses of the proton and delta are in the quenched approximation to lattice gauge theories, when the masses of the pion and rho are fitted. If one recalls the success of valence quark models in describing the static properties of hadrons as well as the success of the OZI rule in decay and scattering processes of hadrons, one may expect that the quenched approximation will give reasonable values for hadron masses with about 10 % errors. However, the results which have been obtained up to now with the standard one-plaquette action and with the Wilson's quark action are not satisfactory (for recent results see ref.[1]): For example, the proton-to-rho mass ratio m_p/m_ρ turned out to be too large. Possible origins for this large discrepancy are finite lattice spacing effects and finite lattice size effects.

In order to reduce finite lattice spacing effects we take a renormalization group (RG) improved lattice SU(3) gauge action

169

$$S = \frac{1}{g^2}\{c_0 \Sigma \text{ Tr (simple plaquette)}$$

$$+ c_1 \Sigma \text{ Tr } (1\times2 \text{ rectangular loop})\} \tag{1}$$

with

$$c_1 = -0.331 , \quad c_0 = 1 - 8c_1 , \tag{2}$$

the form of which has been determined by a block spin renormalization group study[2] and an analysis of instantons on the lattice[3]. In the sum over loops, each oriented loop appears once.

We have obtained several results which support our approach of using a RG improved action for obtaining the continuum limit of lattice theories: First short distance lattice artifacts for the two-point function are reduced with a RG improved action in the two-dimensional $O(3)$ sigma model [4]. Secondly there is no first-order phase transition in SU(5) lattice gauge theory with a RG improved action[5]. This is quite different from the situation with the one-plaquette action. Thirdly we have recently investigated the scaling behavior of the string tension[6]. The results (See Fig. 1) show that asymptotic scaling sets in at $\beta = 6/g^2$ =2.6 ($\xi \approx 5.0$) for the RG improved action, while it does not set in up to $\beta = 6.6$ ($\xi \approx 8.7$) for the one-plaquette action.

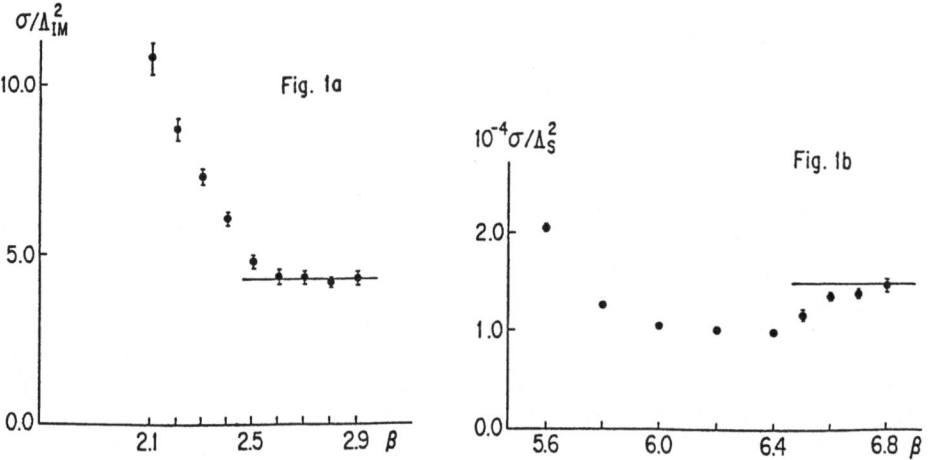

Fig.1: σ / Λ^2 versus ; (a) for the RG improved action and (b) for the one-plaquette action. Λ_{IM} and Λ_S are the scale parameters for the RG improved action and for the one-plaquette action, respectively. The horizontal straight lines both in (a) and (b) correspond to the same value of σ/Λ_S^2.

In order to reduce finite lattice effects we make our calculations on a $16^3 \times 48$ lattice at $\beta=2.4$ ($\beta=6/g^2$). The lattice size and the coupling constant are determined from the results of our previous calculations of hadron masses on a $12^3 \times 24$ lattice[7] and on a $16^3 \times 32$ lattice[8]. (We have increased the number of configurations up to 30 configurations in the case of the $16^3 \times 32$ lattice to improve statistics: The results are essentially the same as those in ref.[8]).

We take Wilson fermions for quarks. We use two operators $\pi_1 = \bar{u}\gamma_5 d$ and $\pi_2 = \bar{u}\gamma_5\gamma_4 d$ for the pion, $\rho_1 = \bar{u}\gamma_i d$ and $\rho_2 = \bar{u}\gamma_i\gamma_4 d$ for the rho, respectively. Here u and d are u-quark and d-quark fields respectively and i=1,2,3. We use non-relativistic operators for the proton and delta as well as $(u^T\gamma_5 C^{-1}d)u$ for the proton and $(u^T\gamma_\mu C^{-1}u)u$ for the delta, where C is the charge conjugation operator. We use two operators for all of them to investigate finite size effects.

Our strategy for extracting masses of hadrons in the quenched approximation is the standard one which has been employed in most previous works: In ref.[7], we have described some details of our method of calculations which we use here. We

Table 1: The ground state masses of the pion, rho, proton and delta at five hopping parameters in lattice unit. The n and r correspond to the non-relativistic and the relativistic operatos, respectively.

K		0.1400	0.1450	0.1500	0.1525	0.1540
π		1.056 4	0.845 4	0.611 4	0.476 5	0.383 6
ρ		1.095 5	0.907 5	0.712 6	0.609 8	0.542 11
P	n	1.722 8	1.421 7	1.099 18	0.918 35	0.811 51
	r	1.719 8	1.419 7	1.094 15	0.912 33	0.795 50
Δ	n	1.748 12	1.458 15	1.161 43	1.004 40	0.922 69
	r	1.747 8	1.457 12	1.156 33	1.014 35	0.944 83

calculate hadron masses (especially the masses of the pion, rho, proton and delta) for 15 gauge configurations, separated by one hundred sweeps after a thermalization of 1000 sweeps. The gauge configurations are generated by a Cabibbo-Marinari algorithm slightly modified for vector processors. We use periodic boundary conditions for both gauge fields and quark fields. The hopping parameters K we take are 0.14, 0.145, 0.15, 0.1525 and 0.154.

The results for the mesons are as follows: The propagators can be excellently fitted to a single hyperbolic-cosine for a wide range of t (t: coordinate in time direction; $0 \leq t \leq 47$ here; the origin of the propagator is t=0). See Fig. 2 for the propagators of π_1 at five hopping parameters together with one-mass fits for $11 \leq t \leq 37$. As a result about 35 points agree with the fitted hyperbolic-cosines within one-standard deviation. The ground state mass at each K is stable for a change of the t-range ($8 \sim 11 \leq t \leq 37 \sim 40$ for the fit. Further when we make a two-mass fit to the propagators at each K for the range $4 \sim 5 \leq t \leq 43 \sim 44$, the ground state mass is stable. The masses and the errors are estimated by a least mean squares method using SALS system here and also hereafter for other particles. The masses determined both from π_1 and π_2 (from ρ_1 and ρ_2) agree with each other within one-standard deviation. Furthermore, the propagators on the $12^3 \times 24$, $16^3 \times 32$ and $16^3 \times 48$ at five K's agree with each other within almost one-standard deviation for the t-region which can be compared. Thus we conclude that we are able to determine the ground state mass of

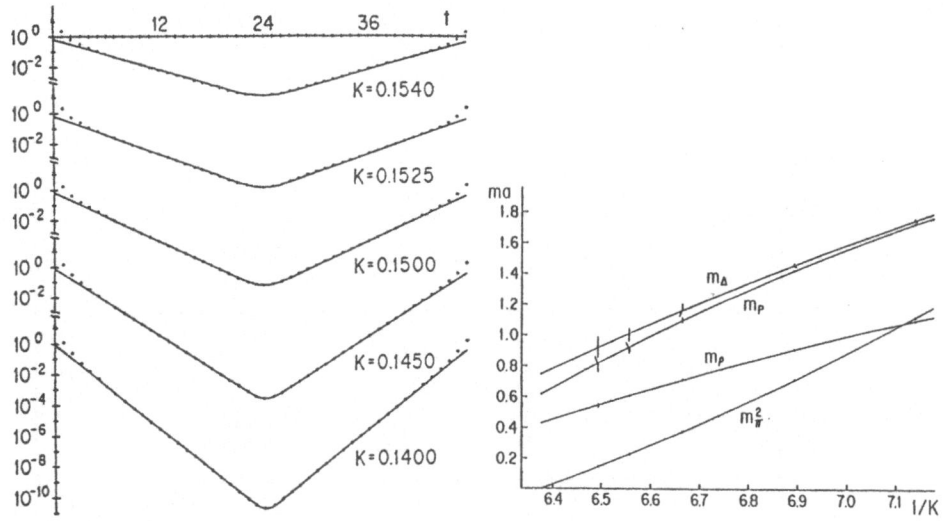

Fig.2: Propagators of π_1 at five hopping parameters with one-mass fit for $11 \leq t \leq 37$.

Fig.3: Masses of pion, rho, proton and delta versus 1/K.

the pion and rho at each K with small errors which is free from finite size effects and free also from the contamination of excited states. See Table 1 for the results of m_π and m_ρ.

We plot the masses of the pion and rho versus 1/K in Fig. 3. We first notice that the linearity between m_π^2 and 1/K as well as that between m_ρ and 1/K are roughly satisfied. However, if we closely look at the 1/K dependence of the masses, we observe that m_π^2 is slightly convex downwords and m_ρ is slightly convex upwards. The results from independent gauge configurations on the $12^3 \times 24$, $16^3 \times 32$ and $16^3 \times 48$ lattices all indicate the same behavior. Therefore we conclude these dependences on 1/K are real and not due to statistical errors. We fit the masses to a quadratic function of 1/K as shown in Fig. 3. This leads to $K_c = 0.1569(2)$ and $m_\rho(K=K_c) = 0.426(15)a^{-1}$, where K_c is the hopping parameter value where the pion mass vanishes and a is the lattice spacing. We do not distinguish K_{phy} (where the ratio m_ρ/m_π becomes the physical value) from K_c, because K_{phy} is within the statistical error for K_c. Inputting the physical ρ mass we have $a^{-1} = 1810(60)$ MeV, which is slightly larger than that obtained previously, because the fitting procedure is different.

Let us now discuss the results for the baryons. As K approaches toward K_c, the fluctuation of propagators at large t becomes large: The statistical errors are larger than 40% for $t \gtrsim 20$ at $K=0.1525$ and for $t \gtrsim 19$ at $K=0.154$ and furthermore some propagators become negative for $t \gtrsim 23$ at $K=0.154$. Therefore we think the data for these t regions are statistically meaningless and consequently do not use them. The propagators of the non-relativistic proton at five hopping parameters are displayed in Fig. 4.

Let us describe in some detail the fitting procedure for the non-relativistic proton. Fitting the propagators to $A_0\exp(-m_0 t)$ for $18 \leq t \leq 24$, we obtain 1.73(1), 1.42(1) and 1.10(2) for m_0 at $K=0.14$, 0.145 and 0.15, respectively. We check that the contamination from excited states for these results is small in the following way: i) Ground state masses remain within one-standard deviation even if we change the fitting range of t to $t_1 \leq t \leq 24$ with $t_1 = 15$ 21. ii) When we fit the propagators to

$$G(t) \sim A_0\exp(-m_0 t) + A_1\exp(-m_1 t) \tag{3}$$

for $t_1 \leq t \leq 24$ with $t_1 = 5, 6$ and 7 the ground state masses are stable and agree with the results obtained by the one-mass fits.

If we would fit the propagators to $A_0\exp(-m_0 t)$ for the range $11 \leq t \leq 16$ on the $16^3 \times 32$ lattice, we would obtain 1.76, 1.47 and 1.14 for m_0 at $K=0.14$, 0.145 and 0.15, respectively. They are larger than those obtained above by two or five standard deviations. Thus we conclude that the temporal linear extention in this work, 48 in lattice unit, is large enough (at least for K $\lesssim 0.15$ at $\beta=2.4$) to obtain ground state masses by one-mass fits, while 32 in lattice unit is not large enough. This is exactly what we have conjectured in ref.[7]. This difficulty in extracting the ground state masses by one-mass fit consists in the fact that the excited states exist·close to the ground states and further that the amplitude A_1 is larger than A_0. This last fact is well known phenomenologically for N(1440). Our results indicate that $\sqrt{A_1}/\sqrt{A_0} \approx$ 2.0. It may be stressed that if some fit would give a good result for the ground state of the proton, but would give either a very large mass for the first excited state or a small amplitude for the first excited state, the fit is not in accord with the experiment. This is one of crucial points to be tested.

Masses and amplitudes for the first excited state by two-mass fits depend on the fitting range of t. This is the notorious problem for two-mass fits. Therefore we are unable to determine excited state masses from the given propagators. Rather we proceed as follows: We check whether the propagators are consistent with that the first excited state (1440 MeV) is 500 MeV heavier than the ground state (940 MeV). We simply assume the mass difference be independent on quark mass (K) and fit the propagators for $7 \lesssim$ $t \lesssim 24$ to eq. (3) with

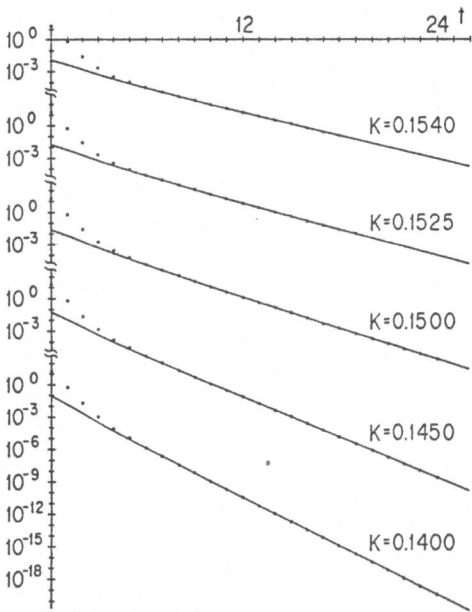

Fig.4: Propagators of proton (non-relativistic operator) with two-mass fits.

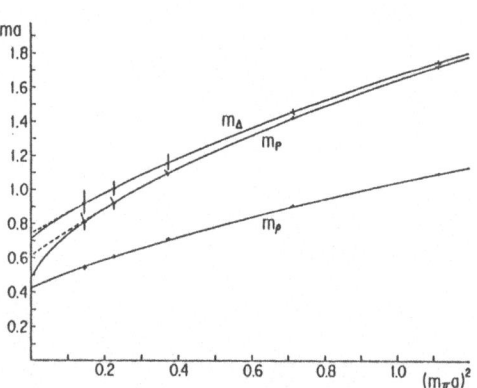

Fig.5: m_Δ, m_p and m_ρ as a functions of m_π^2; the solid curves are the fitts to our data using phenomenological mass formulae with some free parameters, while the dashed curves are the fitts shown in Fig.3 now with slightly different horizontal scale.

174

$$m_1 - m_0 = (1440 - 940)a \qquad (4)$$

fixed. The fits are excellent (see Fig. 4) and the results for the ground state masses completely agree with those obtained by the one-mass fits.

Next let us discuss the propagators at K=0.1525 and K=0.154. Because we cannot obtain reliable values of the propagators for very large t, we have to be satisfied with making a two-mass fit. We fit the propagators to eq. (3) with eq. (4) fixed for $6 \le t \le 19$ at K=0.1525 and for $6 \le t \le 18$ at K=0.154 (see Fig. 4). We obtain 0.92(4) and 0.81(5) for m_0, respectively. These values are slightly smaller than those which would be obtained if we would make one-mass fit for $10 \le t \le 18$.

The ground state masses for other three baryon operators are obtained by similar analyses. See Table 1. The results for the relativistic operators are in excellent agreement with those for the non-relativistic ones at each K. This implies that finite size effects are under control.

We display in Fig. 3 the ground state masses of the baryons at each K. We notice again that although the linearity between the mass and 1/K is roughly satisfied, the mass as a function of 1/K is slightly convex upwards. If we fit the masses of the baryons to quadratic functions of 1/K (See Fig. 2), we obtain m_0 (non-relativistic proton)=1100(90) MeV, m_0 (relativistic proton)= 1080(80) MeV, m_0 (non-relativistic Δ)=1340(120)MeV and m_0(relativistic Δ)=1370(120) MeV at $K=K_c$. The results agree with the physical values with 10~15 % errors.

Although we think the above fitting procedure is a modest one, there is an alternative fit which gives almost exact physical values for the proton mass and the delta mass. We use phenomenological mass formulae which reproduce remarkably the physical hadron masses[9]. The formulae are also used in ref.[10] for comparison with lattice results. They consist of the sum of the quark masses, a constant (different for mesons and baryons) and a term describing the hyperfine spin splitting:

$$M_{baryon} = M_b + \sum_{i=1} m_i + \xi_b \sum_{i>j} \vec{S}_i \cdot \vec{S}_j / m_i m_j ,$$
$$M_{meson} = M_m + \sum_{i=1} m_i + \xi_m \vec{S}_q \cdot \vec{S}_{\bar{q}} / m_q m_{\bar{q}} ,$$

Equating all quark masses and treating M_b, M_m, ξ_b and ξ_n as free parameters, we fit the data to the mass formulae. We obtain m_ρ =900(180) MeV and M_Δ=1290(210) MeV. The fitted curves (solid lines) are displayed in Fig. 5 together with the previous fitted curves (dashed lines) shown in Fig. 3 now with slightly different horizontal scale ($1/K \to m_\pi^2$). The two fits for m_ρ remarkably agree with each other. On the other hand, K dependence of the baryon masses around K_c is more steeper than quadratic functions.

The calculation of hadron masses at K closer to K_c would reveal which of the two fits is closer to the correct one. Of course, the true values of hadron masses can be obtained after including the effect of dynamical quark loops (See ref.[11]) for such an attempt). Our results imply that even in the quenched (valence) approximation we can obtain reasonable values for the masses of flavor non-singlet hadrons with at most 10~15 % errors. This is completely consistent with the success of the valence quark model in describing static properties of hadrons.

This work has been done in collaboration with S. Itoh and T. Yoshié. Details of our data and our analyses will be published elsewhere.

The calculation has been performed with the HITAC S810/10 at KEK. We would like to thank S. Kabe, T. Kaneko, R. Ogasawara and other members of Data Handling Division of KEK for their kind arrangement which made this work possible, and the members of Theory Division, particularly, H. Sugawara and T. Yukawa for their warm hospitality and strong support for this work.

References
[1] K.C. Bowler et al., Nucl. Phys. B240 FS12 (1984) 213;
    A. Billoire, E. Marinari and R. Petronzio, Nucl. Phys. B251
    [FS13] (1985) 141; A. König, K.H. Mütter, K. Schilling and
    J. Smit, Phys. Lett. 157B (1985) 421.
[2] Y. Iwasaki, Nucl. Phys. B258 (1985) 141; preprint UTHEP-118.
[3] Y. Iwasaki and T. Yoshié, Phys. Lett. 131B (1983) 159;
    S. Itoh, Y. Iwasaki and T. Yoshié, Phys. Lett. 147B (1984)
    141.
[4] S. Itoh, Y. Iwasaki and T. Yoshié, Nucl. Phys. B250 (1985)
    312.
[5] S. Itoh, Y. Iwasaki and T. Yoshié, Phys. Rev. Lett. 55
    (1985) 273.
[6] S. Itoh, Y. Iwasaki and T. Yoshié, preprint UTHEP-154.
[7] S. Itoh, Y. Iwasaki, Y. Oyanagi and T. Yoshié, Nucl. Phys.
    B274 (1986) 33.
[8] S. Itoh, Y. Iwasaki and T. Yoshié, Phys. Lett. 167B (1986)
    443.
[9] S. Ono, Phys. Rev. D17 (1978) 888.
[10]K.C. Bowler et al., Phys. Lett. 162B (1985) 354.
[11]M. Fukugita, Y. Oyanagi and A. Ukawa, preprint UTHEP-152;
    F. Fucito, K.J.M. Moriarty, C. Rebbi and S. Solomon, preprint
    BNL 37546.

HIGGS PHASE TRANSITION IN THE (1) AND SU(2) LATTICE HIGGS MODELS*

J. Jersák

Institute of Theoretical Physics E
RWTH Aachen
Physikzentrum
D-5100 Aachen, West Germany

Abstract   I describe some recent Monte Carlo results obtained by the Aachen group, in various collaborations, on the properties of the Higgs phase transition in the lattice Higgs models. We have studied the changes both of the spectrum and of the asymptotic behaviour of the gauge invariant two-point function at the transition. We have also proposed a criterion for confinement in gauge theories with matter fields based on the comparison of the energy of a screened and of a possibly unscreened external charge.

1. Introduction   The lattice Higgs models are studied in the hope that they might elucidate the existence and further nonperturbative aspects of the Higgs mechanism (for a recent review of the progress in this field and for earlier references see ref.[1]). In the case of the U(1) Higgs model with scalar field of charge one the action is

$$S = - \frac{\beta}{2} \sum_{p \in \Lambda} (U_p + U_p^*) - \kappa \sum_{x \in \Lambda} \sum_{\mu=1}^{4} (\Phi_x^* U_{x,\mu} \Phi_{x+\mu} + c.c.)$$

$$+ \lambda \sum_{x \in \Lambda} (\Phi_x^* \Phi_x - 1)^2 + \sum_{x \in \Lambda} \Phi_x^* \Phi_x \ .$$

(1)

The action for the SU(2) model with a doublet scalar field is analogous [2]. In both these models we have concentrated on the properties of the Higgs phase transition (PT). This PT is a lattice analogue of the onset of the Higgs mechanism in the continuum gauge theory according to the quasiclassical approach. A gauge invariant characterization of the region where the Higgs mechanism operates

---

*) Work supported by BMFT and DFG.

(Higgs region) is provided, in the absence of a genuine order parameter [3], by a sensitive dependence of the condensate of the scalar field on κ [2]:

$$\langle \Phi_x^+ \, \Phi_x \rangle \simeq \frac{4}{\lambda} \, \kappa + \text{const.} \tag{2}$$

In the confinement region or in the Coulomb phase, i.e. for $\kappa < \kappa_{PT}$, the values of $\langle \Phi^+\Phi \rangle$ are nearly κ-independent. The Higgs PT is the transition where a nonanalytic change between these two behaviours of $\langle \Phi^+\Phi \rangle$ takes place. This happens on a two-dimensional sheet $\kappa = \kappa_{PT}(\lambda,\beta)$ in the three-dimensional space of couplings λ,β and κ. The Higgs region lies above this sheet, i.e. at $\kappa > \kappa_{PT}$. The sheet does not completely separate the confinement and Higgs regions [4,5] which therefore form one confinement-Higgs phase.

We have systematically investigated the behaviour of various observables at the Higgs PT. This revealed several physical properties of the Higgs PT on the lattice.

2. <u>Higgs boson and gauge vector boson masses</u>   In the vicinity of the Higgs PT the masses both of the gauge vector boson and of the Higgs boson are rather low or even zero in the case of the photon, and depend in different and characteristic ways on κ. In particular the Higgs boson mass has a dip at the Higgs PT in both models.

In the SU(2) model each of the masses has an almost identical dependence on $\kappa-\kappa_{PT}$ for $(\lambda,\beta) = (3,2.25), (0.5,2.25)$ and $(0.5,2.4)$ [6]. Thus in the region we have investigated the lines of constant physics are nearly parallel to the Higgs PT sheet. The masses along these lines vary much less with β than what might have been expected from the asymptotic scaling formula for the pure SU(2) lattice gauge theory.

The Higgs boson and the photon masses in the U(1) model are shown in figs. 1a and 1b, respectively [7]. The photon mass has been obtained both from the plaquette correlations at nonzero momentum [7] (using the method of ref.[8]) and from the static potential determined by means of the Wilson loops [9]. It is vanishing below the Higgs PT, confirming that the free charge phase in the compact U(1) model contains a massless photon, indeed [5].

3. <u>Spectrum in the Coulomb phase of the U(1) model</u>   In the U(1) model the spectrum in the Coulomb phase differs substantially from that in the Higgs region of the confinement-Higgs phase. Unconfined charged particles are encountered and their pairs form bound 'bosonium' states.

In addition to the massless photon we have found also a massive vector state [7] (fig.1c) dominating the correlation function of the operator $\text{Im}\Phi^*U\Phi$ on spatial links. It can be interpreted as a bound state of two charged particles, i.e. a vector bosonium. In the Coulomb phase this state does not mix observably with the photon, whereas in the Higgs region the states contributing to the two vector correlation

178

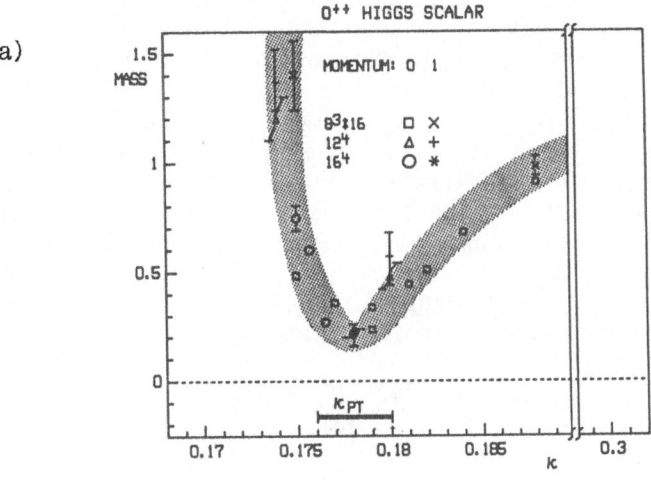

a)

Fig.1. Masses of the Higgs
boson (a), photon (b) and
vector bosonium (c) in the
U(1) model [7]. The width
of the shadowed areas re-
presents the statistical
and systematic errors.

b)

c)

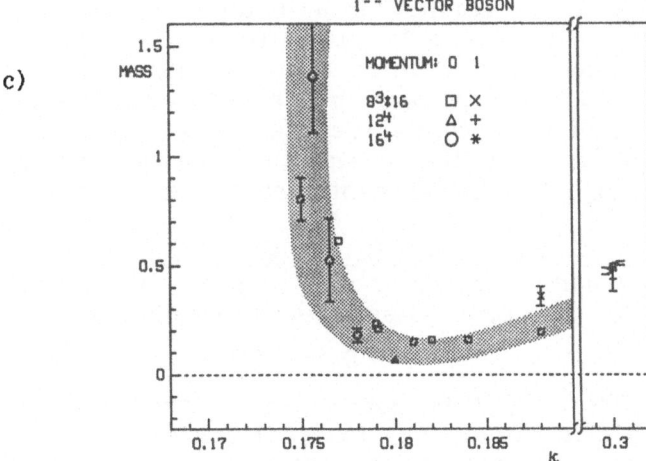

179

functions might be identical. The Higgs boson in the Coulomb phase (fig.1a) may be seen as a scalar bosonium, too. The existence of the free charged particles in the Coulomb phase is supported by the results for the gauge invariant two-point function described below [9]. Its mass m_c grows with decreasing κ and therefore also the bosonium states get heavy in the Coulomb phase. A pair of charged particles interacts both through the gauge and through the Φ^4 couplings, thus the masses of the bosonium states cannot be estimated in some simple way. But the interaction is apparently attractive.

4. The screening energy μ The rate of the decrease of the gauge invariant two-point function

$$G(T,R) = \langle\ \Phi_x^+\ \prod_{\ell \in \Gamma}\ U_\ell\ \Phi_y\ \rangle\ , \qquad |x-y| = T\ , \qquad (3)$$

$$\Gamma = \ \begin{array}{c} y\ \times\!\!-\!\!-\!\!-\!\! \\ \qquad\ \rceil T \\ x\ \times\!\!-\!\!-\!\!-\!\! \\ R \end{array}$$

with growing T defines the 'screening energy' μ,

$$G(T,R)\ \xrightarrow[\substack{T\ \to\ \infty \\ R\ \text{fixed}}]{}\ f_G(R)\ e^{-\mu T}\ . \qquad (4)$$

The screening of an external point-like charge by the dynamical charged field Φ can be realized by several physical mechanisms. In the Coulomb phase the charge can be screened by one unconfined charged particle, in analogy to the hydrogen atom. In the confinement region the screening is due to a confined "constituent" particle, as in the quark model for mesons with two heavy quarks. In the Higgs region the external charge is screened by the condensate of the Φ-field, in analogy to the Debye screening in plasma. In all these cases the screening energy μ is the lowest energy in the dynamical fields in the presence of a screened external charge.

Our results for μ in SU(2) [10] and U(1) [9] models are shown in figs. 2 and 3, respectively. In both models μ is a decreasing function of κ which changes substantially at the Higgs PT. We have found that it is much easier to calculate μ in Monte Carlo simulations than to calculate the Higgs boson mass with a comparable statistical error. Also the asymptotic behaviour according to eq.(4) is achieved for $T \simeq 4$ independently of the value of R.

5. The energy E_q of an external charge   The energy E_q in the dynamical fields in the presence of an external charge is determined by the static potential at large distance,

$$E_q = \frac{1}{2}\ V(\infty)\ . \qquad (5)$$

The energy is normalized so that the ground state energy of the fields without any external charge is zero. Obviously, E_q should equal to μ whenever the external charge must be screened, i.e. in the confinement-Higgs phase. These, and most of the following properties of G(T,R), μ and E_q have been first derived in convergent expansions for the Z(2) Higgs model by Fredenhagen and Marcu [11]. Our data

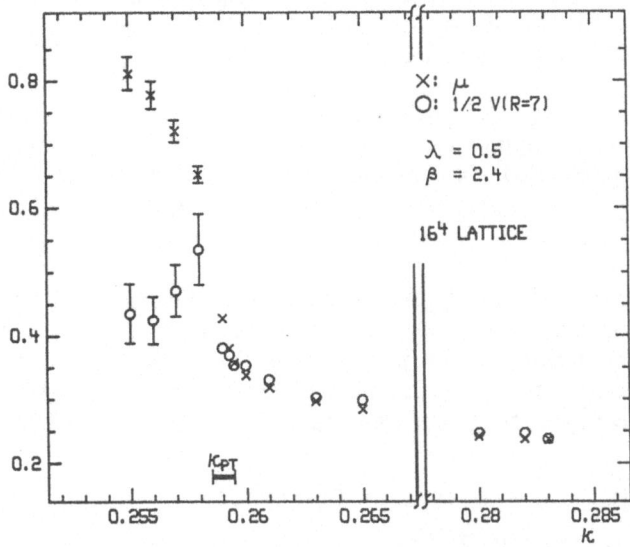

Fig.2. The screening energy μ and $V(R=7)/2$ in the SU(2) model [10].

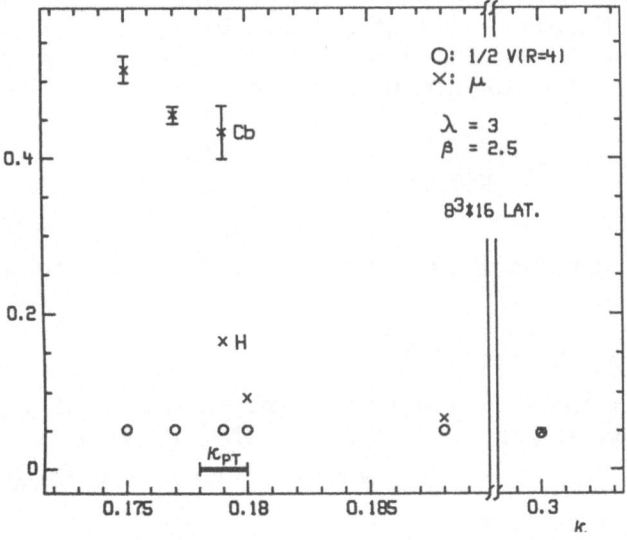

Fig.3. The screening energy μ and the energy E_q of an external charge determined by fits of the potential $V(R)$ in the U(1) model [9]. Cb and H denote the Coulomb and Higgs phase, respectively.

[9,10] for $V(R)/2$ at large R or for E_q obtained from the Coulomb or Yukawa fit to the data for $V(R)$ are shown in figs. 2 and 3. They are consistent with the equality $\mu = E_q$ for $\kappa > \kappa_{pT}$.

For $\kappa < \kappa_{pT}$ in the SU(2) model [10] it was not possible to determine the asymptotic values $V(\infty)$ of $V(R)$ on a 16^4 lattice. The confining potential is screened only at distances greater than the distance $R = 7$, for which we could reliably calculate the values of $V(R)$ shown in fig.2.

6. <u>The mass of the charged particle</u> For $\kappa < \kappa_{pT}$ in the U(1) model in the Coulomb phase, an external charge q can remain unscreened. Then E_q is smaller than μ [9] (fig.3). It is the energy of the lattice Coulomb field of the external charge. The values of the difference $\mu - E_q$ are related to the mass m_c of the unconfined but bound charged particle c,

$$\mu - E_q = m_c - E_b(cq) . \tag{6}$$

The binding energy E_b is in the continuum approximation given by the Bohr formula

$$E_b(cq) \simeq \frac{1}{2} m_c \alpha^2 , \tag{7}$$

α being the coefficient of the $1/R$ term in the potential in the Coulomb phase calculated by means of the Wilson loops. As α is quite small, $\alpha \simeq 0.0385(15)$ [9], the mass m_c of the charged particle in the Coulomb phase is to a good approximation equal to $\mu - E_q$. Thus the values of the quantity $E_q - \mu$ have a fundamental physical meaning.

7. <u>The screening energy criterion for confinement</u> The quantity $\mu - E_q$ is an order parameter distinguishing between phases with and without confinement. The most suitable way to investigate, in the framework of the lattice gauge theories, whether this order parameter vanishes or not is to introduce the ratio

$$\rho_{AC}(T) = \frac{G(T,0)}{W(T,T)^{1/4}} = \frac{\langle \big| T \rangle}{\langle \boxed{} T \rangle^{1/4}} . \tag{8}$$

Here $W(T,T)$ denotes a T×T Wilson loop. This ratio behaves for large T as

$$\rho_{AC}(T) \sim e^{-(\mu - E_q)T} . \tag{9}$$

Our data for the U(1) model and some more general arguments for the ratio (8) suggest that $\mu - E_q \geqslant 0$ [9]. Thus we expect

$$\rho_{AC}^{\infty} = \lim_{T \to \infty} \rho_{AC}(T) \begin{cases} \neq 0 & (\mu - E_q = 0, \text{ confinement}) \\ \\ = 0 & (\mu - E_q > 0, \text{ free charge}) . \end{cases} \tag{10}$$

Our data for $\rho_{AC}(T)$ in the U(1) model are in good agreement with these expectations [9]. On a 16^4 lattice $\rho_{AC}(T)$ behaves differently in the confinement-Higgs and the free charge phases already for rather small T (fig.4). The κ–dependence of the asymptotic value ρ_{AC}^{∞} of $\rho_{AC}(T)$ is shown in fig.5. Instead of $\mu-E_q$ one could alternatively use ρ_{AC}^{∞} as an order parameter, but the physical meaning of its values is less obvious.

8. The vacuum overlap order parameter

Fredenhagen and Marcu [11] proposed a criterion for confinement based on the following idea: One charge of a pair of opposite charges at the distance R is removed by considering the limit $R \to \infty$, $T \sim R$ of G(R,T). This expectation value, properly normalized by the square root of the R×R Wilson loop,

$$\rho_{FM}(R,T) = \frac{\langle \overset{R}{\boxed{}}T \rangle}{\langle \overset{R}{\boxed{}}2T \rangle^{\frac{1}{2}}} \quad , \tag{11}$$

approaches an asymptotic value ρ_{FM}^{∞}. The zero or nonzero values of this parameter mean that the remaining state is either perpendicular to the vacuum or not. In the former case the state is charged whereas in the latter case the original charge has been screened. For more details see the contribution of M. Marcu to this conference.

We have investigated $\rho_{FM}(R,T)$ in the SU(2) model in an earlier work [10]. In the U(1) model [9] the values of $\rho_{FM}(R,R/2)$ turn out to be numerically very close to the values of $\rho_{AC}(T=R)$ shown in fig.4. The values of ρ_{FM}^{∞} are equal within the error bars to ρ_{AC}^{∞} and therefore both quantities are represented by the same symbols in fig.5. Our Monte Carlo results for both models are in good agreement with the expectations for ρ_{FM}^{∞}.

9. The order of the Higgs PT at large λ

The order of the Higgs PT in both models for $\lambda \geqslant O(1)$ and for finite ß remains an open question [12–14]. (For small λ the transition is of 1st order.) Because of the importance of this question for the study of the continuum limit, we are looking for convenient methods of investigation of the order of the Higgs PT on finite lattices. We have noticed in the U(1) model [9] that the screening energy μ can have a significant discontinuity at the Higgs PT (fig.3) even if the local quantities like $\langle \Phi^* U \Phi \rangle$ seem to vary continuously with κ. It is interesting to note that the low and high values of μ at the discontinuity at $\kappa = 0.179$ correspond to large and small fluctuations of $\Phi^* U \Phi$, respectively. As the example of one Monte Carlo run at this κ on an $8^3 \times 16$ lattice shown in fig.6 demonstrates, the corresponding states can be quite long-living.

These phenomena suggest that the Higgs PT in the U(1) model for $\lambda = 3$ and ß = 2.5 might be still of first order. However, we have found [9] that on a 16^4 lattice the discontinuity of μ is smaller than on an $8^3 \times 16$ lattice. Thus a final conclusion on the order is not yet possible. Nevertheless we have learned that the screening energy μ is much more sensitive to the Higgs PT than the local observables are, and should therefore be used for the study of its order.

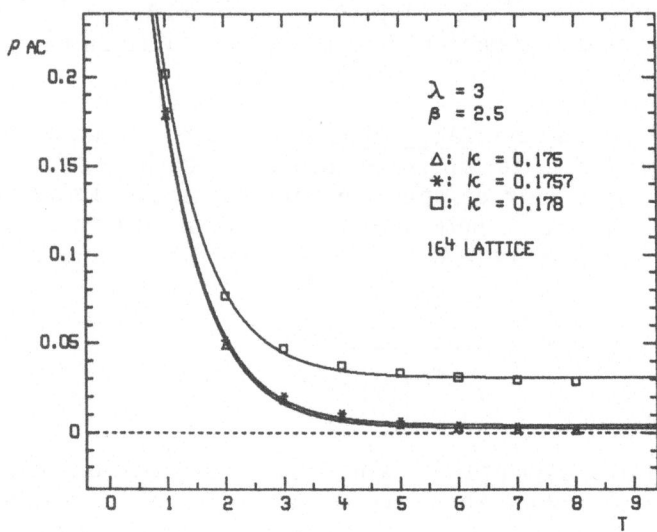

Fig.4. The T-dependence of the ratio $\rho_{AC}(T)$, eq.(8), in the U(1) model [9]. The squares denote data in the Higgs region, the other symbols denote data in the Coulomb phase.

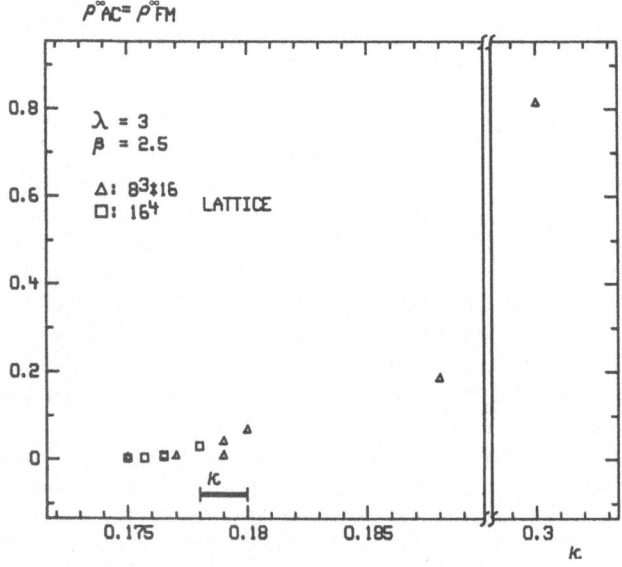

Fig.5. Values of ρ_{AC}^{∞} (eq.10) and ρ_{FM}^{∞} in the U(1) model [9]. Both quantities are equal within the error bars.

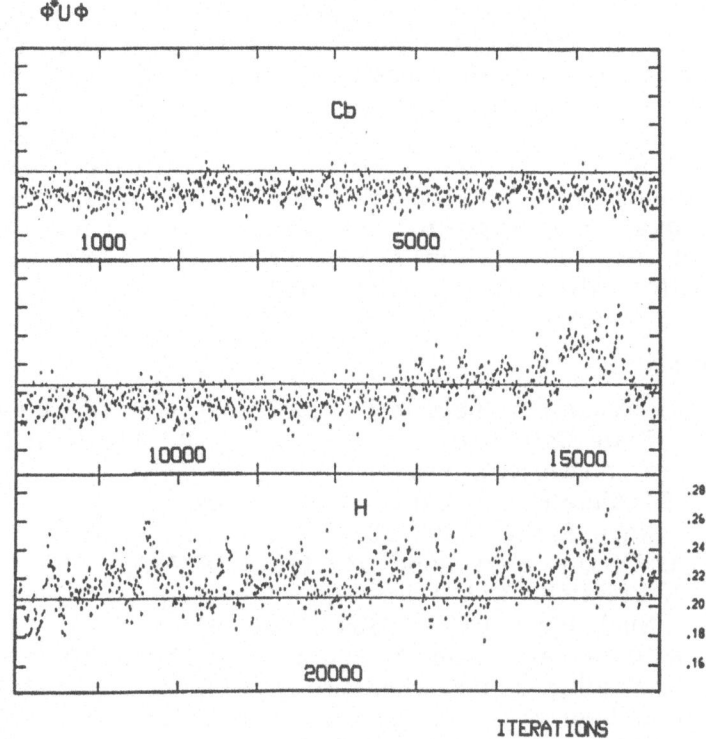

$\Phi^* U\Phi$

Cb

1000   5000

10000   15000

H

.28
.26
.24
.22
.20
.18
.16

20000

ITERATIONS

Fig.6. Fluctuations of $\Phi^* U\Phi$ in one MC run in the U(1) model at
$\kappa = 0.179$ [9]. Long-living states in the Coulomb phase (Cb)
and in the Higgs region (H) correspond to the two values of μ
shown in fig.3 for this κ.

Our preliminary results [15] for the SU(2) model indicate that
for $\lambda = 0.5$ the Higgs PT on a 16^4 lattice is discontinuous at
ß = 2.25, but the discontinuity decreases rapidly when ß changes in
both directions. The data for ß = 2.1 are already consistent with the
second order. It is well possible that the end point of the Higgs PT
line at $\lambda = 0.5$ lies at ß $\simeq$ 2. This end point seems to be a critical
point. These investigations are very difficult and particularly the
finite size effects are not yet under control.

10.   Remark   For completeness I want to mention two further properties of the Higgs PT which have been discussed by other speakers. As
we have heard from Mrs. I-H. Lee, the quenched massless fermion
expectation value $\langle \bar{\psi}\psi \rangle$ is still another interesting observable which
behaves differently in the confinement and Higgs regions of the SU(2)
Higgs model [16]. For small κ its values are large whereas for large κ
they are indistinguishable from zero. As reported by K. Kanaya, the
Higgs PT for $\lambda = 0.5$ changes at high temperatures into a crossover

and shifts slightly to higher κ [17]. The shift corresponds to the occurence of the Kirzhnits-Linde-Weinberg symmetry restoring transition in the Higgs region at high temperatures. The change into a crossover indicates the possibility that this transition is not a PT in the strict mathematical sense.

Acknowledgements I thank H.G. Evertz, V. Grösch, K. Jansen, K. Kanaya, H.A. Kastrup, D.P. Landau, C.B. Lang, T. Neuhaus and J.L. Xu for collaboration and valuable discussions and J. Smižanská-Jersák for help with the manuscript.

References

[1] J. Jersák, in "Lattice Gauge Theory – A Challenge in Large-Scale Computing", ed. B. Bunk, K.H. Mütter and K. Schilling, Plenum Press, 1986.

[2] H. Kühnelt, C.B. Lang and G. Vones, Nucl. Phys. B230 [FS10] (1984) 16.

[3] S. Elitzur, Phys. Rev. D12 (1975) 3978.
J. Fröhlich, G. Morchio and F. Strocchi, Nucl. Phys. B190 [FS3] (1981) 553.

[4] K. Osterwalder and E. Seiler, Ann. Phys. 110 (1978) 440.

[5] E. Fradkin and S. Shenker, Phys. Rev. D19 (1979) 3682.

[6] H.G. Evertz, J. Jersák, C.B. Lang and T. Neuhaus, Phys. Lett. 171B (1986) 271.

[7] H.G. Evertz, K. Jansen, J. Jersák, C.B. Lang and T. Neuhaus, Aachen preprint PITHA 86/16.

[8] B. Berg and C. Panagiotakopoulos, Phys. Rev. Lett. 52 (1984) 94.

[9] H.G. Evertz, V. Grösch, K. Jansen, J. Jersák, H.A. Kastrup and T. Neuhaus, Aachen preprint PITHA 86/15.

[10] H.G. Evertz, V. Grösch, J. Jersák, H.A. Kastrup, D.P. Landau, T. Neuhaus and J.-L. Xu, Phys. Lett. 175B (1986) 335.

[11] K. Fredenhagen and M. Marcu, Commun. Math. Phys. 92 (1983) 81.
K. Fredenhagen and M. Marcu, Phys. Rev. Lett. 56 (1986) 223.
M. Marcu, "Lattice Gauge Theory – A Challenge in Large-Scale Computing", ed. B. Bunk, K.H. Mütter and K. Schilling, Plenum Press, 1986.

[12] K. Jansen, J. Jersák, C.B. Lang, T. Neuhaus and G. Vones, Nucl. Phys. B265 [FS15] (1986) 129.

[13] W. Langguth and I. Montvay, Phys. Lett. 165B (1985) 135.
W. Langguth, I. Montvay and P. Weisz, Nucl. Phys. B277 (1986) 11.

[14] J. Jersák, C.B. Lang, T. Neuhaus and G. Vones, Phys. Rev. D32 (1985) 2761.

[15] H.G. Evertz, J. Jersák, D.P. Landau, T. Neuhaus and J.-L. Xu, (in preparation).

[16] I-H. Lee and J. Shigemitsu, Phys. Lett. 178B (1986) 93.

[17] H.G. Evertz, J. Jersák and K. Kanaya, Aachen preprint PITHA 86/14.

THERMAL GREEN'S FUNCTIONS OF THE ENERGY-MOMENTUM TENSOR AND TRANSPORT COEFFICIENTS OF THE SU(3) YANG-MILLS GAS

F. Karsch

Theory Division, CERN, 1211 Geneva 23, Switzerland

H.W. Wyld

University of Illinois at Urbana-Champaign, Dept. of Physics, Loomis Laboratory, Urbana, IL 61801

Forthcoming heavy ion experiments[1] will probe the existence of a quark-gluon plasma at high temperatures. The equilibrium properties of the plasma phase have been studied in lattice Monte Carlo simulations[2]. However, in a heavy ion experiment, a hot QCD plasma will not be produced in equilibrium and moreover will cool down rapidly during the subsequent expansion. To judge, for instance, the temperature decrease during the expansion phase of the plasma, an understanding of dissipative effects will be essential.

Transport coefficients are important ingredients in the description of dissipative effects in an expanding quark-gluon plasma using hydrodynamic equations of motion[3-6]. For instance, the change in energy density ε is determined by

$$\frac{d\varepsilon}{d\tau} + \frac{1}{\tau}(\varepsilon + p) = \left(\frac{4}{3}\eta + \zeta\right)/\tau^2 \qquad (1)$$

with p denoting the pressure and $\eta(\xi)$ the shear (bulk) viscosity. Transport coefficients like η or ξ can be calculated by means of Monte Carlo simulations. This, however, requires the analytic continuation of thermal correlation functions of the energy-momentum tensor to real time[4,7], which in general is difficult as the knowledge of the spectral decomposition of the thermal Green's functions[8-10] is needed. We will avoid this problem by using an ansatz for the spectral density function involving a few parameters. These are determined from a fit of the Monte Carlo data for the thermal Green's functions.

In the following, we will concentrate on a discussion of the calculation of the shear viscosity η and present first results from a Monte Carlo simulation on a $8^3 \times 4$ lattice for pure SU(3) gauge theory[11].

From a Kubo type formula, we find for the shear viscosity[9]

$$\eta = - \int d^3x' \int_{-\infty}^{t} dt_1 \, e^{\varepsilon(t_1-t)} \int_{-\infty}^{t_1} dt' \langle T_{12}(x,t), T_{12}(x',t') \rangle_{ret} \quad (2)$$

Here the limit $\varepsilon \to +0$ has to be taken, and $T_{\mu\nu}$ denotes the energy-momentum tensor, which in terms of the field strength tensor $F_{\mu\nu}$, is given by

$$T_{\mu\nu} = 2 \, Tr \left(F_{\mu\sigma} F_{\nu\sigma} - \frac{1}{4} \delta_{\mu\nu} F_{\sigma\delta} F_{\sigma\delta} \right) \quad (3)$$

The retarded Green's function $\langle A(x,t)B(x',t') \rangle_{ret}$ is defined as

$$\langle A(x,t), B(x',t') \rangle_{ret} = -i\theta(t'-t) \langle [A(x,t), B(x',t')] \rangle_0 \quad (4)$$

with $\langle ... \rangle_0$ denoting the usual thermal expectation value of the correlation function with respect to the thermal equilibrium distribution. The correlation functions appearing in Eq. (2) represent correlations in real time. In a non-perturbative lattice calculation on a Euclidean lattice, we obtain instead thermal Green's functions. The determination of the retarded Green's functions in Eq. (2) thus requires an analytic continuation of the thermal Green's functions

$$G_\beta(x,\tau) = -i < A(x,\tau) B(0,0)>_0 \; , \; 0 < \tau < \beta = 1/T \quad (5)$$

The formalism to do this has recently been discussed by Hosoya et al.[8] [see also Ref. 10)]. Knowing the thermal Green's function $G_\beta(x,\tau)$, we can determine the Fourier transform $G^\beta(p,\omega_n)$ at the discrete set of frequencies $\omega_n = 2\pi n/\beta$ with $n = 0, \pm 1, \ldots$. In terms of the spectral function $\rho(p,\omega)$, G^β is given as

$$G^\beta(p,\omega) = \int d\omega \; \frac{\rho(p,\omega)}{i\omega_n - \omega} \quad (6)$$

Once the spectral function ρ is known, the analytic continuation is immediate as the retarded Green's function has the same spectral representation as G^β with $i\omega_n$ replaced by $p_0 + i\varepsilon$.

Thus the aim of a non-perturbative Monte Carlo calculation of transport coefficients should be to extract from thermal correlation functions of suitable components of the energy-momentum tensor the spectral function $\rho(p,\omega)$, and use it as an input for the analytic continuation. In practice, however, we are far away from this ideal situation. On lattices with a rather small number of sites in the temporal direction, the thermal correlation functions can only be determined for a few points. A more pragmatic way to proceed is to start with a reasonable ansatz for the spectral density which contains a few parameters. Using the Fourier transform of Eq. (6) gives then an ansatz for the thermal correlation function $G_\beta(x,\tau)$ which can be used to fit the Monte Carlo data.

As an ansatz for the spectral density $\rho(\omega)$ [only the zero momentum part contributes in Eq. (2)], we use[11]

$$\rho(\omega) = \frac{A(1-e^{-\beta m})}{\pi} \left[\frac{\gamma}{(m-\omega)^2 + \gamma^2} - \frac{\gamma}{(m+\omega)^2 + \gamma^2} \right] \quad (7)$$

which is motivated by the spectral density for a free bosonic field and attempts to include the effect of interactions through the parameter γ. The above ansatz for $\rho(\omega)$ has three free parameters A, m and γ which have to be determined from fits of Monte Carlo data with the analytic form obtained from Eq. (7), i.e.,

$$G_\beta(\tau) = -i \int d\omega \, \frac{e^{\beta\omega}}{e^{\beta\omega} - 1} \, e^{-\omega\tau} \rho(\omega) \quad (8)$$

Once these parameters are determined we get the shear viscosity η as

$$\eta = 2A(1-e^{-\beta m}) \frac{2\gamma m}{(\gamma^2 + m^2)^2} \quad (9)$$

or in terms of dimensionless lattice parameters $A_\ell = Aa^5$, $m_\ell = ma$ and $\gamma_\ell = \gamma a$

$$\eta/T^3 = N_\tau^3 \frac{2A_\ell}{m_\ell} (1 - e^{-N_\tau m_\ell}) \frac{2\gamma_\ell/m_\ell}{(1 + (\gamma_\ell/m_\ell)^2)^2} \quad (10)$$

where we have used $\beta \equiv 1/T = N_\tau a$, a being the lattice spacing and N_τ the number of lattice sites in temporal direction of the lattice.

To test the applicability of this approach, we have performed a Monte Carlo simulation on a $8^3 \times 4$ lattice. To evaluate η we have to measure thermal correlations of off-diagonal spacelike components of the energy-momentum tensor $T_{\mu\nu}$. Here we take advantage of the cubic symmetry of space components of $T_{\mu\nu}$[11] which allows us to express these correlation functions in terms of diagonal elements,

$$\langle T_{12}(x) T_{12}(y) \rangle = \frac{1}{2} \left[\langle T_{11}(x) T_{11}(y) \rangle - \langle T_{11}(x) T_{22}(y) \rangle \right] \quad (11)$$

The diagonal components of $T_{\mu\nu}$ represent the energy (T_{00}) and pressure (T_{ii}) in thermal equilibrium. Thus we can obtain $T_{\mu\mu}$ as a suitable derivative of the partition function of the SU(3) Yang-Mills system[12]. This way we also obtain the correct trace anomaly for the energy-momentum tensor[13,14]. As is usual in thermodynamic calculations on the lattice we have neglected the $O(g^2)$ terms in the definition of $T_{\mu\mu}$ in our present calculation. As in calculations of T_{00} we expect the influence of the neglected terms to be small. Thus we use for the diagonal components of the energy-momentum tensor

$$T_{\mu\mu}(x) = \frac{2}{g^2} \left[-\sum_{\nu \neq \mu} \text{Tr} U_{x,\mu\nu} + \sum_{\substack{\xi, \nu \neq \mu \\ \xi > \nu}} \text{Tr} U_{x,\xi\nu} \right] \quad (12)$$

On a $8^3 \times 4$ lattice, we can measure thermal correlation functions at distance $\tau = 0,1,2$. As we have to determine three parameters to extract η from the Monte Carlo data, calculations on such a small lattice can only serve as a test of the procedure used. We have performed simulations at various values of the bare coupling β in the confined and deconfined region. It turned out that the correlation functions drop very fast, making a measurement at distance $\tau = 2$ difficult even with runs of 180 000 iterations[11]. In Fig. 1 we show results at $6/g^2 = 5.6$, which is close to the deconfinement transition ($6/g_c^2 = 5.68$) in the confined region. Figure 1a shows fits of $G(\tau)$ for fixed $m\beta = 11.6$ and various values of m/γ whereas in Fig. 1b we show fits for fixed $m/\gamma = 0.02$ and various values of $m\beta$. The third parameter, A, has been fixed to be identical to $G(0)$. Clearly, larger lattices will be necessary to do a better fit.

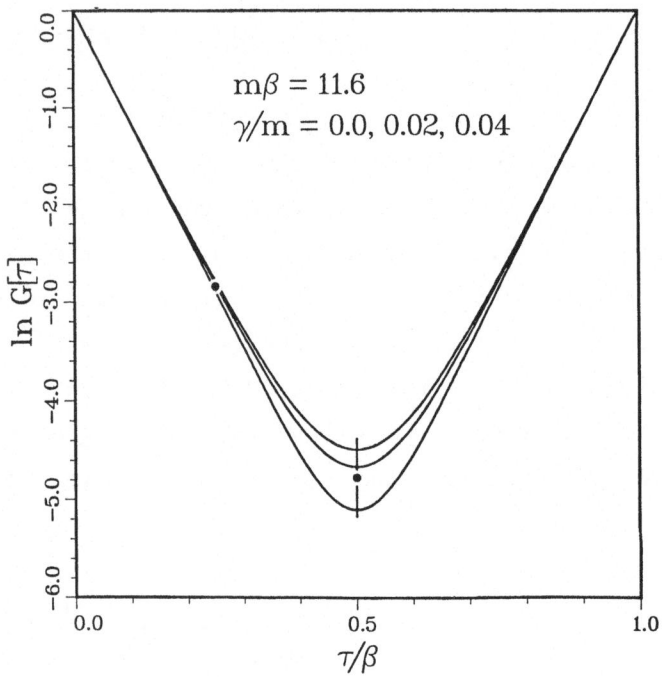

Fig. 1a Thermal Green's function $G_\beta(\tau)$ versus τ for $m\beta = 11.6$ and various values of the interaction parameter $\gamma/m = 0.0$, 0.02 and 0.04. Also shown are Monte Carlo data from simulations at $6/g^2 = 5.6$ for the shear viscosity correlation function.

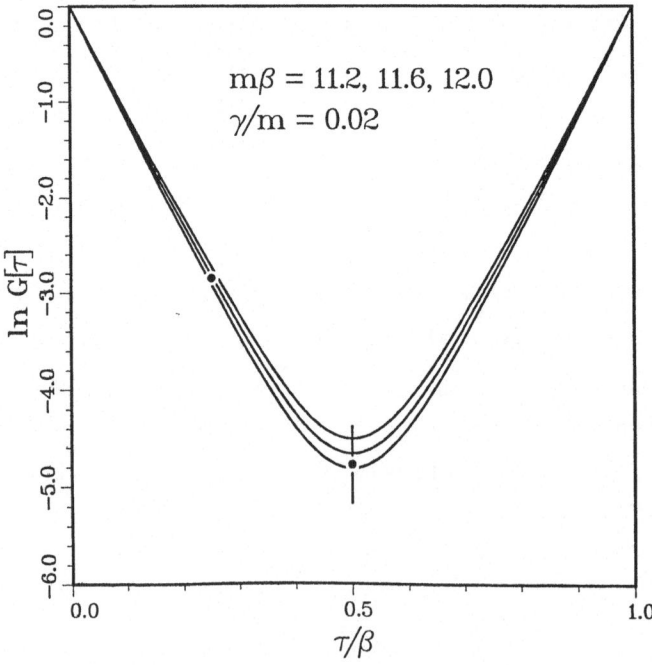

Fig. 1b Same as Fig. 1a, however, for fixed $\gamma/m = 0.02$ and $m\beta = 11.2$, 11.6 and 12.0.

We see that while the mass parameter seems to be well determined (ma = 2.5±0.5), the value of the interaction parameter γ/m is difficult to determine and we can only give an upper bound for it

$$\gamma/m < 0.05 \tag{13}$$

Similar results hold slightly above the deconfinement transition[11]. From our present calculations, we thus can give only quite loose bounds for η/T^3 for temperatures close to T_c

$$0 \le \eta/T^3 \le 3.5 \tag{14}$$

It would be of considerable interest to get better bounds on η/T^3 as this would give indications for the applicability of hydrodynamical models for the expansion of a quark-glon plasma[6]. To achieve this, it seems that other methods have to be tested which allow us to improve the signal to noise ratio in our data. Like in glueball calculations where similar correlation functions are studied, it may be useful to develop source methods for the energy momentum tensor.

ACKNOWLEDGEMENTS

This work was partly supported by the grant NSF-PHY-82-01948. F.K. would like to thank M. Gyulassy for a very helpful discussion.

REFERENCES

1. M.A. Faessler, CERN preprint, CERN-EP/86-102 (July 1986).

2. F. Karsch, in: Lattice Gauge Theory - A Challenge in Large-Scale Computing, Eds. B. Bunk, K.H. Mütter and K. Schilling, Nato ASI Series B:Physics Vol. 140 (1986).

3. L.P. Kadanoff and P.C. Martin, Ann. Phys., 24:419 (1963).

4. A. Hosoya and K. Kajantie, Nucl. Phys., B250:666 (1985).

5. S. Gavin, Nucl. Phys., A435:826 (1985).

6. P. Danielewicz and M. Gyulassy, Phys. Rev., D31:53 (1985).

7. R. Horsely and W. Schoenmaker, Kaiserslautern preprints (September 1985).

8. A. Hosoya, M. Sakayami and M. Takao, Ann. Phys., 154:229 (1984).

9. D.N. Zubarev, "Non-Equilibrium Statistical Thermodynamics", Plenum Press, New York, 1974.

10. G. Baym and N.D. Mermin, J. Math. Phys., 2:232 (1961).

11. F. Karsch and H.W. Wyld, Illinois preprint, ILL-(TH)-86-51 (1986).

12. J. Engels, F. Karsch, I. Montvay and H. Satz, Nucl. Phys., B205:545 (1982).

13. F. Fucito and B. Svetitsky, Phys. Lett., 131B:165 (1983).

14. We thank A. Gocksch for a discussion on this point.

RENORMALIZATION FLOW STRUCTURE IN LATTICE QED

C.B. Lang

Institut f. theoretische Physik
Universität Graz
A-8010 Graz, Austria

INTRODUCTION

In my talk I want to discuss results regarding the critical structure of U(1) lattice gauge theory obtained in two independent studies. The emphasis is put on the study of the flow structure in coupling constant space[1]. Further results come from a collaboration with Claudio Rebbi[2] where we studied the critical exponents in the even and the odd sector of the renormalization group transformations.

There have been related investigations of the pure gauge U(1) system also with help of Monte Carlo renormalization group (MCRG) methods by other authors[3,4], one of them presented at this conference[5]. At the Wuppertal conference last year there has been a review talk by Gupta[6] covering applications to gauge theories in general.

In this presentation I first remind you on the underlying ideas of Real Space Renormalization Group (RSRG) and briefly discuss the methods involved in our study. Then the situation of pure, compact U(1) gauge theory on the lattice is reviewed. In the last two parts the results for the flow structure under scale transformations are given and the values obtained for the leading critical exponents are presented. In the conclusion the possible critical structure is discussed.

REAL SPACE RENORMALIZATION GROUP

In the formulation of a quantum field theory on an euclidean space-time lattice[7,8] the final aim is to study the system near a phase transition of higher order. There the correlation length measured in lattice units a diverges and the long distance properties of the system should become independent on the lattice structure. The physical properties of the continuum theory to be constructed are then identified with these, lattice regularization independent, properties. The principle underlying the RSRG approach is to study the behaviour of such a system under scale changing transformations in real space (as contrasted to momentum space). The system may be characterized by its coupling constants β_α entering in the action

$$S = \sum_\alpha \beta_\alpha\, S_\alpha(U) \quad , \quad S_\alpha = \sum_\alpha S_{\alpha,x}(U) \tag{1}$$

where $S_{\alpha,x}$ denote possible local contributions to the action; for lattice gauge theories they may assume the form of characters of closed loops of link variables or products thereof. The system may also be characterized by the whole entity of its observables, the expectation values of all possible gauge invariant objects.

The configurations of link variables $\{U\}$ of a theory given by (1) are distributed according to the Gibbs measure

$$\exp(-S(\beta,U))\; d\mu(U) \quad , \tag{2}$$

where $d\mu(U)$ denotes the Haar integration measure of the link variables. A scale changing transformation may be introduced by reducing the set of variables $\{U\}$ to a smaller set $\{U'\}$. For a real space scale factor of $s=2$ the number of variables is reduced by a factor 2^d, i.e. 16 for $d=4$. Technically this is achieved by mapping each configuration of U-variables to one of U'-variables with the probability measure

$$P(U',U)\; d\mu(U') \quad . \tag{3}$$

This measure is normalized to unity under integration over all U' variables. The resulting set of configurations $\{U'\}$ is now distributed according

$$\exp(-S'(\beta',U'))\; d\mu(U') \quad , \tag{4}$$

where

$$\exp(-S'(\beta',U')) = \int \exp(-S(\beta,U))\; P(U',U)\; d\mu(U) \quad . \tag{5}$$

Whereas this transformation is straightforward in configuration space the corresponding transformation of the action $S \to S'$, where

$$S'(\beta',U') = \sum_\alpha \beta'_\alpha\, S_\alpha(U') \tag{6}$$

is the "blocked action", is a priori unknown. Only in a few simple cases of spin models it may be determined explicitly. Usually one has to rely on (a) truncations in the set of coupling constants from an infinite number to a few, and (b) approximation of the effect of the RG transformation with help of pertubation expansions[8].

This "block spin transformation" (BST) may be performed sucessively, always leading from configurations after n BSTs and action $S^{(n)}$ to configurations after n+1 BSTs with action $S^{(n+1)}$. At each BST the correlation length measured in lattice units is reduced by a factor 1/s. Obviously there are only two possible fixed point (FP) values :
(a) the original ensemble of configurations represented a situation with finite correlation length ξ, then it will decrease and vanish in the limit n→∞;
(b) the original ensemble of configurations represented a critical situation with infinite correlation length, then it has to stay infinite under successive BSTs.
The second FP is the one interesting for the continuum limit; its domain of attraction defines the universality class. The position of the FP may depend on the specific form of the BST but the physical properties like the number of relevant couplings and the critical exponents are independent on the form of the BST.

At the various levels of blocking one may determine various observables and infer from them information on the flow structure and the critical properties.

Expectation values of operators
From the values $\langle S_i^{(n)} \rangle$ we find the position of the corresponding ensemble of configurations in the "space of operators" Σ; there is a surjective mapping from the space B of coupling parameters onto Σ, i.e. each set of couplings constants (each action S) corresponds to a unique set of operator values.

Conditional expectation values
We may measure operators in a conditional way, i.e. defining

$$\langle \tilde{S}_i \rangle_{\{U\}} = \left\langle \frac{\int dU_0 \; S_i(U_0,\{U\}) \; \exp[-\sum_\alpha \tilde{\beta}_\alpha \, S_\alpha(U_0,\{U\})]}{\int dU_0 \; \exp[-\sum_\alpha \tilde{\beta}_\alpha \, S_\alpha(U_0,\{U\})]} \right\rangle_{\{U\}} \cdot (7)$$

Here one determines first the operator in a frozen background configuration integrating explicitly over the local variable U_0, then one sums over the whole ensemble of configurations $\{U\}$. It may be shown[9] that

$$\langle \tilde{S} \rangle = \langle S \rangle \quad \text{if and only if} \quad \tilde{\beta} = \beta \quad , \tag{8}$$

i.e. even if we do not know the values of the coupling parameters β we may guess them and check the quality of this guess with help of the above relation. Swendsen[10] suggested to employ this relation to iteratively improve on that estimate of β. This technique was used successfully within quantum field theory to study the universality class of Φ^4-theory[11]. We

197

thus may determine the renormalized couplings and trace the flow in coupling parameter space B: $\beta^{(n)} \to \beta^{(n+1)} \to \ldots \to \beta^*$. Recently further methods have been proposed in order to determine the couplings from the ensembles of configurations[12] but they have not yet been applied to our problem.

Correlations

Derivatives of expectation values of observables may be obtained from correlations between observables on lattice configurations after n and m block steps.

$$\frac{\partial \langle S\{n\} \rangle}{\partial \beta\{m\}} = \langle S\{n\} \, S\{m\} \rangle - \langle S\{n\} \rangle \langle S\{m\} \rangle \quad , \tag{9}$$

and from these one may obtain approximations to the linearized BST[13],

$$T_{jk}^{(n+1,n)} \equiv \frac{\partial \beta_j^{(n+1)}}{\partial \beta_k^{(n)}} \quad , \tag{10}$$

$$\frac{\partial S_i^{(n+1)}}{\partial \beta_k^{(n)}} = \sum_j \frac{\partial S_i^{(n+1)}}{\partial \beta_j^{(n+1)}} \frac{\partial \beta_j^{(n+1)}}{\partial \beta_k^{(n)}} \quad . \tag{11}$$

At a fixed point of the BST one expects $T^{(n+1,n)} \to T^*$ (for $n \to \infty$). The largest eigenvalues of T^* are related to the leading critical exponents through

$$y_E = \log \lambda_E / \log s \quad , $$
$$\nu \equiv 1/y_E \quad , \tag{12}$$

$$y_0 = \log \lambda_0 / \log s \quad , $$
$$\delta \equiv y_0 / (d - y_0) \quad , \tag{13}$$

where the scale factor of the BST s=2 and the dimension d=4 in our case. The linearized BST $T^{(n+1,n)}$ factorizes into an even and an odd sector corresponding to the even and odd operators. For that reason the leading even and odd critical exponents may be studied independently and the corresponding eigenvalues of T have been denoted by λ_E and λ_0.

$T^{(n+1,n)}$ is obtained by solving (11); this equation involves only a finite number of operators and therefore the determination of the eigenvalues of the truncated T involves two sources of systematic errors: from the inversion of the truncated matrix and from the truncation itself. In a recent discussion of these problems[14] it turned out that in a systematic expansion for the leading eigenvalue these two errors cancel in the first order correction term. A certain chance to estimate the systematic errors is to compare the results for the diagonalization for different stages of truncation.

198

BST AND OPERATORS FOR U(1) LATTICE GAUGE THEORY

U(1) is the most elementary Lie group that can be used to construct a quantum field theory. Yet, when formulated on a lattice, the U(1) system becomes one of the most intriguing and less understood quantum gauge models. The pure U(1) gauge system, while trivial in the continuum, is non-trivial on the lattice because of the non-linearities and consequent self-couplings inherent in the formulation. The lack of asymptotic freedom makes the weak coupling domain less interesting from the point of view of establishing a continuum limit. Non-trivial topological excitations, singular in the continuum but regular on the lattice, produce an interesting phase structure.

It has been established[15] that, for sufficiently weak coupling, the U(1) quantum lattice gauge system is in a Coulombic phase, characterized by long range correlations. Early numerical calculations[16], based on the Monte Carlo technique, gave indication of a phase transition at a coupling parameter $\beta \equiv 1/g^2 \approx 1$ for the system defined with Wilson's action[7]. Further Monte Carlo studies[17] of the scaling properties led to the picture of a rather straightforward second order phase transition, with definite exponents. These conclusions were subsequently challenged by the results of more extensive computations, which gave evidence for metastabilities and discontinuities to be associated with a first order phase transition[18]. Evidence for a tricritical point in an extended space of couplings was also produced[19]. However, a rather peculiar dynamical behaviour of the topological excitations present in the model[20] interferes with the stochastic evolution of the Monte Carlo simulation and adds some degree of uncertainty to the interpretation of the observed metastabilities and related critical phenomena.

The general U(1) lattice action (1) lives in a space of infinitely many different interaction terms S_α and their coupling constants β_α. If there is no external field these terms have to be even in the link field variables, i.e. invariant under the complex conjugation $U \leftrightarrow U^*$. Any point in the multidimensional parameterspace B of couplings corresponds to a specific action S. A straightforward extension of the original Wilson action[7] is to consider a term like $Re(U_p^2)$ in the so-called mixed action,

$$-S_P = \beta \, Re \, U_P + \gamma \, Re \, U_P^2 \quad , \tag{14}$$

or modify the single plaquette action in some other way like e.g. Villain's form[21]

$$\exp(-S_P) = \sum_{n=-\infty}^{\infty} \exp[\frac{\beta}{2} - (\phi - 2n\pi)^2] \quad , \tag{15}$$

and both cases will be discussed below.

In the presented work[1-2] the critical behaviour of the U(1) lattice quantum gauge system was studied with the discussed real space Monte Carlo RG methods. The extensive simulations were performed for various forms of the action for lattice size 16^4, which is then renormalized to 8^4 and 4^4. Other MCRG studies of the U(1) transition have been presented in the literature[3-6]; our results add several new elements of information, especially by determining the flow structure in a space of 10 couplings[1] and through the implementation of two independent blocking procedures and the consideration of odd observables[2]. Further investigations of this interesting model, not based on MCRG, can be found in refs.22-23.

All BST's considered have scale factor s=2. In ref.1 a form suggested by Swendsen[24] has been chosen. The sites x' of the blocked lattice are those sites of the large lattice which have even coordinates n_x, n_y, n_z, and n_t. The links U' are constructed from the sum of 7 paths of length 2 and 4 between such sites, in the form

$$U'_{x',\mu} = V_{x',\mu} / |V_{x',\mu}| \quad ,$$

$$V_{x',\mu} = U_{x,\mu} U_{x+\mu,\mu} + \alpha \sum_{\substack{\pm\nu \\ \nu\perp\mu}} U_{x,\nu} U_{x+\nu,\mu} U^*_{x+\nu+\mu,\mu} U^*_{x+2\mu,\nu} \quad . \tag{16}$$

A gauge transformation on the larger lattice corresponds to a gauge transformation on the blocked lattice, thus the necessary properties of gauge invariance are guaranteed. The parameter α is a free parameter and could be used to optimize the BST; in ref.1 $\alpha=1$ and in ref.2 $\alpha=1/2$ throughout the analysis. Where V'=0 we chose U'=1 as a tiebreaker in ref.1, whereas we added a small random number to V' in ref.2.

In ref.2 we also considered another BST related to Wilson's early proposal[25] where the new site on the block lattice is situated at the center of a 2^4 hypercube of the larger lattice. The new link variable is constructed by summing over all 8 links connecting the corresponding hypercubes. In this approach one first has to fix the gauge completely within the hypercube. This was done resorting to a Coulomb like gauge condition on all vertices of the hypercube. The sum is normalized to unit modulus in order to stay within the group manifold U(1).

In this presentation I put the emphasis on the results of ref.1 where 15 observables were considered. They may be written as real parts of products of link fields along the boundary of geometric objects, i.e.

S_1	: single plaquette	
S_2	: double plaquette	
S_3	: bent double plaquette	for charge 1 representation
S_4	: twisted bent double plaquette	
S_5	: planar 2×2 loop	
S_{6-10}	: like S_{1-5} but for charge 2 representation	
S_{10-15}	: like S_{1-5} but for charge 3 representation	

The representation for charge n amounts to taking the real part of the n-th power of the variables. The multiplicity of these operators (number of geometric inequivalent occurrences on the lattice) is 6, 18, 72, 24 and 12 times the number of sites for S_1 to S_5 and the higher operators respectively. The action includes of course all occurrences of the loops; in the later presentation of results the expectation values of the operators are normalized to lie $\epsilon[0,1]$, i.e. they are normalized by their multiplicities.

The reason to restrict ourselves to this set of simple loop operators is that they are straightforward and quickly to construct and they are likely to include important contributions to the FP-action. In ref.2 we also included the even operators:

$-\partial S_{P,V}/\partial \beta$, the internal energy per plaquette for the Villain action,
the total number of monopole string bits,
the product $\mathrm{Re}\, U_{P1}\, \mathrm{Re}\, U_{P2}$ for parallel plaquettes P1 and P2, facing each other on opposite sides of a 3-cube, and
$\mathrm{Re}\, U_{3\times 1}$, the 3×1 Wilson loop, wrapped around a 3-cube.

Furthermore there we considered operators odd under $U \leftrightarrow U^*$

$$R_1 \equiv \sum_{x\,\epsilon\,Q_{\mu\nu}} \sin \theta_{x,\mu\nu} \quad , \qquad (17)$$

and

$$R_2 \equiv \sum_{x\,\epsilon\,Q_{\mu\nu}} \bar{\theta}_{x,\mu\nu} \qquad (\bar{\theta}_{x,\mu\nu} \epsilon\, [-\pi,\pi]), \qquad (18)$$

where the domains $Q_{\mu\nu}$ are partitions of the hypercube defined as follows: one takes any quarter of the (μ,ν) plane and translates it over all possible displacements in the two orthogonal directions. The union of such quarter planes forms an $8^2\times 16^2$ parallelepipedon which constitutes $Q_{\mu\nu}$.

RESULTS ON THE RENORMALIZATION FLOW

Details of the results and the statisticscan be found in ref.1. The system was simulated by standard MC techniques on lattices of size 16^4 at various values of the couplings $\beta\equiv\beta_1$ and $\gamma\equiv\beta_6$ and the Villain coupling β_V:

for $\gamma=0.15$ at $\beta=0.912$
for $\gamma=0.$ at $\beta=0.97, 1.01, 1.0103, 1.0105, 1.02, 1.05$
for $\gamma=-0.15$ at 1.121
for $\gamma=-0.20$ at $\beta=1.158, 1.16$
for $\beta_V=0.643, 0.644, 0.645$

The idea was to study the behaviour close to the phase transition (PT), possibly at the PT. It turned out that there is a discontinuity in the internal energy at all PTs in that range. The typical lifetime of the metastable states is, however, very long (several 10000 configurations for the Wilson action at $\gamma=0$). We therefore considered only run sequences in a definite state, i.e.

without tunneling, for the analysis. For the Wilson action the PT lies at $\beta-1.0105(3)$ and no tunneling was observed for either sequence, the hot one (originating from a hot start configuration) studied over more than 50000 configurations and the cold one (from an ordered start) over more than 40000 configurations, although both sequences had identical β.

Fig.1 shows the flow in a subspace of observables (S_1, S_6) and coupling constants $(\beta_1 \equiv \beta, \beta_6 \equiv \gamma)$. These four run sequences originate from starts at the PT on the hot branch of the PT. One finds a clear flow towards the hot, trivial fixed point (FP) at $\beta_\alpha = 0$. A remarkable feature is that, although the flow started at very different points in the (β, γ) plane the values approach each other quickly and apparently follow a common behaviour already after just one BST.

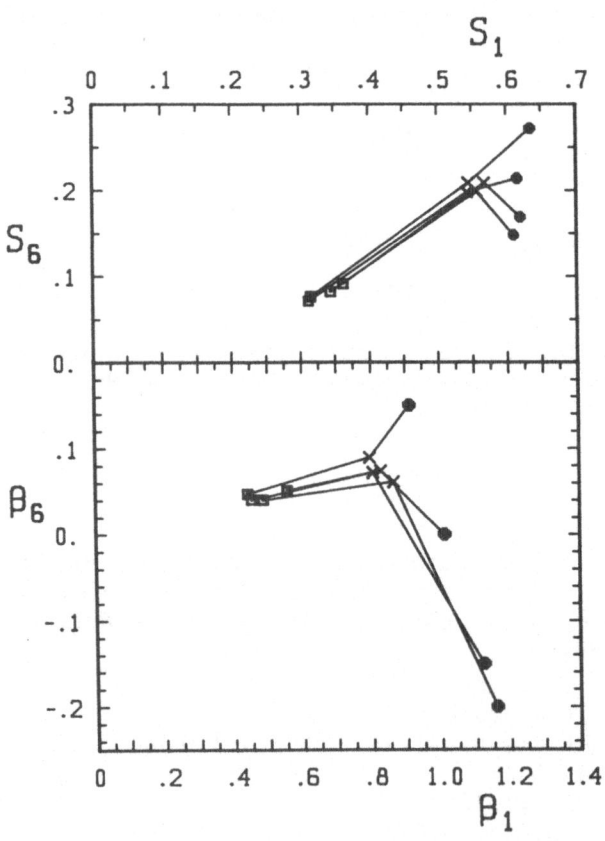

Figure 1

In fig.2 the observables S_2-S_{11} are plotted versus S_1, as obtained after one and two BSTs on lattices of size 8^4 and 4^4. For sake of the presentation the values of the observables have been shifted by multiples of 0.2 in the figure. In this figure the start

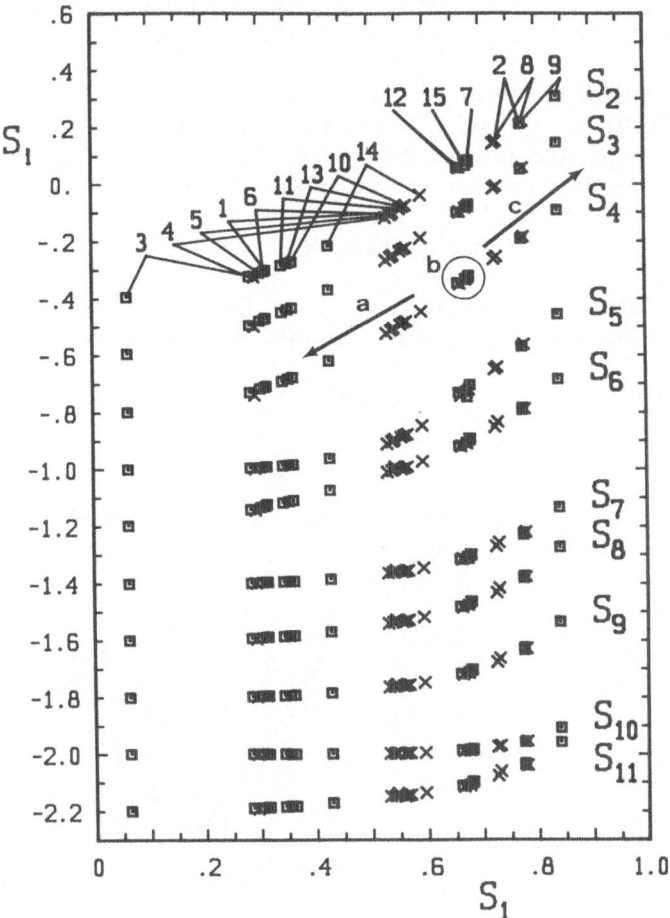

Figure 2

points on the 16⁴ lattice have been omitted and the renormalization flow in operator space is thus more clearly visible.

The observed flow behaviour may be grouped into three categories (cf. fig. 2):

a) Starting on the hot side of the PT <u>and</u> at the PT on the hot branch the flow is always towards smaller values of observables and coupling constants, i.e. towards the trivial, hot FP.

b) Starting at the PT, but on the cold branch we observed three run sequences (all at $\gamma \lesssim 0$) whith a FP behaviour: the flow approaches a FP in the first BST and stays there in the second BST. Remarkable enough the values of the 15 observables remain constant within the first two decimals at this point from 8⁴ to 4⁴.

(c) Starting above the PT one observes a flow away from the PT without any signal of saturation. This indicates that if there is a line of FPs it is not within the range of values studied.

The results for the coupling constants $\beta_1-\beta_{10}$, determined as discussed in sect.2, confirm this picture. Figs.3 show some of our results; obviously the statistical uncertainty is larger there.

Our results indicate the existence of a FP close to $(\beta_1 \equiv \beta,\ \beta_2,\ \beta_6 \equiv \gamma)\ \approx\ (0.97(7),\ -0.10(5),\ 0.11(2))$. That this FP is reached even from starting points below the Wilson line shows that there can be no tricritical point in the $(\beta,\ \gamma)$ plane between $\gamma=0$ and $\gamma \approx -0.20$. A puyyling feature is that the flow runs away from this FP in the hot phase, even when starting at the PT.

A possible interpretation of this behaviour is that the PT is really of second order at least for $\gamma \leq 0$. The observed two-state

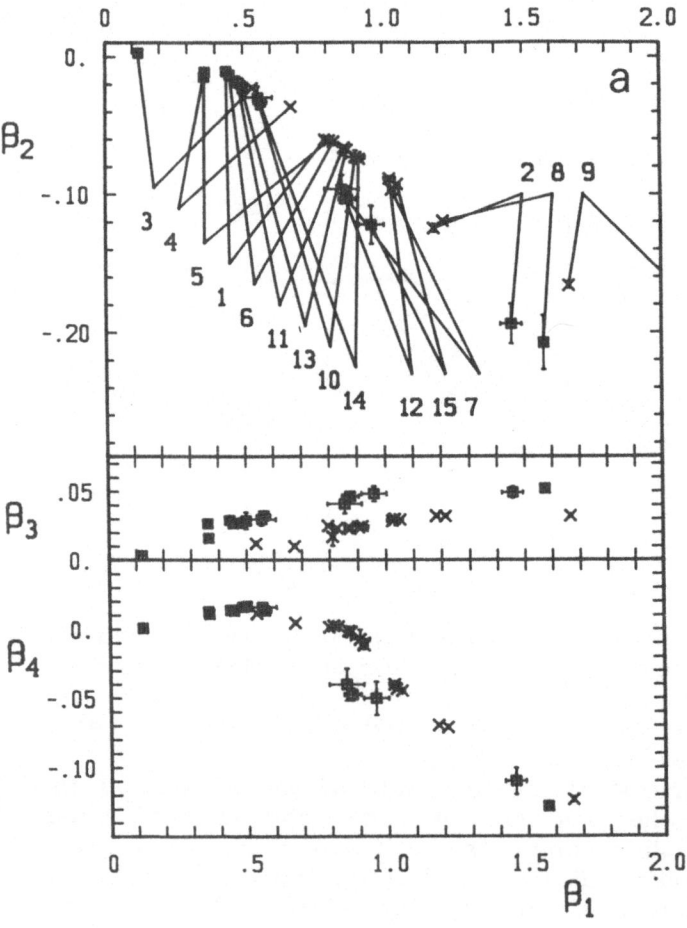

Figure 3a

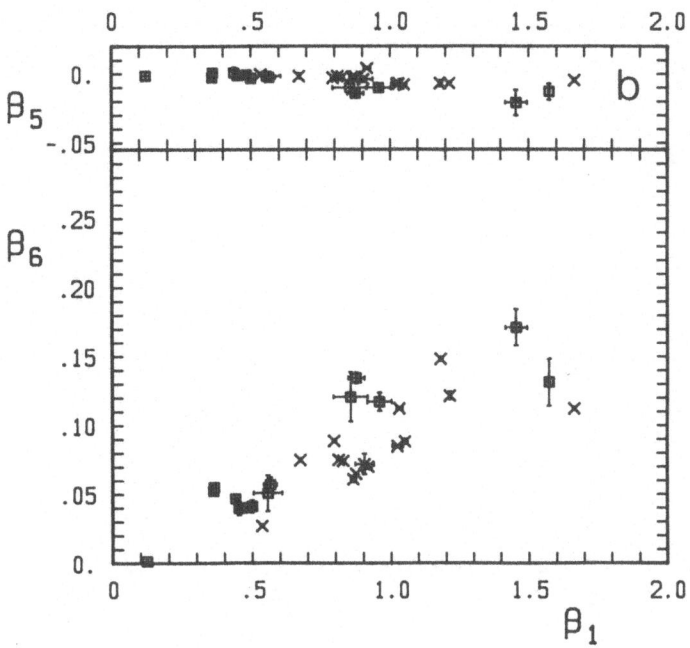

Figure 3b

signals then would have to be explained by spurious finite size effects, possibly related to the occurrence of monopole loops closed only due to the periodic boundary conditions[2,20]. The tricritical point would then be situated at values of $\gamma>0$, maybe between $\gamma=0$ and $\gamma=0.15$.

At this conference there has been another suggestion[5] proposing that the observed FP is really a discontinuity FP[26]. In this case the tricritical FP would be expected to lie at values γ below the considered range and the PT of first order in the considered domain of coupling constants. This picture is consistent with the observed flow and a possibility to decide between the two alternatives is to study the critical exponents which will be done in the next section.

RESULTS ON THE CRITICAL EXPONENTS

In fig.4 we plot the largest eigenvalue of the linearized BST against the operator S_1. Since this value of λ comes from correlations involving operators after n and n+1 BSTs the vertical bar runs from the values of S_1 after n+1 BSTs to that after n BSTs. We do not give the errors of the eigenvalues here; they are typically of the order of 5% of the value itself and are due to a small varia-

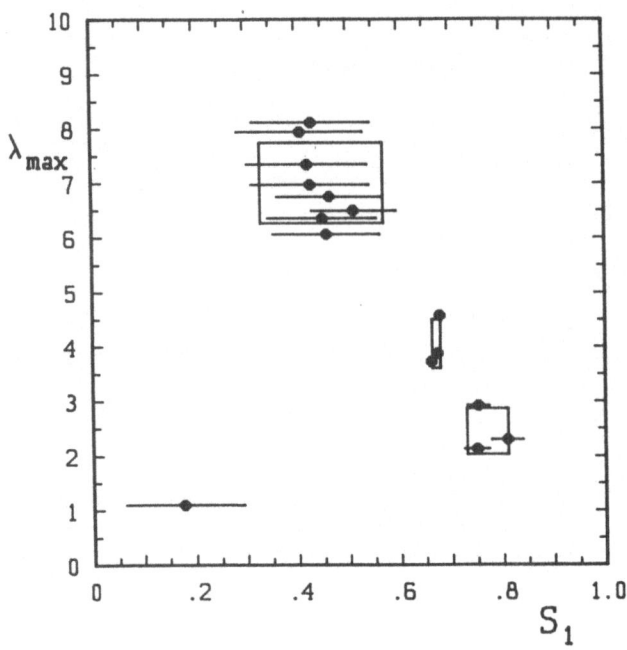

Figure 4

tion of the result with regard to different truncations. We find that once one has included operators S_{1-6} in the analysis the resulting largest eigenvalue remains stable up to consideration of all 15 operators.

One finds a dependence $\lambda(S_1)$ which is compatible with a smooth curve running through 4 at the critical FP. If we continue to interpret the values away from the FP still along the lines of (12) we expect that the ß-function becomes flater towards smaller g (larger β_1). The bending of λ towards 0 for small S_1 can be understood: in a strong coupling expansion we expect $\lambda \approx \beta_1$, i.e. vanishing at the origin.

In fig.4 it becomes obvious that the large eigenvalues 6-8 come from ensembles of configurations deep in the hot phase, well away from the critical surface. They have no relation to the critical properties of the FP. On the other hand they may indicate the domain of influence of a possible TCP, and the large eigenvalue might be associated with the tricritical exponent.

These numbers agree with the results in ref.2 where the system was simulated at the PT of the Villain action (15). There two different types of BSTs were studied and we found a good agreement of the resulting eigenvalues with each other and with the results in fig.4. As concerns the odd sector we considered only two operators, which are strongly correlated. We find a leading eigenvalue 2.6(2) for the hot run sequences and 4.1(7) (corresponding to a critical exponent $\delta=0.53(7)$) for the runs on the cold side of the PT. The overall behaviour is consistent with the flow structure discussed.

CONCLUDING REMARKS

The results on the critical exponents of the even sector agree, where applicable, with independent MCRG determinations along these lines[3,4]. There is a disagreement with results[5] presented at this conference, where a operator matching technique produced an eigenvalue around 16 which is the value expected at a discontinuity FP. This disagreement still has to be resolved by a careful anlysis of both methods under the circumstances of U(1) lattice gauge theory.

Based on the results discussed here[1,2] the flow structure is as follows (cf. fig.5). If one starts somewhere in the hot phase or even at the PT on the hot branch the flow quickly approaches a common RT after one BST and stays on that RT after further BSTs (type (a) runs in the tables). The RT is unique in the space of measured operators Σ and connects a critical FP with the trivial FP at $\beta = 0$.

Along this RT the leading eigenvalue of the linearized BST assumes values around 8; since this value is not observed at a FP it may not be related to a critical exponent (which would be

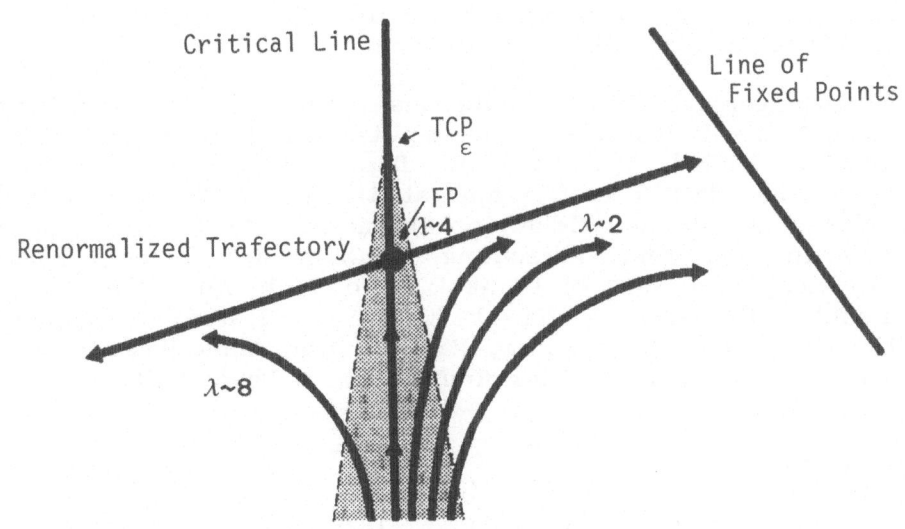

Figure 5

$\nu \dot= 0.33(2)$; such a value agrees with earlier determinations of ν that have involved mainly extrapolations of quantities measured on the hot side of the PT, like e.g. the string tension[22,27,18].) It seems conceivable that the RT enters the domain of influence of the TCP soon after having left the critical FP. Then one might consider this value being the leading tricritical exponent belonging to the TCP.

Run sequences starting at the PT on the cold branch (at values of $\gamma \lesssim 0$ in our study) show the existence of a critical FP where the set of measured operators stays invariant under the second BST from 8^4 to 4^4. This position is compatible with the values for ß and γ obtained in ref.4. Only one relevant eigenvalue was observed within the limits of our statistics. The critical exponent comes out as $\nu = 0.50(4)$ which is the value for a gaussian theory of non-interacting bosons.

Throughout the cold phase we expect a massless particle state: the photon. There is Monte Carlo evidence for such a particle; it was observed in correlations[23] of the odd operator Im U_p. Two possible structures of FPs may be imagined: (i) there may be a line of FPs leading from large ß to the FP on the critical surface, or (ii) there may be a line of FP at larger ß but without connection to the FP on the critical surface. In case (i) if one starts somewhere in the cold phase repeated BSTs should lead one towards this line of FPs and eventually one should converge to some point on this line. The position of this point would depend on the start couplings. The closer one starts to the critical surface, the closer should the saturation point on the fixed line be to the FP. Along the line there would be a marginal operator, i.e. the next to leading eigenvalue would assume the value 1. The leading eigenvalue should be the same all along the line of FPs. In case (ii) starting points away from the critical point would under BSTs lead towards the line of FPs, which may be situated at infinite ß. The flow structure observed supports this second picture. No saturation of the flow within the first two BSTs is observed.

Acknowledgement: The computations have been performed at the Florida State University Supercomputer Research Institute which is partially funded by the U.S. Dept. of Energy Contract No. DE-FC05-85ER250000 (ref.1) and at the CDC 7600 and CYBER 875 machines at BNL and CERN (ref.2); support in part by Fonds zur Förderung der Wiss. Forschung in Österreich, project P5965P is acknowledged. I want to thank Claudio Rebbi for collaboration and valuable discussions and I want to gratefully acknowledge discussions with Ken Bowler, Rajan Gupta, Anna Hasenfratz, Peter Hasenfratz, Jiri Jersák, Doug Toussaint and David Wallace.

REFERENCES

1.   C.B. Lang, Phys. Rev. Lett. 57:1828 (1986); C.B. Lang, Preprint FSU-SCRI-58 (1986), to appear in Nucl. Phys. B.

2. C.B. Lang and C. Rebbi, BNL preprint (Oct.1986).
3. R. Gupta, M.A. Novotny and R. Cordery, Phys. Lett. 171B:86 (1986).
4. A.N. Burkitt, Nucl. Phys. B270[FS16]: 575 (1986).
5. K. Decker, A. Hasenfratz and P. Hasenfratz, talk presented at this conference.
6. R. Gupta, Proceedings of the Wuppertal Conference Lattice Gauge Theories: A Challenge in Large Scale Computing. Eds. K.-H. Mütter and K. Schilling (Plenum Press, New York: 1986).
7. K.G. Wilson, Phys. Rev. D10:2445 (1974).
8. K.G. Wilson and J. Kogut, Phys. Reports 12C:76 (1974).
9. H.B. Callen, Phys. Lett. 4B:161 (1963); R.L. Dobrushin, Theor. Prob. Appl. 13:387 (1969); D.E. Lanford III and D. Ruelle, Commun. Math. Phys. 13:194 (1969).
10. R.H. Swendsen, Phys. Rev. Lett. 52:1165 (1984).
11. C.B. Lang, Phys. Lett. 155B:399 (1985); Nucl. Phys. B265: 630 (1986).
12. R. Gupta and R. Cordery, Phys. Lett. 105A:415 (1984); G. Bhanot, Phys. Lett. 154B:63 (1985); K.M. Bitar,preprint FSU-SCRI-85-7 (1985).
13. S.K. Ma, Phys. Rev. Lett. 37:461 (1976); R.H. Swendsen, Phys. Rev. Lett. 62:859 (1979); R.H. Swendsen, in Real Space Renormalization, T.W. Burkhardt and J.M.J. van Leeuwen, Topics in Current Physics 30 (Springer: Berlin 1982).
14. R. Shankar, R. Gupta and G. Murthy, Phys. Rev. Lett. 55:1812 (1985).
15. A.H. Guth, Phys. Rev. D21:2291 (1980); J. Fröhlich and T. Spencer, Commun. Math. Phys. 83:411 (1982).
16. M. Creutz, L. Jacobs and C. Rebbi, Phys. Rev. D20:1915 (1979).
17. B. Lautrup and M. Nauenberg, Phys. Lett. 95B:63 (1980).
18. J. Jersák, T. Neuhaus and P.M. Zerwas, Phys. Lett. 133B:103 (1983).
19. H.G. Evertz, J. Jersák, T. Neuhaus and P.M. Zerwas, Nucl. Phys. B251[FS13]: 279 (1985); J. Jersák, T. Neuhaus and P.M. Zerwas, Nucl. Phys. B251[FS13]: 299 (1985).
20. J.S. Barber, Phys. Lett. 147B:330 (1984); J.S. Barber, R.E. Shrock and R. Schrader, Phys. Lett. 152B:221 (1985); V. Grösch, K. Jansen, J. Jersák, C.B. Lang, T. Neuhaus and C. Rebbi, Phys. Lett. 162B:171 (1985).
21. J. Villain, J. de Phys. 36:581 (1975).
22. T.A. DeGrand and D. Toussaint, Phys. Rev. D22:2478 (1980); G. Bhanot, Nucl. Phys. B205[FS5]:168 (1982); G. Bhanot, Phys. Rev. D24:461 (1981).
23. B. Berg and C. Panagiotakopoulos, Phys. Rev. Lett. 52:94 (1984).
24. R.H. Swendsen, Phys. Rev. Lett. 47:1775 (1981).
25. K.G. Wilson in Recent Developments in Gauge Theories (Cargese 1979), Eds. G. 't Hooft et al. (Plenum, New York: 1980).
26. B. Nienhuis and M. Nauenberg, Phys. Rev. Lett. 35:477 (1975).
27. D.G. Caldi, Nucl. Phys. B220[FS8]: 48 (1983).

THE GLUON IS MASSIVE: A LATTICE CALCULATION
OF THE GLUON PROPAGATOR IN THE LANDAU GAUGE

J. E. Mandula
U.S. Department of Energy
Washington, D.C. 20545
and Department of Physics
Washington University
St. Louis, Missouri 63130

M. Ogilvie
Department of Physics
Washington University
St. Louis, Missouri 63130

ABSTRACT

A Monte Carlo calculation of the gluon propagator in the Landau gauge in SU(3) lattice gauge theory is described. The results of calculations at $\beta = 5.6$ (200 $4^3 \times 8$ lattices), $\beta = 5.8$ (400 $4^3 \times 10$ lattices), and $\beta = 6.0$ (100 $4^3 \times 8$ lattices) indicate that the gluon propagator resembles a massive particle propagator with asymptotic mass near 600 MeV.

This talk describes a calculation of the gluon propagator in the Landau gauge in SU(3) Wilson lattice gauge theory.[1] Our goal in this calculation is the use of lattice gauge Monte Carlo techniques to estimate a function which, while not directly measurable, is one of the most fundamental quantities in a Yang-Mills theory. In this way, lattice gauge techniques can make contact with continuum approaches to gauge theories. The gluon propagator will be calculated from first principles in QCD with the same level of confidence as other quantities on the lattice. The techniques we use can be employed to calculate other propagators and gauge dependent quantities of interest in field theories. These calculations are important not because they directly verify QCD or show that it is inadequate, but because they offer a window into the way that QCD "works" with theoretical and phenomenological implications.

The Monte Carlo calculations that we have performed have been carried out on small lattices (4 sites in each spatial direction). The values of β (5.6-6.0) are also small, below the onset of two loop asymptotic scaling. Thus the results will require confirmation from studies on larger lattices before being regarded as definitive. Our initial efforts have focused on the gluon propagator at zero temperature, but we have also obtained preliminary results for the gluon propagator near the deconfinement temperature.

The principal result of this investigation is that at the largest available distances the gluon propagator in the Landau gauge falls off exponentially like a massive particle propagator. The mass is about 600 MeV. Although the bare gluon propagator has a singularity at $q^2 = 0$, in the dressed propagator this singularity has moved up to a finite mass. This is an instance of a purely dynamical Higgs mechanism, without an explicit Higgs field.

The possibility of dynamical mass generation without an explicit scalar field was first pointed out by Schwinger,[2] before the discovery of the conventional Higgs mechanism. He gave an explicit example of this phenomenon, two-dimensional massless quantum electrodynamics. The possible occurrence of a dynamical Higgs mechanism in four-dimensional gauge theories has been studied by Jackiw and Johnson[3] and by Cornwall and Norton.[4] All of the above work is concerned with gauge fields coupled to massless fermions, whereas we work in the pure gauge sector. Our results provide the first evidence for the existence of a dynamical Higgs mechanism in a four-dimensional gauge theory from a direct simulation of the theory.

There is indirect theoretical evidence from other lattice gauge studies that supports the idea that the gluon is massive, although this evidence has not always been interpreted in terms of an effective gluon mass. The fact that as $N \to \infty$ there survives a linear potential between adjoint sources in SU(N) gauge theory ("the adjoint string doesn't break")[5] suggests that, in the sense of a mechanical mass, not only is the gluon massive, but that as $N \to \infty$ its mass becomes infinite.

Bernard has performed lattice Monte Carlo studies of the existence and breaking of an adjoint string in SU(2) gauge theory.[6] As those studies note, the association of a gluon mass with the distance at which the adjoint string breaks is ambiguous, because the measured mass value includes an unknown binding energy to the adjoint source.

The gluon propagator, and especially its infrared behavior, has been the focus of several discussions that aim at explaining confinement or address other fundamental properties of QCD. Arguments have been made, based on truncations of the Schwinger-Dyson equations and Ward identities, that there is a $1/q^4$ infrared singularity in the gluon propagator, and models of confinement based on that singularity have been constructed.[7] Different

212

approximations to the Schwinger-Dyson equations suggest a $1/q^2 \ln^2 q^2$ IR singularity,[8] and, by studying a gauge invariant function somewhat related to the axial gauge propagator, it has been argued that the gluon is effectively massive.[9] Of course, perturbation theory gives a $1/q^2$ behavior. Of course, the observed behavior of the gluon propagator will be gauge-dependent, and the extraction of gauge-invariant information non-trivial. DeTar *et al.*[10] have argued that in a lattice version of the axial auge, the color electromagnetic field propagator has an asymptotic decay governed by the mass of what they call "light-heavy glueballs." These states consist of a gluons bound to a "hidden" octet sources, which are a feature of the axial gauge. There is a binding energy uncertainty, like that which makes ambiguous the extraction of the gluon mass from the gauge-invariant calculation of the adjoint string breaking energy.[6] In order to understand the significance of the gauge variant calculations we have attempted, Monte Carlo simulation should be used to deterimne the behavior of the gluon propagator in gauges which lack hidden sources.

The interpretation of the gluon mass may be clearer at high temperature. Because gluons themselves carry color electric charge, the creation of a gluon plasma above the deconfinement temperature should lead to color Debye screening. Our preliminary results indicate a weak temperature dependence on the gluon mass, with no indication of dramatic behavior at the deconfinement transition. There is another aspect of finite temperature behavior we can observe. Current ideas about the deconfinement transition[11] stress the similarity between the role of the temporal component of the gauge field in the deconfinement transition and the fundamental scalar field in a Higgs model. We have seen strong evidence of this similarity in the behavior of the A_o propagator. Above the deconfinement temperature, this propagator undergoes a sudden and uniform increase, as if the A_o field had developed a vacuum expectation value.

In the rest of this note we will summarize the calculation: the specification of a lattice Landau gauge, our method for calculating in that gauge with the proper functional measure, the results and their implications. We also make a brief remark about the problem of Gribov copies.

The association between lattice and continuum variables is not unique. We make a maximally local choice for the lattice gauge potential $A_\mu(n)$ $= (U_\mu(n) - U_\mu^\dagger(n))_{traceless}/2ia$. We implement the Landau gauge by maximizing the quantity

$$\text{Re } Tr\sum_\mu (U_\mu(n) + U_\mu^\dagger(n - \mu)) \tag{1}$$

at each site.[12] This implies that the maximally local difference form of the Landau gauge condition, $\sum_\mu (A_\mu(n) - A_\mu(n - \mu)) = 0$, holds.

It is very difficult to do Monte Carlo calculations in a fixed gauge by using a Metropolis algorithm on trial updates that preserve the gauge condition, starting from a configuration that satisfies it. Sweep-to-sweep correlations between lattices updated in this way can be much greater than in the usual non-gauge-constrained Monte Carlo method, so that many more sweeps are needed between each pair of statistically independent lattices; this is known to occur in the case of discrete gauge groups.[13] Also, it would be necessary, for each Monte Carlo trial, to calculate the Faddeev-Popov determinant, making the calculation as computationally intensive as the inclusion of dynamical fermions.

To avoid these problems, we update the lattice using the conventional Metropolis Monte Carlo algorithm without any gauge constraint. We then independently gauge transform each lattice into the Landau gauge.[14] This gives us an ensemble of Landau gauge lattices, correctly distributed according to the Wilson action and Faddeev-Popov determinant, without having to explicitly compute a Faddeev-Popov determinant.

The basis for this algorithm lies in the Faddeev-Popov analysis, which we briefly review.[15] The Faddeev-Popov determinant Δ corresponding to the gauge condition $f(U) = 0$ is defined by

$$\Delta_f (U) \int dg \ \delta(f(U^g)) = 1 \qquad (2)$$

where U^g is the gauge transform of U and the integration is over all gauge transformations g. $\Delta_f (U)$ is gauge invariant by construction. Multiplying the gauge invariant functional integration measure DU by 1 in the form of Eq. (2), interchanging orders of integration, changing variables $U \rightarrow U^{g^{-1}}$, and noting that both DU and $\Delta_f (U)$ are gauge invariant, gives the functional integral subject to the gauge constraint:

$$\int DU = \int DU \ \Delta_f (U)\delta(f(U)) \qquad (3)$$

where we have conventionally taken the group volume to be 1.

The expected value of any function of link variables $\Phi(U)$, gauge invariant or local or not, is given by

$$<\Phi(U)>_f = \frac{1}{Z} \int DU \ \Delta_f (U)\delta(f(U))\Phi(U) \qquad (4)$$

where the normalization Z is the same integral without the $\Phi(U)$ factor. We reverse the Faddeev-Popov analysis by integrating $\int dg$ (on which nothing depends), interchanging the order of integration, and changing variables $U \rightarrow U^g$. We can do the dg integration because of the δ function, using Eq. (3). Only the value of $\Phi(U^g)$ at that value of g which maps U into the gauge $f = 0$ is needed. The result is

$$<\Phi(U)>_f = \frac{1}{Z} \int DU \ \Phi(U^{g(U)}) \qquad (5)$$

where $g(U)$ is that gauge transformation for which $f(U^{g(U)}) = 0$.

Equation (5) is the statement of the algorithm. In words it says that the expected value of any quantity $\Phi(U)$ in the gauge $f(U) = 0$ may be calculated from an ensemble of configurations weighted without gauge constraint $(\int DU)$, but that each configuration should be gauge transformed into the $f(U) = 0$ gauge before evaluating Φ.

In the foregoing analysis, we have tacitly assumed that the condition $f(U) = 0$ specified the gauge uniquely. However, it is known that many differential conditions, such as $\partial_\mu A_\mu = 0$ in the continuum, are subject to Gribov ambiguities.[16] This may be so with our lattice Landau gauge condition as well.[17] We have observed that Gribov copies do occur with a finite difference gauge condition. However, because the gauge condition used in this paper is a global maximization rather than a difference condition, the Gribov ambiguity may be absent. We have performed a variety of simple tests to check that our gauge-fixing procedure is reasonably robust. For example, we have checked that improving our gauge-fixing figure-of-merit by a factor of 2 does not sigificantly change the propagators. This does not, however, resolve the Gribov problem. It is possible to look for Gribov copies by applying the gauge-fixing procedure to field configurations related by a gauge transformation. The condition that non-trivial infinitesimal gauge transformations maintain the gauge condition is also subject to numerical analysis. Such investigations may give some insight into the Gribov problem.

Our method for transforming a lattice into the Landau gauge consists of sweeping through the lattice, applying at each site the gauge transformations that maximizes the quantity Eq. (1). This procedure is somewhat delicate, since a gauge transformation at one site upsets the gauge condition at the surrounding sites. We know of no proof that the procedure converges. Empirically, however, the method seems to work very well.

We have used this method to compute the gluon propagator in SU(3) pure Yang-Mills theory for $\beta = 5.6$ and 6.0 on a $4^3 \times 8$ lattices and for $\beta = 5.8$ on a $4^3 \times 10$ lattice, all with periodic boundary conditions. Simulations were performed on two machines, a Ridge-32 and Cray-XMP, with different acceptance rates, numbers of hits, and sweeps between each measurement, as well as slightly different update procedures. The data were combined only in the final analysis. We used 200 lattices for $\beta = 5.6$, 400 lattices for $\beta = 5.8$, and 100 lattices for $\beta = 6.0$. Each lattice was separated by 10-30 sweeps, with 10-20 hits per sweep. We fixed the gauge well enough so that $< Tr(\partial_\mu A_\mu)^2_{\text{lattice}} >$ was always less than 0.01, and often less than 0.0005. To verify that our results are stable to better gauge fixing, we reran the $\beta = 5.6$ case, taking 100 lattices, but spending twice the time gauge fixing, with no significant change in the gluon propagator.

We examined the time displacement correlations of the gauge potential summed over all sites on each fixed t (space-like) hyperplane $A_\mu(t) = \Sigma A_\mu(t,\vec{x})$. In the Landau gauge with periodic boundary conditions, $A_o(t)$ should be a constant, and we measure $Tr < A_o(t)A_o(0)>$ as a function of t to monitor that we have adequately fixed the gauge. The dynamical content of the Landau gauge potential is expressed in $\sum\limits_{i=1,2,3} Tr < A_i(t)A_i(0)>$. Note that A_i couples the vacuum to the color octet sector. Both the $<A_i(t)A_i(0)>$ and the $<A_o(t)A_o(0)>$ correlations are shown in the Figures 1. The errors are statistical errors on the measured propagators from each ensemble, including corrections for residual sweep to sweep correlation in the few cases these were detectable.

To guide the eye, note that the free propagator for mass m, summed over spatial sites, is

$$\Delta_m(n_t) = C \cosh(aM \mid n_t - N_t/2 \mid) \tag{6}$$

where $\sinh(aM/2) = am/2$, C is a normalization constant, and N_t is the total number of sites in the t direction. As the mass moves toward zero, the propagator becomes progressively flatter. For a canonically normalized field, C diverges as m goes to zero, showing that the massless canonical propagator does not exist on a periodic lattice. An interacting theory with massless particles will surely show behavior reflecting the production of all numbers of particles at threshold, since the assumption of an isolated pole at $m = 0$ is invalid.

Far from being constant, the actual propagators $<A_i A_i>$ in Figs. 1 look like massive particle propagators. In fact, using Eq. (6) we can compute an effective mass governing the falloff from each hyperplane to the next. These are given in Table 1. The errors are estimated by binning the data and are statistical only.

A striking aspect of Table 1 is that for each β value the effective mass grows with separation. Such behavior is only possible if the spectral function describing the gauge potential propagator is not positive definite. Otherwise, the effective mass must fall monotonically. A rise is entirely possible in the Landau gauge, which has ghosts.

To estimate finite size effects and possible scaling violations, we express the effective masses from Table 1 in physical units, using the values of the inverse lattice spacing determined by Barkai, Moriarty, and Rebbi from the string tension ($\sqrt{\sigma} = 0.42$ GeV) on large lattices at each of our values of β.[18] Note that in physical units, the $\beta = 6.0$ lattice is smallest, so that finite size effects are greatest, while the $\beta = 5.6$ lattice is coarsest.

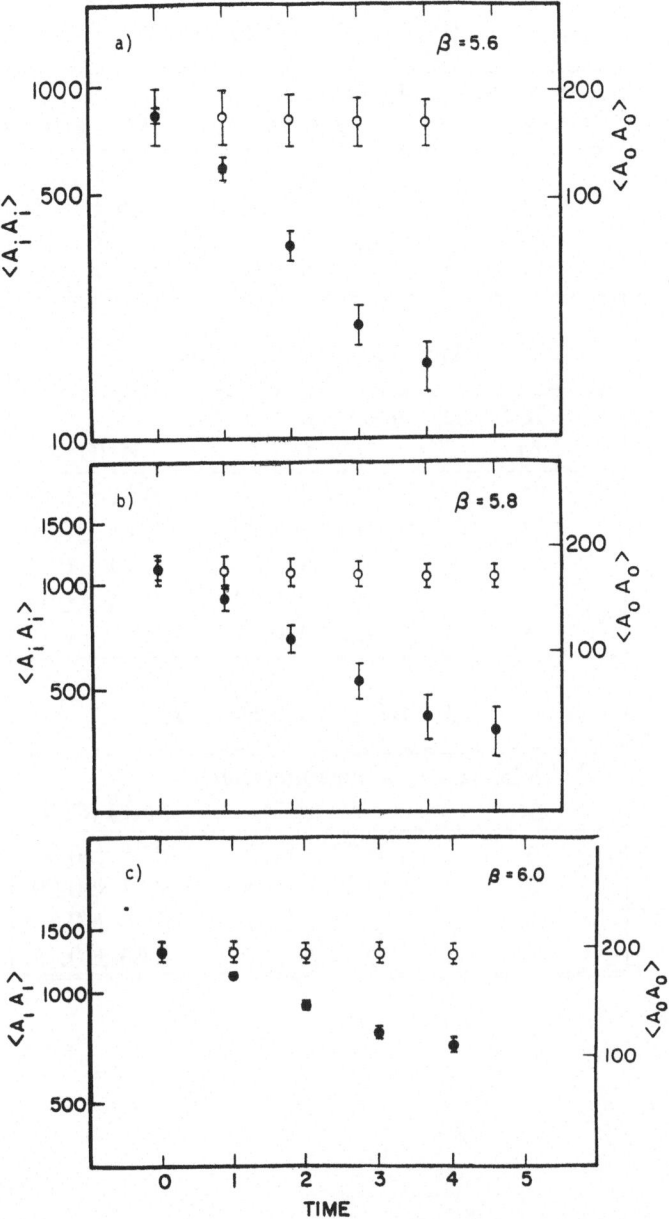

Figure 1: Spatial and temporal gluon propagators as a function of t. Errors are statistical only in the individual data points. a) $\beta = 5.6$ on $4^3 \times 8$ lattice. b) $\beta = 5.8$ on $4^3 \times 10$ lattice. c) $\beta = 6.0$ on $4^3 \times 8$ lattice.

Table 1.

Correlation functions versus separation. Errors are statistical only.

$$\beta = 5.6 \ a^{-1} = 0.80 \ \text{GeV}$$

$4^3 \times 8$ lattice	200 measurements	
Δt	$M \ (a^{-1})$	M (GeV)
1	0.39 ± 0.03	0.32 ± 0.03
2	0.57 ± 0.06	0.46 ± 0.05
3	0.69 ± 0.12	0.55 ± 0.10
4	0.75 ± 0.21	0.60 ± 0.17

$$\beta = 5.8 \ a^{-1} = 1.27 \ \text{GeV}$$

$4^3 \times 10$ lattice	400 measurements	
Δt	$M \ (a^{-1})$	M (GeV)
1	0.24 ± 0.03	0.31 ± 0.04
2	0.33 ± 0.05	0.41 ± 0.06
3	0.38 ± 0.07	0.49 ± 0.09
4	0.42 ± 0.09	0.54 ± 0.11
5	0.44 ± 0.10	0.56 ± 0.13

$$\beta = 6.0 \ a^{-1} = 1.69 \ \text{GeV}$$

$4^3 \times 8$ lattice	100 measurements	
Δt	$M \ (a^{-1})$	$M(t)$ (GeV)
1	0.23 ± 0.02	0.38 ± 0.04
2	0.30 ± 0.05	0.51 ± 0.09
3	0.35 ± 0.08	0.60 ± 0.13
4	0.39 ± 0.09	0.65 ± 0.15

In Figure 2, we plot the $<A_i A_i>$ propagators versus separation in physical units for all three values of β, each normalized to one at zero separation. As may be seen from the figure, the agreement with dynamical scaling is good.[19] Note that the range of physical distance is different for each value of β. The cosh form of the propagator results in the last point for each β value being above the envelope.

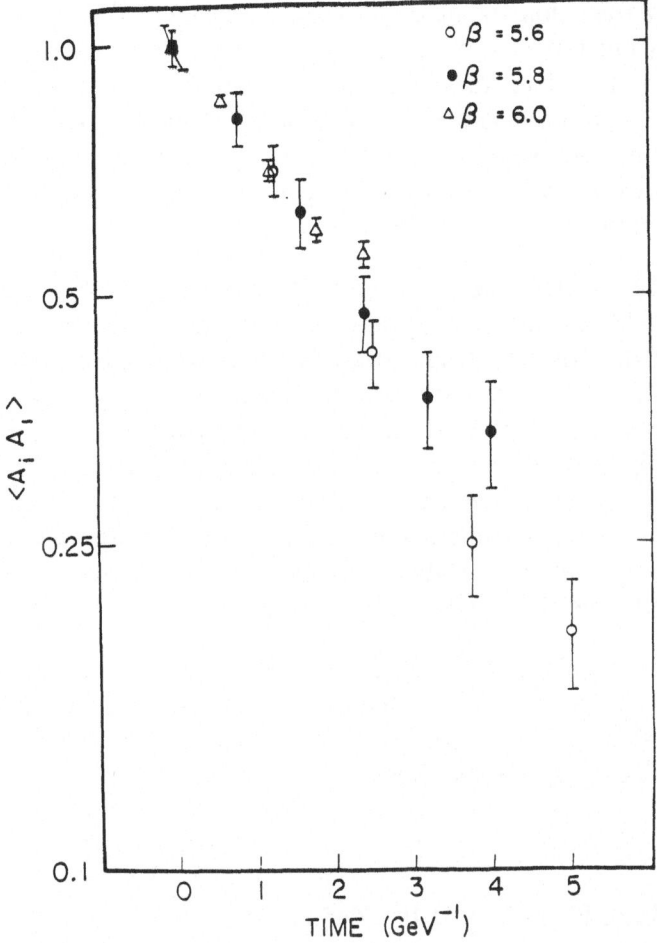

Figure 2: The $<A_i\,A_i>$ propagators, all normalized to 1 at zero separation.

The calculation has a remarkably high intrinsic "signal to noise" ratio. Even at separations of 4 or 5 lattice spacings, the value of the propagator is reasonably well determined with only 100 lattices. This should be compared to calculations of glueballs of comparable mass without using spacially distributed sources. In these calculations, with hundreds or thousands of times as many lattices, the computed correlations are lost in statistical noise beyond 2 spacings.

The calculations reported here have been performed on small lattices and at fairly small β. In the range of β from 5.6 to 6.0, significant deviations from two-loop asymptotic behavior are known to occur, although dynamical scaling seems to hold.[19] A further concern is deconfinement in the spatial directions.

The deconfining transition for pure QCD occurs at about $\beta = 5.7$ for a lattice size of 4. Thus our range of β values cross the transition region, although no profound jump in the gluon mass was seen. Simulations on larger lattices are clearly desirable, and are underway. Further work on the behavior of the gluon propagator at finite temperature is also underway, focusing on behavior bear the deconfinement transition. We are concentrating particularly on the appearance of symmetry-breaking behavior.

From the behavior of the gluon propagator reported here, it appears that in a pure Yang-Mills theory, a dynamical Higgs phenomenon occurs. Our best estimate of the effective gluon mass, as determined at large distances, is about 600 MeV, with finite size effects, possible scaling violations, and statistical uncertainties of at least $\pm 25\%$. In analogy with the concept of a constituent quark mass, it may be useful to think of the mass in the gluon propagator as a constituent gluon mass. The massiveness of the gluon may be connected to the apparent suppression of many-gluon intermediate states in J/ψ and   decay, and the relative absence of mixing between the lowest quark model states and those with gluonic excitations.

The authors would like to thank Drs. R. Gupta, G. Guralnik, M. B. Halpern, A. J. G. Hey, P. K. Williams, A. Warnock, and C. Zemach for valuable discussions. This work has been supported in part by the U.S. Department of Energy and by the Research Corporation.

REFERENCES

1. K. G. Wilson, Phys. Rev. D **10**, 2445 (1974).

2. J. Schwinger, Phys. Rev. **125**, 397 (1962); Phys. Rev. **128**, 2425 (1962).

3. R. Jackiw, Phys. Rev. D **8**, 2386 (1973).

4. J. Cornwall and R. Norton, Phys. Rev. D **8**, 3339 (1973).

5. J. Greensite and M. B. Halpern, Phys. Rev. D **27**, 2545 (1983).

6. C. Bernard, Phys. Lett. **108B**, 431 (1982); Nucl. Phys. **B219**, 341 (1983).

7. M. Baker, J. S. Ball, and F. Zachariasen, Nucl. Phys. **B186**, 531, 560 (1981); S. Mandelstam, Phys. Rev. D **20**, 3223 (1979); H. Pagels, Phys. Rev. D 14, 2747 (1976); **15**, 2991 (1977); J. M. Cornwall, Phys. Rev. D **22**, 1452 (1980).

8. R. Delbourgo, J. Phys. G **5**, 603 (1979); E. J. Gardnere, J. Phys. G **9**, 139 (1983).

9. J. M. Cornwall, Phys. Rev. D **26**, 1453 (1982); Nucl. Phys. **B157**, 392 (1979).

10. C. DeTar, J. E. King, S. P. Li, and L. McLerran, Nucl. Phys. **B249**, 621, 644 (1985).

11. B. Svetitsky, Physics Reports **132**, 1-53 (1986).

12. K. G. Wilson, in *Recent Developments in Gauge Theories* (Proc. NATO Advance Study Institute, Cargese, 1979), ed. G. 't Hooft *et al.*, Plenum, New York, 1980.

13. M. Creutz, private communication.

14. This approach was also used by Wilson in Ref. 14.

15. L. Faddeev and V. Popov, Phys. Lett. **25B**, 29 (1967).

16. V. N. Gribov, Nucl. Phys. **B139**, 1 (1978); I. M. Singer, Comm. Math. Phys. **60**, 7 (1978).

17. B. Sharpe, J. Math. Phys. **25**, 3324 (1984).

18. D. Barkai, K. J. M. Moriarty, and C. Rebbi, Phys. Rev. D **30**, 1293 (1984).

19. See, for example, the articles by A. Hasenfratz, P. Mackenzie, R. Gupta, and A. Patel in *Gauge Theory on a Lattice: 1984*, C. Zachos, W. Celmaster, E. Kovacs, and D. Sivers, ed., U.S. Government Publication CONF-8404119, and references therein.

NEW DEVELOPMENTS ON ORDER PARAMETERS

AND RELATED PROBLEMS

Mihail Marcu

II. Institut für Theoretische Physik
Luruper Chaussee 149
2000 Hamburg 50, West Germany

ABSTRACT

The status of order parameters in lattice gauge theories with matter fields is reviewed, together with several related problems. Then the flux correlations order parameter is discussed in detail. An analysis of its finite distance behaviour leads to the definition of a characteristic length scale r_D. This is possible both at zero and at finite temperature. It is argued that r_D is the screening length for dynamical charge fluctuations (Debye screening). In general, the length defined from the exponential decay of Polyakov loop correlations is different from r_D.

1. INTRODUCTION

In this talk I will try to present the current status of the order parameters (OP) proposed some time ago by Klaus Fredenhagen and myself.

Typically a lattice gauge theory (LGT) with matter fields has three regions: a free charge phase, a screening (Higgs) region and a confinement region /1, 2/. In an actual model some of these regions may be absent. If the matter fields are in the fundamental representation of the gauge group, the Higgs and the confinement regions are not separate phases.

For a long time after LGT's were invented, no OP was known except in the limiting cases of pure matter and pure gauge theories /3/. In ref. /4/ we proposed three different OP's. In fact the vacuum overlap OP (VOOP) had been proposed as early as 1982 /5/.

The VOOP has the most rigorous foundation of our OP's. It is based on the following ideas for constructing a charged state /4, 6, 7/: separate a charge-anticharge pair, regularize the energy, then send the anticharge to infinity, and check whether the resulting state is indeed charged. The quantity testing whether this candidate for a charged state is indeed charged is its scalar product with the vacuum (the "vacuum overlap"). In the limit of the pure matter theory the VOOP goes into the square of the usual local order parameter (e.g. magnetization). In the Higgs region the VOOP is the "Higgs expectation value", defined properly, in a gauge invariant way.

The construction of charged states /4/ that led to the VOOP was also meant as an example for the general analysis of particle states in theories with a local gauge symmetry /8/. This general analysis also suggested to investigate correlations of electric fluxes associated with the gauge group center, in order to understand those properties of the vacuum which are responsible for the existence or absence of charged states. For theories with a mass gap, a useful quantity for studying these correlations is the flux correlations OP (FCOP), defined in ref. /4/. In the limit of the pure gauge theory, the FCOP goes into the 't Hooft loop /9/ (the dual of Wilson loop in the case of selfdual theories). As opposed to the VOOP, the FCOP is well-defined even at finite temperature.

Our third OP compares the charge of the candidate for a charged state to that of the vacuum. We therefore call it the charge measurement OP (CMOP). In ref. /4/ this comparison was possible although we did not succeed in regularizing the cahrge operator itself. In this talk the CMOP will not be discussed further.

A discussion of the three OP's in terms of general properties of particle states in gauge theories can be found in ref. /10/. Here the accent will lie on the ideas underlying the OP's and their practical use in LGT's.

All our OP's are limits of local quantities. Before the limit, these quantities contain valuable additional information. In pure gauge theories, knowledge of the finite Wilson loops allows us to determine the quark-antiquark potential. In ref. /7/ it was shown that in the confinement region of LGT's with matter fields, the VOOP before the limit can be used to define a length scale. It was argued that this is the scale where quark fragmentation sets in. In this talk it will be shown that the FCOP also allows us to define a length scale, this time in the Higgs region. Furthermore, at finite temperature this length scale may be defined for all values of the coupling constants. It will be argued that it can be interpreted as the screening length for dynamical charge fluctuations (Debye screening).

224

All other OP's for LGT's with matter fields that have been
proposed in the literature and that correctly reproduce the known
phase diagrams, are closely related to the VOOP. This will be discussed in section 2, in a brief survey on the work done on the VOOP
and related problems. In section 3 the FCOP will be presented and
the model of a plasma for the vacuum or for a thermal equilibrium
state will be discussed.

For simplicity, the discussion will be given an terms of a LGT
with a scalar matter field in the fundamental representation of the
gauge group. The action is

$$ S = -\beta \sum_p \chi(U(p)) - \kappa \sum_l 2 \, \text{Re}(\varphi^\dagger U \varphi)(l) + \sum_x V(\varphi(x)) \qquad (1.1) $$

Here p, l, x are the plaquettes, links and sites of a d-dimensional
Euclidean lattice, U are the gauge fields, φ are the matter fields,
and χ is some character that contains the fundamental character
as an irreducible component. The time-zero field operators will be
denoted by $\hat{U}(\underline{l})$ and $\hat{\varphi}(\underline{x})$. In general, underlined quantities will
denote geometrical objects in d-1 (space) dimensions.

2. THE VACUUM OVERLAP ORDER PARAMETER AND RELATED PROBLEMS

A candidate for a charges state is obtained by considering
the following sequence of norm-one vectors /4, 6, 7/:

$$ |\underline{x},\underline{y},n\rangle = \frac{1}{\|\cdot\|} \sum_{a,b} \hat{\varphi}_a(\underline{x}) \, \hat{\varphi}_b^+(\underline{y}) \, \hat{T}^n \, \hat{U}(\underline{L}) |0\rangle \qquad (2.1) $$

where a,b are gauge group indices, $\underline{L}$ is a spatial path connecting
$\underline{x}$ and $\underline{y}$, $\hat{U}(\underline{L})$ is the path ordered product of $\hat{U}$'s along $\underline{L}$ and $\hat{T}$ is
the transfer matrix (derived in the usual way in the temporal
gauge). For simplicity, let $\underline{L}$ be along a coordinate axis. The
Euclidean time translation by n achieves the regularization of
the energy for the charge-anticharge pair: the relative weight of
the low-energy components of $|\underline{x},\underline{y},n\rangle$ increases with n. Actually
for an Abelian theory we can extend the methods used in /4/ to
prove that the energy of $|\underline{x},\underline{y},n\rangle$ stays bounded in the limit
$|\underline{x}-\underline{y}| \to \infty$, $n \to \infty$ provided n grows at least linearly with $|\underline{x}-\underline{y}|$
(this statement is true for the whole phase diagram). Unfortunately
the proof cannot be easily generalized for non-Abelian theories.

Let us denote by $\rho(r,n)$ the vacuum projection of $|\underline{x},\underline{y},n\rangle$.
Up to a minor modification /6, 7/, the expression of $\rho(r,n)$ in
terms of expectation values in the d-dimensional Euclidean path-

integral formulation of the model is:

$$\rho(r,n) = \langle 0 | \underline{x}, \underline{y}, n \rangle = \frac{\langle \;\; n \rangle}{\langle \;\; 2n \rangle^{1/2}} \tag{2.2}$$

Here $r = |\underline{x}-\underline{y}|$, and the symbolic notation means a gauge invariant path ordered product of the gauge fields along the open string and the closed loop resepctively.

The VOOP is the $r, n \to \infty$, $n \gtrsim r$ limit of $\rho(r,n)$. In the free charge phase the limit should be zero, since a charged state is orthogonal to the vacuum. In the Higgs and confinement regions the limit should be nonzero. Thus the criterion for existence of charged states reads:

$$\lim_{\substack{r,n \to \infty \\ r \lesssim n}} \rho(r,n) = \begin{cases} = 0 & \text{free charge} \\[2ex] \neq 0 & \text{confinement, Higgs} \end{cases} \tag{2.3}$$

For Abelian gauge groups and scalar matter fields, eq. (2.3) follows immediately from Griffith inequalities. In models with Fermions or a non-Abelian gauge group, (2.3) has not been proven, with the exception of the massless Schwinger model (in the continuum!) /11/.

The construction of charged states in Z_2 theories, which was our starting point in /4/, can be easily generalized only for models having a convergent expansion in the free charge region. Unfortunately this excludes most cases of real interest.

In the Z_2 model even more can be rigorously established. Barata and Fredenhagen proved that the charged states are really particle states /12/.

A totally different construction of the charged states in Z_2 theories was given by Szlachányi /13, 14/. He formulates in a gauge invariant way the old idea /2/ that the charged fields can be constructed using the matter fields in an appropriately chosen gauge. Then he succeeds in finding a suitable class of gauges and goes on to construct the charged states. The resulting charged sector is identical to that obtained by our construction. Moreover, his proofs

226

also rely heavily on the convergent expansion.

For QED we expect our charged state to be similar to the physical electron in the Coulomb gauge, i. e. the bare electron dressed with the electric field of a pointlike external source. The same is true for Szlachányi's construction, if he does it in a rotationally invariant way. In ref. /14/ he gives a method to generalize to arbitrary compact gauge groups the gauge invariant definition of expectation values of products of time-zero Coulomb-gauge matter fields.

A similar construction using the Landau gauge was proposed by Kennedy and King /15/. Borgs and Nill /16/ showed that, due to problems with Gribov copies, this construction in fact works only for the noncompact scalar lattice QED.

In the Higgs phase the Szlachányi and Kennedy-King constructions define the "Higgs expectation value" in a gauge invariant way. The same is true for the VOOP. Numerically, these different definitions will not be identical. However, this is not necessary for a correct discussion of the Higgs mechanism.

As opposed to the VOOP, the ideas of Szlachányi and Kennedy-King do not seem to work in the confinement region.

The behaviour of $\wp(r,n)$ for finite r and n was discussed in detail in ref. /7/. In the free charge phase and in the Higgs region, $\wp(r,\infty)$ decays exponentially with r to the asymptotic value of the VOOP. From this decay we can determine the charged particle mass and the Higgs mass respectively. This is however difficult numerically, since it is difficult to compute Euclidean correlations of zero momentum charged states. In the confinement region, $\wp(r,\infty)$ decays exponentially with the perturbative charged particle mass for small values of r, while for large values it shoots up again to the asymptotic value of the VOOP. The scale of the crossover is the scale where fragmentation of the charge-anticharge pair sets on. The asymptotic value of the VOOP can be estimated more easily using the Bricmont-Fröhlich parameter /17/, which is equal to the VOOP in the Higgs-confinement phase. In the free charge phase, the Bricmont-Fröhlich parameter actually tests the existence of a hydrogen atom /7, 4/.

The VOOP has been investigated numerically in a variety of Monte Carlo simulations. The most accurate study was possible in the Z_2 model in 4 dimensions /7, 18/, where the VOOP was used to give strong numerical evidence for a line of second order phase transitions with mean field exponents separating the free charge phase from the Higgs region. The Aachen-Georgia group have studied the VOOP in the SU(2) Higgs model in 4 dimensions /19/. This case is much more difficult numerically than the Z_2 case. Within their accuracy, the predictions of /7/ were verified. This is very important, since the theoretical arguments are much weaker for non-Abelian gauge groups. The VOOP has also been computed in the 4-dimensional U(1) Higgs model. For matter fields with unit charge, the Aachen group has investigated the Higgs-free charge transition /20/. They have some evidence that the transition is first order. The case of doubly charged matter fields was simulated by Azcoiti and Tarancón /21/. Alessandrini et al. /22/ analized it using mean field methods. The U(1) numerical studies also confirm the theoretical predictions for $\varrho(r,n)$.

The VOOP cannot be defined at a finite temperature T since the lattice is finite in Euclidean time. At finite T the theory effectively decouples (see e.g. /23/) into a pure matter theory with magnetic field and a LGT with matter fields, both in d-1 dimensions and at zero temperature. If we consider the paths of eq. (2.2) as lying in a specelike plane, $\varrho(r,n)$ will define the VOOP for the d-1 dimensional LGT with matter fields. It can be used to study e.g. the "symmetry restauration" transition, which was repeatedly discussed in this conference.

3. THE FLUX CORRELATION ORDER PARAMETER AND THE SCREENING OF DYNAMICAL CHARGE FLUCTUATIONS

The general analysis of charged states in a <u>massive</u> gauge theory with matter fields /8/ leads to the following picture. A charged state can be constructed from a charge-anticharge pair by sending the anticharge to infinity. Things can always be arranged such that asymptotically the lines of electric flux starting at the charge localization point and going to infinity are inside a cone of given solid angle. The total charge can be determined by measuring the electric flux through an arbitrarily large closed surface (Gauss' law). However, since the asymptotic direction of the cone axis is not an observable, it is not possible to measure a nontrivial electric flux through a solid angle of less than 4π. This latter property prevents the existence of charged states carrying an additive charge (Swieca's theorem) /24/. On the other hand, states carrying a multiplicative charge may exist if electric fluxes in different directions are correlated strongly enough.

This general picture is not contradicted by any LGT study in the literature. In a LGT Gauss' law holds only for the center $\mathcal{C}$ of the gauge group. Thus a charged state may only carry the quantum numbers of $\mathcal{C}$. In all known examples of a massive free charge phase (the photon is massive) $\mathcal{C}$ is a finite Abelian group, and therefore the charge is multiplicative rather than additive.

Before defining the FCOP some additional notation must be introduced. Let $\hat{\mathcal{E}}_V(\underline{l})$ be the left translation operator with the element V of the gauge group. As in the case of the gauge fields, it is useful to consider $\underline{l}$ as an orient ed link. If $\underline{l}$ and $\underline{l}'$ differ only by their orientation, then $\hat{\mathcal{E}}_V(\underline{l}') = \hat{\mathcal{E}}_{V^+}(\underline{l})$. For a set $\underline{\mathcal{L}}$ of oriented links and an element C of $\mathcal{C}$, let us define $\hat{\mathcal{E}}_C(\underline{\mathcal{L}}) = \prod_{\underline{l} \in \underline{\mathcal{L}}} \hat{\mathcal{E}}_C(\underline{l})$. Then

$$\langle 0 | \hat{\mathcal{E}}_C(\underline{\mathcal{L}}) | 0 \rangle = \left\langle \prod_{\underline{l} \in \underline{\mathcal{L}}} \exp \beta \{ \chi(C U(p_{\underline{l}})) - \chi(U(p_{\underline{l}})) \} \right\rangle \quad (3.1)$$

Here $p_{\underline{l}}$ is a time-like plaquette between the time-zero and time-one hyperplanes such that $\underline{l}$ is the part of $\partial p_{\underline{l}}$ contained in the time-zero hyperplane.

Let Λ be a cube in d−1 dimensions with side length r. Let $\partial^* \Lambda = \underline{\mathcal{L}} \cup \underline{\mathcal{R}}$ be the decomposition of the coboundary of Λ into a left and right "hemisphere" (all geometrical considerations are in d−1 dimensions; on the dual lattice, $\underline{\mathcal{L}}$ and $\underline{\mathcal{R}}$ are d−2 dimensional open surfaces with the same boundary). Denote by $\underline{\mathcal{M}}$ the minimal set of spatial links such that $\partial^* \underline{\mathcal{L}} = \partial^* \underline{\mathcal{R}} = \partial^* \underline{\mathcal{M}}$ (minimal surface on the dual lattice).

Assume now that we are in a massive free charge phase and we try to bring a charge inside Λ such that the electric flux coming in from infinity is localized in a cone. Since the asymptotic direction of the cone axis is not observable, there are vacuum fluctuations that delocalize the electric flux. The probability for the vacuum to contain large closed lines of electric flux is relatively high. As a consequence the vacuum expectation values of the electric flux operators $\hat{\mathcal{E}}_C(\underline{\mathcal{L}})$ and $\hat{\mathcal{E}}_C(\underline{\mathcal{R}})$ $(C \in \mathcal{C})$ are strongly correlated (it only makes sense to consider fluxes associated with the gauge group center). If on the other hand we are in a massive phase without charged states, the asymptotic direction of the cone could be observed by measuring the electric flux through appropriate portions of $\partial^* \Lambda$. There are no vacuum fluctuations delocalizing the electric flux. This also means that in the vacuum the fluxes through $\underline{\mathcal{L}}$ and $\underline{\mathcal{R}}$ are only weakly correlated.

Since the electric fluxes are multiplicative, the quantity to be considered for a study of the flux correlations is:

$$f_C(\underline{\Lambda}) = \frac{\langle 0 | \hat{\mathcal{E}}_C(\underline{\mathcal{L}}) | 0 \rangle \, \langle 0 | \hat{\mathcal{E}}_C(\underline{\mathcal{R}}) | 0 \rangle}{\langle 0 | \hat{\mathcal{E}}_C(\partial^* \underline{\Lambda}) | 0 \rangle} \quad (3.2)$$

If our intuitive picture of the role played by vacuum fluctuations is correct, in a massive free charge phase the main contributions to (3.2) will come from the closed lines of electric flux passing through both $\underline{\mathcal{L}}$ and $\underline{\mathcal{R}}$ (remember that $\hat{\mathcal{L}}_c(\underline{\mathcal{L}})$ and $\hat{U}(\underline{L})$ do not commute if $\underline{L}$ passes through $\underline{\mathcal{L}}$). The other vacuum fluctuations play a similar role in the numerator and denominator and we expect them to cancel up to a contribution at the "perimeter" $\partial^*\underline{\mathcal{L}} = \partial^*\underline{\mathcal{R}} = \partial^*\underline{\mathcal{M}}$. The closed lines passing through $\underline{\mathcal{L}}$ and $\underline{\mathcal{R}}$ can be classified according to their intersection with the minimal surface $\underline{\mathcal{M}}$. This immediately suggests that (3.2) decays with the area of $\underline{\mathcal{M}}$. In a phase without free charges this area effect is absent and (3.2) will decay with the perimeter of $\underline{\mathcal{M}}$. To sum up,

$$
f_c(\underline{\Lambda}) \underset{\underline{\Lambda} \to \infty}{\sim}
\begin{cases}
e^{-c_1|\underline{\mathcal{M}}|} & \text{free charge} \\[2em]
e^{-c_2|\partial^*\underline{\mathcal{M}}|} & \text{confinement, Higgs}
\end{cases}
\tag{3.3}
$$

We call $f_c(\underline{\Lambda})$ the FCOP.

By Gauss' law, the denominator of (3.2) is nothing else than the vacuum expectation value of the operator measuring the charge inside $\underline{\Lambda}$. In the limit of the pure gauge theory it becomes 1. In this limit the numerator of (3.2) converges towards the square of the 't Hooft loop /9/, which is known to have the behaviour (3.3).

For the Z_2 theory (3.3) was proven using convergent expansion techniques /4/.

In the d=3 Z_2 theory the FCOP is dual to the VOOP. In particular, in the Higgs region the FCOP behaves differently for small and for large r, since it is dual to the VOOP in the confinement region. This situation is not restricted to d=3. For arbitrary dimensions, the 0th order contribution to $f_{-1}(\underline{\Lambda})$ in the Higgs region expansion is

$$
f_{-1}(\underline{\Lambda}) = \frac{\left(e^{-c_2|\partial^*\underline{\mathcal{M}}|-2\beta|\underline{\mathcal{L}}|} + e^{-2\beta|\underline{\mathcal{M}}|-\varkappa|\underline{\Lambda}|} \right)^2}{e^{-2\beta|\partial^*\underline{\Lambda}|} + e^{-2\varkappa|\underline{\Lambda}|}}
\tag{3.4}
$$

Thus for large values of β and $\varkappa$,

$$
f_{-1}(\underline{\Lambda}) \sim
\begin{cases}
e^{-2\beta|\underline{\mathcal{M}}|} & r \ll r_D \\[1em]
e^{-c_2|\partial^*\underline{\mathcal{M}}|} & r \gg r_D
\end{cases}
\tag{3.5}
$$

where $r_D = 2d\beta/\varkappa$.

The scale r_D can be interpreted in the framework of a plasma model for the Higgs phase. The vacuum is a condensate of charged particles. For distances smaller than r_D, the flux correlations are strong. If by some fluctuation a charged particle came inside Δ, a charge measurement could be performed by measuring the electric flux through $\partial^*\Delta$, provided the linear size of Δ is smaller than r_D. Thus r_D is the scale at which the fluctuations of the dynamical charge are screened. Usually this is called the Debye screening length /25/.

In Plasma Electrodynamics the screening length r_{pot} for the potential (i.e. for the energy of an external charge-anticharge pair) is identical to r_D. This is shown by using Gauss' law to relate the charge density and the potential /25/. However, a similar argument is not possible if Gauss' law holds only in multiplicative form, as in our case. A 0th order computation of r_{pot} in the Higgs region of the Z_2 model results in:

$$r_{pot} = \frac{1}{4(d-1)\beta + h} \qquad (3.6)$$

In an electrodynamical plasma $r_D \to \infty$ as $\beta \to \infty$ (zero gauge coupling). The same is true in the Z_2 model for r_D but not for r_{pot}.

The FCOP can be considered also at finite temperatures T, since eq. (3.2) involves only time-zero operators. The Debye screening length r_D can now be defined in the whole phase diagram, not only in the Higgs region. We expect the following qualitative behaviour:

a) If at T = 0 we are in the free charge phase, at finite T we will have a plasma of charged particles. r_D diverges as $T \to 0$ and monotonically decreases as T grows. As $T \to \infty$, r_D settles to a value around one lattice spacing.

b) If at T = 0 we are in the Higgs region and $r_D > 1$, r_D will monotonically decrease with increasing T. The asymptotic value is again of the order of 1.

c) If at T = 0 we are in the confinement region, then we expect r_D to be very small as T increases towards the deconfinement transition T_d. At T_d there is no phase transition, but a crossover from the confinement regime to a charged particle plasma regime. r_D will rise sharply as T reaches T_d from below, and then decrease as described in a) - b) if T is increased further.

For the Z_2 model simple arguments can be given that support the finite T behaviour of r_D described here /26/. We also plan to do a Monte Carlo investigation of the FCOP. In general the FCOP is difficult to simulate for large values of β, as seen from eq. (3.1). However, for the Z_2 case a duality transformation can be used such that in the new model one has to compute expectation values of products of ± 1.

The decay of Polyakov loop correlations can be used to compute r_{pot} at finite T. There is again no similarity at all between r_D and r_{pot}. However, it is quite plausible that the deconfinement peak of r_D coincides with the rapid rize in the value of one Polyakov loop. Unfortunately we have no good theoretical argument to support this belief.

Eq. (3.4) suggests that one can use the denominator of (3.2) alone in order to define r_D and to investigate the deconfinement transition.

The study of the FCOP is just at its beginning. Reliable methods to compute it in various models have to be developed. The theoretical foundations are also not as strong as for the VOOP. Maybe different quantities can be defined that have essentially the same meaning but are easier to calculate. It would be interesting to analyse the role played by flux correlations in massless phases. Last but not least, it would be interesting to see how far the plasma analogy can be pushed.

REFERENCES

1 K. Osterwalder and E. Seiler, Ann. Phys. 110 (1978) 440

2 E. Fradkin and S. Shenker, Phys. Rev. D19 (1979) 3682

3 K.G. Wilson, Phys. Rev. D10 (1974) 2445

4 K. Fredenhagen and M. Marcu, Commun. Math. Phys. 92 (1983) 81

5 K. Fredenhagen, Talk presented at the Colloquium in honour of
       Prof. R. Haag at the occasion of his 60th birthday,
       Hamburg, November 15, 1982, appeared as Freiburg Univer-
       sity preprint THEP 82/9

6   K. Fredenhagen and M. Marcu, Phys. Rev. Lett. 56 (1986) 223

7   M. Marcu, in "Lattice gauge theory - a challenge in large-scale computing", B. Bunk, K.H. Mütter and K. Schilling editors, Plenum (1986) p. 267

8   D. Buchholz and K. Fredenhagen, Commun. Math. Phys. 84 (1982) 1

9   G. 't Hooft, Nucl. Phys. B153 (1979) 141

10 K. Fredenhagen, Lectures given at the Erice Summer School on "Fundamental Problems of Gauge Field Theory" (1985), DESY preprint 85-120

11 J.L. Alonso and A. Tarancón, Phys. Lett. 165B (1985) 167

12 J. Barata and K. Fredenhagen, "Charged particles in Z_2 gauge theories", to be published

13 K. Szlachányi, Phys. Lett. 147B (1984) 335

14 K. Szlachányi, Budapest Central Research Institute of Physics preprint KFKI-1986-35/A

15 T. Kennedy and C. King, Phys. Lett. 55 (1985) 776 and Commun. Math. Phys. 104 (1986) 327

16 C. Borgs and F. Nill, Nucl. Phys. B270 (FS16) (1986) 92

17 J. Bricmont and J. Fröhlich, Phys. Lett. 122B (1983) 73

18 T. Filk, K. Fredenhagen and M. Marcu, Phys. Lett. 169B (1986) 405

19 H.G. Evertz et al., Phys. Lett. 175B (1986) 335

20 J. Jersák, this conference, Aachen group, in preparation

21 V. Azcoiti and A. Tarancón, Phys. Lett. 176B (1986) 153

22 V. Alessandrini et al., Zaragoza preprint DFTUZ-86.9

23 B. Svetitzky and L.G. Yaffe, Nucl. Phys. B210 (FS6) (1982) 423

24 J.A. Swieca, Phys. Rev. D13 (1976) 312, D. Buchholz and K. Fredenhagen, Nucl. Phys. B154 (1979) 226

25 A.F. Alexandrov, L.S. Bogdankevich and A.A. Rukhadze, "Principles of Plasma Electrodynamics", Springer (1984)

26 K. Fredenhagen and M. Marcu, in preparation

THE LATTICE-REGULARIZED STANDARD HIGGS MODEL

I. Montvay

Deutsches Elektronen-Synchrotron DESY
D-2 Hamburg, FRG

ABSTRACT

Some recent non-perturbative investigations of the lattice-regularized $SU(2)$ Higgs model with a scalar doublet field are reviewed.

INTRODUCTION

The lattice regularization makes possible to define and study quantum field theories non-perturbatively. The main emphasis in recent years' numerical Monte Carlo investigations was put on quantum chromodynamics or on related gauge models with only fermionic matter fields. The obvious motivation is the basic property of asymptotic freedom, which implies strongly interacting infrared physics and hence an immediate need for non-perturbative methods. The necessity of non-perturbative investigations in the standard $SU(2) \otimes U(1)$ electroweak model is not so obvious. Although the $SU(2)$ gauge coupling is also asymptotically free, the Higgs mechanism cuts off the low energy growth of the coupling at a scale where it is still rather weak. The $U(1)$ gauge coupling and the scalar self-coupling in the Higgs sector are not asymptotically free, therefore have a tendency to be weak in the infrared. The electromagnetic $U(1)$ coupling is, indeed, known to be weak at low energies. However, the Higgs sector is not yet known phenomenologically, therefore it can, in principle, be a source of non-perturbative effects. The question of the large cut-off behaviour of non-asymptotically free couplings is always non-perturbative. These two latter points make, in my opinion, the non-perturbative study of the electroweak sector interesting (perhaps also unavoidable).

For the question of the large cut-off (or small lattice spacing) behaviour numerical Monte Carlo calculations are not very well suited: not even the best existing computer is able to treat correlation lengths much more than 10 lattice units. A reliable study of four-dimensional gauge field systems with correlation lengths in the order of 100 lattice spacings seems to be still far away. Nevertheless, there are interesting unsolved problems in the standard electroweak model which can be studies with present computers:

1. The simplest model for a strongly interacting Higgs sector is the standard Higgs model [1] even if it implies a relatively low cut-off.

2. The phase transition between the confining- and Higgs-phase is, both at zero and non-zero temperatures, a non-perturbative phenomenon.

3. The previous two points imply that in the standard Higgs sector there is an upper and a lower limit for the Higgs to W-mass ratio $R_{HW} \equiv m_H/m_W$. The precise value of R_{HW}^{max} and R_{HW}^{min} can presumably be obtained only by combining analytical and numerical work in lattice regularization.

4. There is a confining ("composite model") interpretation of the standard electroweak model [2], which is not yet ruled out by experiment. This would be a genuine strongly interacting theory, in many respects very similar to QCD (in some other aspects, however, different from it).

EXPANSIONS AT THE PARAMETER SPACE BOUNDARY

Because of the limited applicability of numerical methods, analytic expansions are very important to support and supplement the numerical work. The general strategy of analytic expansions is to reduce the number of coupling parameters by sending some of them to the boundary of the coupling parameter space. The small parameter in the expansion is the distance from the boundary in some appropriately chosen metric. In the standard Higgs model there are three couplings: the scalar self-coupling λ, the gauge coupling g (or $\beta \equiv 4g^{-2}$) and the hopping parameter κ standing in lattice regularization for the mass parameter of the scalar field. Possible expansions in the standard Higgs model are:

  i.) strong gauge coupling expansion (SGCE) at $\beta = 0$ [3];
 ii.) hopping parameter expansion (HPE) at $\kappa = 0$;
iii.) strong self-coupling expansion (SSCE) at $\lambda = \infty$ [4];
 iv.) weak gauge coupling expansion (WGCE) at $\beta = \infty$ [5];
  v.) inverse hopping parameter expansion (IHPE) at $\kappa = \infty$;
 vi.) weak self-coupling expansion (WSCE) at $\lambda = 0$.

The latter two (IHPE and WSCE) were, to my knowledge, not yet studied in the standard Higgs model. The difficulty with WSCE is that the Higgs phase is singular for $\lambda \to 0$. At the intersections one can also define double- or triple-expansions. (For instance, combined WGCE and SSCE at the line $\lambda = \beta = \infty$.)

Here we shall discuss in some detail WGCE. In this case the expectation values at an arbitrary point (λ, β, κ) of the bare parameter space are expressed in terms of a series of expectation values at the point $(\lambda, \beta = \infty, \kappa_0)$ with vanishing gauge coupling. This is achieved by performing the integration over the gauge field variables in perturbation theory, thereby explicitly displaying the dependence on the gauge field propagators and vertices. In Ref. [5] the generating function of connected expectation values of some gauge invariant composite fields was considered in WGCE. For the study of renormalization and large cut-off behaviour it is better to expand the gauge dependent generating functions in renormalizable gauges. In order to write down the WGCE master formula for the gauge dependent Green's functions, it is useful to introduce a shorthand notation for index repetitions:

$$(f.)_\nu^n \equiv f_{\nu_1} f_{\nu_2} \cdots f_{\nu_n} \qquad \sum_{[\nu]_n} (f.)_\nu^n \equiv \sum_{\nu_1 \cdots \nu_n} f_{\nu_1} f_{\nu_2} \cdots f_{\nu_n} \qquad (1)$$

In the master formula the gauge field expectation values $< \cdots >_{\alpha g}^c$ have to be expressed by gauge field propagators and vertices in the same way as in pure gauge perturbation theory. (α is the gauge parameter.) The occuring expectation values in the ϕ^4 model at $g^2 = 0$ contain, in addition to the original scalar fields σ_{0x} and π_{rx}, $(r = 1, 2, 3)$, also the bilinear composite fields

$$s_{x\mu} = 2(\sigma_{0x}\sigma_{0x+\hat\mu} + \pi_{rx}\pi_{rx+\hat\mu})$$

$$u_{rx\mu} = 2i(\pi_{rx}\sigma_{0x+\hat\mu} - \sigma_{0x}\pi_{rx+\hat\mu} + \epsilon_{rst}\pi_{sx}\pi_{tx+\hat\mu}) \qquad (2)$$

The derivatives of the generating function $W[h, i]$ with respect to $h \equiv (h_{0x}, h_{rx})$ and $i_{rx\mu}$ give the connected Green's functions of the scalar fields (σ_{0x}, π_{rx}) and, respectively, of the gauge field $A_{rx\mu}$. Let us first introduce the notations

$$C[i]_{[rx\mu]_m [y\nu]_n}^{mn} \equiv \frac{1}{2^{m+3n}m!n!} (i. + i\kappa u.)_{rx\mu}^m (1 - \kappa s.)_{y\nu}^n$$

$$A_{[rx\mu]_m [y\nu]_n}^{(\alpha)mn} \equiv \left\langle (A.)_{rx\mu}^m (A_s.A_s.)_{y\nu}^n \right\rangle_{\alpha g}^c \qquad (3)$$

The result for the generating function of connected Green's functions W is in this notation (applying the trick (1) twice):

$$W[h, i]_{\lambda\beta\kappa}^\alpha = \sum_{LMN} \sum_{[X]_L [RY]_M [Z\lambda]_N} \sum_K \sum_{[m]_K [n]_K} \sum_{[[rx\mu]_m [y\nu]_n]_K}$$

$$\frac{(h_0.)_X^L (h.)_{RY}^M (\kappa - \kappa_0)^N}{K!L!M!N!} \left(g^{m+2n} A_{..}^{(\alpha)mn} \right)_{[rx\mu]_m [y\nu]_n}^K \cdot$$

$$\cdot \left\langle (\sigma_0.)_X^L (\pi.)_{RY}^M (s.)_{Z\lambda}^N \left(C[i]_{..}^{mn} \right)_{[rx\mu]_m [y\nu]_n}^K \right\rangle_{\lambda\kappa_0}^c \qquad (4)$$

In the connected ϕ^4 expectation value $< \cdots >_{\lambda\kappa_0}^c$ the contents of the parentheses have to be considered as single entities in the definition of connectedness. The same applies also to $(A_{sy\nu}A_{sy\nu})$ in the connected gauge field expectation value of Eq. (3). The most useful way to apply WGCE is to choose the expansion point (λ, κ_0) in the vicinity of the critical line of the ϕ^4 model. In this case it is possible to obtain enough information about the behaviour of the ϕ^4 Green's functions by assuming the triviality of the continuum limit in ϕ^4 [6,7]. By combining the hopping parameter expansion ("high temperature expansion" in the terminology of statistical physics) and the Callan-Symanzik renormalization group equations one can, for instance,

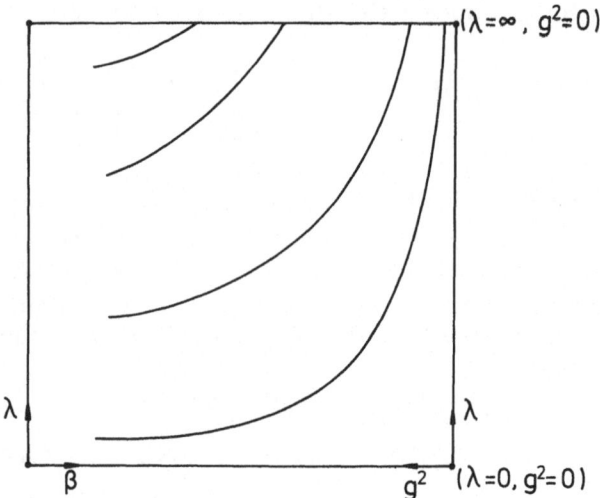

Fig. 1. The qualitative picture of CCP's in the standard Higgs model projected on the (g^2, λ) plane. The small g^2 behaviour is the result of WGCE. The extension to larger g^2 is a guess supported by some approximate numerical Monte Carlo calculations at $\lambda = \infty$, $\beta = 2 - 3$ [9,10]. Note that in reality there is a two-parameter family of CCP's, but here only a one-parameter subset is shown for simplicity.

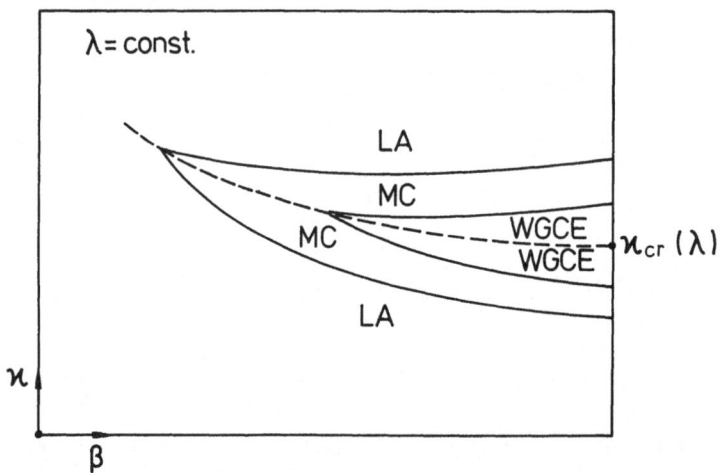

Fig. 2. The schematic lay-out of the regions where interesting Monte Carlo calculations can be done (MC) and where WGCE can be expected to give a good approximation (WGCE). The uninteresting region of dominant lattice artifacts is denoted by LA. The confining-Higgs phase transition is at the dashed line. The whole picture is for $\lambda = const.$.

238

determine the curves of constant physics of ϕ^4 everywhere in the bare parameter space [7]. The knowledge of the non-perturbative dynamics of pure ϕ^4 is an input in WGCE. In what follows, we shall assume the triviality of the continuum limit in ϕ^4, but if this would finally turn out to be wrong, WGCE could still be a useful expansion in order to determine the perturbation due to a weak $SU(2)$ gauge coupling.

CURVES OF CONSTANT PHYSICS

Before going to the Higgs model, let us first formulate how to obtain differential equations for the curves of constant physics in the general case. Let us consider a lattice quantum field theory with n bare couplings $g_1, g_2, \ldots, g_n$. In order to define the CCP's one has to keep $(n-1)$ independent physical quantities $F_2, F_3, \ldots, F_n$ constant (we are assuming here that the number of relevant couplings is n):

$$F_j(g_1, \ldots, g_n) = F_{j0} = const. \quad (j = 2, \ldots, n) \tag{5}$$

The CCP's are characterized by the constant values F_{j0}. The points of a singled out CCP can be parametrized, for instance, by the first bare coupling g_1: $g_j = g_j(g_1)$ $(j = 2, \ldots, n)$. In this case we have

$$\frac{dg_j(g_1)}{dg_1} = \frac{\det_{n-1}^{[1,j]}\left(\frac{\partial F}{\partial g}\right)}{\det_{n-1}^{[1,1]}\left(\frac{\partial F}{\partial g}\right)} \tag{6}$$

Here $\det_{n-1}^{[i,k]}\left(\frac{\partial F}{\partial g}\right)$ denotes the $(n-1) \times (n-1)$ subdeterminant of the $n \times n$ derivative matrix $\frac{\partial F}{\partial g}$ belonging to the matrix element $\frac{\partial F_i}{\partial g_k}$.

Another possibility is to parametrize the points of a CCP by the value of some reference physical quantity F_1. (In practical cases F_1 is usually some physical mass in lattice units.) In this case the differential equations for $g_i(F_1)$ $(i = 1, \ldots, n)$ are:

$$\frac{dg_i(F_1)}{dF_1} = \frac{\det_{n-1}^{[1,i]}\left(\frac{\partial F}{\partial g}\right)}{\det_n\left(\frac{\partial F}{\partial g}\right)} \tag{7}$$

where $\det_n(\cdots)$ is the $n \times n$ determinant of the derivative matrix.

Sometimes it is also useful to consider curves in subspaces of the bare parameter space which belong to constant values of an appropriately smaller number of physical quantities. These "curves of partially constant physics" (CPCP's) are defined by fixing $(n-k)$ physical quantities $F_2, \ldots, F_{n-k+1}$ and $(k-1)$ bare parameters $g_{n-k+2}, \ldots, g_n$. The differential equations for CPCP's have the same form as Eqs. (6-7). For simplicity, let us consider here only the case with $n = 3$ bare parameters (as we have in the standard Higgs model) and look at the plane with constant bare coupling g_3. Keeping the value of some physical quantity $F_2(g_1, g_2, g_3) = F_{20}$ fixed and parametrizing the points of the curve by the reference quantity F_1, the differential equation for the function $g_2(F_1)$ is:

$$
\frac{dg_2(F_1)}{dF_1} = \left(\frac{-\frac{\partial F_2}{\partial g_1}}{\frac{\partial F_1}{\partial g_1}\frac{\partial F_2}{\partial g_2} - \frac{\partial F_1}{\partial g_2}\frac{\partial F_2}{\partial g_1}} \right)_{g_2 = g_2(F_1, g_2, g_3)}
\tag{8}
$$

As an example in the standard Higgs model, one can take $g_1 = \kappa$, $g_2 = \lambda$, $g_3 = g^2$ and $F_1 = \mu_W$ (the W-mass), $F_2 = R_{HW} \equiv m_H/m_W$ (the ratio of Higgs- to W-mass). In this case Eq. (8) gives the curves with constant Higgs- to W-mass ratio in the $g^2 = const.$ planes.

As discussed before, in WGCE the expectation values at a point (λ, g^2, κ) with weak bare gauge coupling and arbitrary bare scalar self-coupling are given in WGCE by explicit gauge propagators and vertices and by scalar blobs representing expectation values in the pure ϕ^4 model at $g^2 = 0$ and (λ, κ_0). We are assuming that the continuum limit in ϕ^4 is trivial therefore, if the point (λ, κ_0) is close to the critical line, the renormalized ϕ^4 coupling λ_r in the pure ϕ^4 model is small and the renormalized ϕ^4 Green's functions can be well approximated by a low order perturbative expansion in λ_r. In $\tilde{\text{W}}\text{GCE}$ we need unrenormalized $g^2 = 0$ expectation values, which are obtained from the renormalized ones by multiplying with the wave function renormalization factor of the scalar field Z_r and with the multiplicative renormalization factors of the composite fields Z_s, Z_u.

Since the $g^2 = 0$ expectation values in WGCE are given in terms of the pure ϕ^4 renormalized coupling λ_r, it is natural to parametrize the points of the bare parameter space, instead of (λ, g^2, κ), by (λ_r, g^2, κ). (In this case the ϕ^4 Z-factors have to be considered also as functions of λ_r and κ: $Z_{r,s,u} = Z_{r,s,u}(\lambda_r, \kappa)$.) In this way the problem of determining the CCP's for small bare gauge coupling is reduced by WGCE to the problem of finding the CCP's, with $\lambda_r = const.$, in the ϕ^4 model at $g^2 = 0$. The renormalization is also decomposed in two steps: after going to the renormalized variables at $g^2 = 0$, λ_r is considered as one of the bare parameters for WGCE in the Higgs model. The renormalized quantities of the Higgs model are introduced in WGCE in the same way as in ordinary perturbation theory. (In what follows only 1-loop graphs will be considered. The renormalization of WGCE up to higher loops have to be investigated in the future.) The CCP's in the Higgs model can be defined by the requirement that the renormalized ϕ^4 coupling λ_R and renormalized gauge coupling squared g_R^2 be constant. (Note that capital R denotes renormalized quantities in the Higgs model, whereas small r is reserved for the renormalized quantities at $g^2 = 0$.) As a parameter along the CCP's, one can take the renormalized ϕ-mass squared μ_R^2 (or $\tau \equiv \log \mu_R^{-1}$). The differential equations corresponding to Eq. (7) are, in the large cut-off (small μ_R^2) limit up to leading order:

$$\frac{d\lambda_r(\tau)}{d\tau} = -\frac{9}{16\pi^2}\lambda_r g^2 + \cdots \qquad \frac{dg^2(\tau)}{d\tau} = -\frac{43}{48\pi^2}g^4 + \cdots \qquad (9)$$

The dots stand here for higher orders in λ_r and g^2. In what follows the higher order terms will be neglected, although an estimate of their importance can only be given after a more detailed study of multiloop renormalization of WGCE. The solution of the asymptotic equations is:

$$g^2(\tau) = \left[g_0^{-2} + \frac{43}{48\pi^2}(\tau - \tau_0)\right]^{-1} \qquad \lambda_r(\tau) = \lambda_{r0}\left[1 + \frac{43g_0^2}{48\pi^2}(\tau - \tau_0)\right]^{-\frac{27}{43}} \qquad (10)$$

Here g_0^2 and λ_{r0} are the initial values at $\tau = \tau_0$.

Since both $g^2(\tau)$ and $\lambda_r(\tau)$ tend to zero for $\tau \to \infty$, WGCE is an asymptotically free expansion. This does not, however, mean that for $\tau \to \infty$ a non-trivial continuum limit exists. The reason is that on the (τ, λ_r) plane not every point is possible. The triviality of the continuum limit of ϕ^4 implies that the CCP's in ϕ^4 with $\lambda_r = const.$ are ending at $\lambda = \infty$ near the critical point $\kappa_{cr}(\lambda = \infty)$ for some finite cut-off [7]. This means that on the (τ, λ_r) plane there is a limiting curve and the allowed points are below this (at smaller values of τ and λ_r). To obtain the exact shape of the limiting curve is a non-perturbative problem in the four-component $O(4)$-symmetric ϕ^4 model. In order to have a rough guess about its qualitative behaviour, one can take for large τ the position of the "Landau-pole" in one-loop perturbation theory:

$$\lambda_r(\tau)_{max} \simeq \frac{\pi^2}{6\tau} \qquad (11)$$

The intersection of the curve $\lambda_r(\tau)$ in Eq. (10) with $\lambda_r(\tau)_{max}$ determines the maximal cut-off τ_{max} which belongs to the Higgs model CCP given by $(g^2(\tau), \lambda_r(\tau))$. If the maximal cut-off is required to be the Planck mass ($\tau \simeq 20$), this crude estimate for a CCP with gauge coupling $g_R^2 = 0.5$, roughly equal to the physical value in the standard electroweak model, gives λ_R about a factor of 2 larger than a one-loop perturbative calculation [8] would give. In order to obtain this "Planck mass cut-off upper limit" more precisely for λ_R (or for the Higgs mass to W-mass ratio $R_{HW} \equiv m_H/m_W$), we need a careful non-perturbative study of the 4-component ϕ^4 model to pin down the limiting curve $\lambda_r(\tau)_{max}$. In addition, a non-perturbative investigation in the Higgs model itself is also necessary for obtaining the non-asymptotic form of $\lambda_r(\tau)$.

The above asymptotic estimates determine the qualitative behaviour of the CCP's for small g^2 in the (λ, g^2) plane (assuming that the higher order contributions can be neglected, indeed). Since every CCP with non-zero λ_R and g_R^2 is ending at a finite cut-off in the $\lambda = \infty$ plane near the phase transition line, the only possibility to reach an infinite cut-off is to put $\lambda_R = g_R = 0$. Therefore, the continuum limit in the Higgs model on the critical line at $g^2 = 0$ is trivial, both in the confining- and Higgs-phase. The triviality of the $g^2 = 0$ continuum limit in the Higgs phase was recently concluded also in Ref. [11] on the basis of perturbation theory near the Gaussian fixed point ($\lambda = 0, g^2 = 0, \kappa = 1/8$). Perturbation theory is, however, not applicable for large bare self-coupling λ, therefore Ref. [11] did not exclude the possibility of a non-trivial fixed point in the combined gauge-scalar system at $g^2 = 0$ and large λ. In any case, the non-trivial λ-independent continuum limit conjectured in Ref. [9] is not possible. The behaviour of the CCP's near $\beta = \infty$ is different from the picture suggested there. The only open possibility for a search of a non-trivial continuum limit in the standard Higgs model is to go inside the bare parameter space to points where also the gauge coupling is non-perturbative. However, even if such a fixed point would exist, it would not necessarily be adequate for the description of the standard electroweak physics.

The framework of WGCE is obviously more general than the specific case of the standard $SU(2)$ Higgs model. It would certainly be interesting to consider in the future more general Higgs models, too. In particular, as one can see from Eq. (10), there is an interesting class of models, where the Callan-Symanzik β-function coefficients are such that the power of the squared brackets in $\lambda_r(\tau)$ is, instead of $\frac{-27}{43}$, equal to -1. In this case the leading asymptotic behaviour of $\lambda_r(\tau)$ coincides with the asymptotics of the limiting curve in Eq. (11). The question of a possible non-trivial continuum limit at $\lambda = \infty$ is then decided on the next-to-leading order level. In any case, even if the strict continuum limit would turn out to be trivial, such models are interesting, because they can easily allow for very large cut-off's in a wide range of physical situations. A simple example of a model with a τ^{-1} leading λ_r-behaviour is an $SU(2)$ Higgs model with 1 scalar doublet and 4 vector-like spin-$\frac{1}{2}$ fermion doublets. Namely, in this case the coefficient of the g^4 term in Eq. (9) is equal to $-27/(48\pi^2)$. Because of the vector-like fermions, Yukawa-couplings are forbidden. In cases with Yukawa-couplings and chiral fermions (as in the standard model) the appropriate lattice formulation has to be constructed first, and similar questions can be asked only afterwards.

MONTE CARLO CALCULATIONS AT WEAK GAUGE COUPLING

The SU(2) coupling in the standard electroweak sector is weak: at the W-mass scale we have $g_R^2 \simeq 0.5$ (or $\alpha_{SU(2)} \equiv g_R^2/(4\pi) \simeq 0.04$). Since the Monte Carlo calculations are done in a region where the W-mass in lattice units am_W is of the order 1, this implies for the bare gauge coupling a value of about $\beta \simeq 8$. A numerical Monte Carlo calculation in this β region is feasible and it can yield interesting information about the non-perturbative behaviour in the strong self-coupling regime [12]. Such calculations are complementary to the information one can obtain from WGCE. In general, WGCE can be expected to give a good approximation for small g^2, in the immediate vicinity of the phase transition surface, where the correlation lengths are large. Somewhat farther away from the critical surface one can do Monte Carlo calculations. Optimistically we can expect that the MC and WGCE regions touch. In reality there may be some no-man's-land inbetween, where the correlation lengths are too large for a numerical investigation but not large enough to make the couplings small enough for a low order WGCE. The phase transition line is given in the small g^2 region by WGCE as $\kappa_{cr}(\lambda, g^2) = \kappa_{cr}(\lambda, 0)\{1 + g^2 0.04358\ldots\}$.

The Monte Carlo calculation at $\beta = 8$ was restricted in Ref. [12] to two points at $\lambda = 1.0$. To explore the λ-dependence in the large bare self-coupling region, the same calculation has been repeated recently on 12^4 lattice in a few other points at $\beta = 8$ [13]. The points were selected in such a way that the link expectation value L was nearly constant. This is known from previous studies [9] to minimize the λ-dependence for a given β. The new Monte Carlo results show that the renormalized gauge coupling is, within errors, the same in the three points: $\alpha_{SU(2)}^{(R=3)} = 0.051 - 0.052$. The Higgs- to W-mass ratio R_{HW} is monotonously increasing with λ: from $R_{HW} = 4.6(5)$ at $(\lambda = 0.1, \kappa = 0.177)$ to $R_{HW} = 7.5(8)$ at $(\lambda = \infty, \kappa = 0.370)$. The value of R_{HW} at $\lambda = \infty$ can be considered as a first rough estimate for an upper limit [14] of the Higgs mass. Of course, our 12^4 lattice is not big enough for $\lambda = \infty$, we need many other points to check the λ- and β-dependence of R_{HW} etc.

REFERENCES

[1] M. E. Peskin, in *Proc. of the 1985 Int. Symp. on Lepton and Photon Interactions at High Energies*, Kyoto;
M. K. Gaillard, in *Proceedings of the 1985 Theoretical Advanced Study Institute*, Yale University
[2] L. F. Abbott, E. Farhi, Phys. Lett. **B101** (1981) 69;
Nucl. Phys. **B189** (1981) 547;
M. Claudson, E. Farhi, R. L. Jaffe, MIT preprint CTP#1331, 1986
[3] U. Wolff, DESY preprint 86-085, 1986
[4] K. Decker, I. Montvay, P. Weisz, Nucl. Phys. **B268** (1986) 362
[5] I. Montvay, Phys. Lett. **B172** (1986) 71
[6] E. Brezin, J. C. Le Guillou, J. Zinn-Justin, in *Phase transitions and critical phenomena,* ed. C. Domb, M. S. Green (Academic Press, London, 1976) vol. 6, p. 125
[7] M. Lüscher, P. Weisz, to be published
[8] N. Cabibbo, L. Maiani, G. Parisi, R. Petronzio, Nucl. Phys. **B158** (1979) 295;
D. J. E. Callaway, Nucl. Phys. **B233** (1984) 189;
M. A. Beg, C. Panagiotakopoulos, A. Sirlin, Phys. Rev. Lett. **52** (1984) 883
[9] I. Montvay, Nucl. Phys. **B269** (1986) 170
[10] D. Callaway, R. Petronzio, Nucl. Phys. **B267** (1986) 253
[11] A. Hasenfratz, P. Hasenfratz, Tallahassee preprint FSU-SCRI-86-30
[12] W. Langguth, I. Montvay, P. Weisz, Nucl. Phys. **B277** (1986) 11
[13] W. Langguth, I. Montvay, in preparation
[14] R. Dashen, H. Neuberger, Phys. Rev. Lett. **50** (1983) 1897

CONTINUUM SYMMETRY RESTORATION IN LATTICE MODELS WITH STAGGERED

FERMIONS

A. Morel

Service de Physique Théorique
CEN-SACLAY
91191 Gif-sur-Yvette Cedex, France

ABSTRACT

This talk is a report on results obtained by T.Jolicoeur, R.Lacaze, B.Petersson and the author: staggered fermions can be consistently interpreted as flavoured quarks in the continuum limit of asymptotically free theories on the lattice. This statement is supported by analytical results for the Gross-Neveu model at large N and for a QCD two point function, and by a numerical simulation of SU(2) quenched QCD.

1.INTRODUCTION

Latticizing a continuum field theory breaks many of its symmetries. The lattice action, however, still has some remnant symmetries, including a discrete subgroup of the original group, and it is hoped that they are sufficient to insure continuum symmetry restoration as the cut-off is removed. In the case of an asymptotically free theory such as QCD, the continuum limit is eventually reached as the lattice coupling g^2 goes to zero, when the correlation length in units of the lattice spacing goes to infinity, hopefully washing out all the (short distance) artefacts of the regularization. As an example, evidence for rotational symmetry restoration in the static quark potential has been reported in the past[1].

When fermions are incorporated, the well-known degeneracy problem comes in, and the chiral symmetry of the continuum QCD action also has to be broken on the lattice, at least under the general assumptions of the Nielsen-Ninomyia theorem [2]. These assumptions are fulfilled by the two most popular ways of putting quarks on the lattice, namely using Wilson or Kogut-Susskind (staggered) fermions. In the former case, the vector part of the chiral group is preserved, and the condition for the axial part to be realized à la Nambu-Goldstone is set by parameter adjustement. In the latter case, here under study, the chiral group should be *dynamically* restored in the continuum limit.

Within the staggered fermion formalism in (even) D dimensions, the 2^D-fold degeneracy of the naive discretization is reduced to a $2^{[D/2]}$-fold degeneracy by throwing away all but one of the $2^{[D/2]}$ Dirac components. One is left with one pair $(x_i(x), \bar{x}_i(x))$ of Grassmann variables per site and per colour, and this leads to $C = 2^{[D/2]}$ species of fermions in the free spectrum, to be interpreted as flavours [3]. The states of the interacting theory are classified according to the irreducible representations of the *lattice* symmetry group [4], and their interpretation according to the *continuum* group for C flavours (broken by the discretization) requires the knowledge of the embedding of the lattice group into its continuum extension. For this purpose, an explicit construction of the flavour degrees of freedom is needed, and the consistency of a construction suggested by the free fermion case with the dynamics of the interacting theory has to be checked: is the flavour symmetry of the classical limit of the action a symmetry of the continuum limit ?

The purpose of this talk is to discuss flavour symmetry restoration in the continuum limit of lattice models, including QCD, formulated with staggered fermions. It is based on work done in common with T.Jolicoeur and B.Petersson[5], and also on papers by Jolicoeur[6] and by Lacaze[7].

In section 2, two flavour formalisms in the free case are reviewed and compared, and their apparently different symmetry patterns are elucidated. The effect of interactions is first studied on a lattice version of the Gross-Neveu Model for N species of staggered fermions (section 3); it is shown at the first non trivial order in 1/N that the continuum limit of the 4-point function is fully consistent with that of a 2N-species continuum theory. In section 4, we construct a QCD gauge invariant two-fermion function, which coincides with the quark propagator in the free case, and we prove that it is C-flavour invariant in the continuum limit. Section 5 contains results obtained by Lacaze for SU(2) quenched QCD. He shows that flavour symmetry (in fact *chiral* symmetry à la Nambu-Goldstone) is being restored in the coupling region where scaling behaviour starts to show up. As the present outline of the talk already states the main outputs of these investigations, there will be no conclusion.

2.FLAVOUR FORMULATIONS OF STAGGERED FERMIONS

Here we recall the basics of two distinct formulations of the flavour interpretation of lattice fermion degeneracy in the free case. We are dealing with the Kogut-Susskind action, defined on a D-dimensional hypercubical lattice with sites x, link directions $\hat{\mu}$, and lattice spacing a (taken as unity unless specified):

$$S_{ks} = -\frac{1}{2} \sum_{x,\mu} \alpha_\mu(x) [\bar{\chi}(x)\chi(x+\hat{\mu}) - \bar{\chi}(x+\hat{\mu})\chi(x)] ,$$

with the choice

$$\alpha_\mu(x) = (-)^{x_1 + \ldots + x_{\mu-1}} . \tag{1}$$

If needed, a colour index i=1,...,N is understood, or specified when necessary. Such an action is expected to describe $C = 2^{[D/2]}$ Dirac quarks, but neither these flavour indices nor the Dirac components are apparent yet.

The first formulation leads to "local quarks" $q^{\alpha a}(y)$[8,9], described as follows:

$$q^{\alpha a}(y) = \frac{1}{c\sqrt{c}} \sum_\eta \Gamma_\eta^{\alpha a} \chi(2y+\eta) \tag{2}$$

where the sites are labelled by

$$x_\mu = 2y_\mu + \eta_\mu \quad \text{with} \quad \eta_\mu = 0 \text{ or } 1 \text{ and } y_\mu \text{ integer,}$$

and

$$\Gamma_\eta = \gamma_1^{\eta_1} \ldots \gamma_D^{\eta_D} .$$

We use conventions where $\gamma_\mu^\dagger = \gamma_\mu$, $\{\gamma_\mu,\gamma_\nu\} = 2\delta_{\mu\nu}$. The indices α and a are interpreted respectively as Dirac and flavour indices, as suggested by the rewriting

$$S^q_{ks} = \sum_{y,\mu} \left[\bar{q}(y)(\gamma_\mu \otimes 1)\Delta_\mu q(y) + \bar{q}(y)(\gamma_5 \otimes \gamma_\mu^* \gamma_5^*)\delta_\mu q(y) \right] , \qquad (3)$$

where we defined the first and second lattice derivatives:

$$\Delta_\mu f(y) = \frac{1}{4a} (f(y+\hat{\mu}a) - f(y-\hat{\mu}a)) ,$$

$$\delta_\mu f(y) = \frac{1}{4a^2} (f(y+\hat{\mu}a) + f(y-\hat{\mu}a) - 2f(y)) .$$

Flavour symmetry breaking is explicit in the second term of Eq.(3).

In the second formulation[10], one divides the Brillouin zone into C^2 pieces, writing the momentum k_μ as

$$k_\mu = \frac{p_\mu}{2} + (\pi_A)_\mu , \qquad (\pi_A)_\mu = 0 \text{ or } \pi, \quad A=1,\ldots,C^2,$$

and identifying as (Dirac × Flavour) components the quantities

$$\tilde{\phi}_A(p) = \tilde{\chi}(\frac{p}{2} + \pi_A) ,$$

where $\tilde{\chi}(k)$ is the Fourier transform of $\chi(x)$ (likewise for $\bar{\tilde{\phi}}_A$ associated with $\bar{\tilde{\chi}}$). A suitable unitary transformation $\tilde{\phi} = V \tilde{\psi}$ in the C^2-dimensional space spanned by the $\tilde{\phi}_A$'s then gives the same action the form

$$S^\psi_{ks} = \int \frac{d^D p}{(2\pi)^D} \bar{\tilde{\psi}} \left[i(\gamma_\mu \otimes 1) \sin \frac{p_\mu}{2} \right] \tilde{\psi}(p) , \qquad (4)$$

with the apparently paradoxical feature that the *full* chiral group SU(C) ⊗ SU(C) is preserved. There is no paradox however, because of the non-local relationship between the ψ's and the χ's introduced by cutting the Brillouin zone sharply. In fact, the $\tilde{\psi}(p)$'s and the $\tilde{q}(p)$'s are related through the *p-dependent* unitary transformation[9]

$$\tilde{\psi}(p) = w(p) \; \tilde{q}(p)$$

$$w(p) = \exp\left\{-i \sum_\mu p_\mu a(\gamma_\mu \gamma_5 \otimes \gamma_\mu^* \gamma_5^*)\right\} . \tag{5}$$

Examination of the effect, on the local (q) or momentum space (ψ) quarks, of the lattice symmetry transformations leads to the result [4,5] that the two flavour constructions are consistent with each other. Amongst the generators of the chiral group exhibited by the form (4), only one, $T_A = \gamma_5 \otimes \gamma_5^*$, generates *local* transformations on the χ's and is a symmetry of the form (3) of the action. Note for further reference that the identity of the two descriptions in the classical limit is clearly exhibited by Eq.(5): the basis transformation w(p) tends to unity, $1 \otimes 1$, when $a \to 0$ at fixed physical momenta p. The relevant question concerns the fate of these properties in the continuum limit of the interacting case.

3. FLAVOUR SYMMETRY RESTORATION IN THE GROSS-NEVEU MODEL

We will here sketch the method and give the results obtained [5] for a particular lattice version of the large N, 2-dimensional Gross-Neveu model[11]. We recall that this model has in common with QCD that it is asymptotically free; so its (known) properties can be investigated from the lattice. The interaction we add to the kinetic term S_{ks} (Eq.(1)) is[12]

$$S = \frac{g^2}{32} \sum_{\text{plaquettes}} (\bar{\chi}\chi(0) + \bar{\chi}\chi(1) + \bar{\chi}\chi(2) + \bar{\chi}\chi(3))^2, \tag{6}$$

where the labels 0,1,2,3 denote the corners of each plaquette. A "colour" index $i=1,\ldots,N$ is understood $(\bar{\chi}\chi = \sum_1^N \bar{\chi}_i \chi_i)$, corresponding to a global U(N) symmetry, in addition to its discrete symmetries identical to those of the kinetic term. The classical limit of S is $(\bar{q}q)^2$ in terms of the local quarks q of Eq.(2). At finite spacing however it also involves flavour breaking terms. We have studied the continuum limit of the 4-quark amplitude

$$A = \langle q^{\alpha a} \bar{q}^{\beta b} q^{\alpha' a'} \bar{q}^{\beta' b'} \rangle \; , \tag{7}$$

in the large N limit, at first non trivial order (1 loop) and fixed fermion mass. Exactly as in the continuum theory, the fermions indeed receive a mass m_F (dimensional transmutation), expressed by[13]

$$m_F = g\langle\sigma\rangle \propto a^{-1} \exp\left[-\frac{\pi}{2g^2 N}\right] \tag{8}$$

in the weak lattice coupling limit ($g^2 N \to 0$) associated with the continuum limit ($a \to 0$). In Eq.(8), $\langle\sigma\rangle$ denotes the vacuum expectation value of the auxiliary field σ_P conjugate to the plaquette fermionic operator $(\bar{\chi}\chi)_P$ which appears squared in Eq.(6):

$$\exp S = \int [\pi d\sigma_P] \exp\left\{ -\frac{1}{2} \sum_P \left(\sigma_P^2 + \frac{g}{2} \sigma_P (\bar{\chi}\chi)_P \right) \right\} \; .$$

In terms of σ_P, χ, $\bar{\chi}$, the action is quadratic in $\chi, \bar{\chi}$ and after a shift $\sigma_P \to \sigma_P + m_F/g$ one gets a free massive quark propagator which reads in the q basis

$$\Delta_q = \frac{a}{2\left(\sum_\alpha \sin^2 p_\alpha /2 + m_F^2 a^2\right)} \times \mathcal{N}$$

$$\mathcal{N} = \left\{ i\sum_\alpha \sin p_\alpha (\gamma_\alpha \otimes 1) + 2m_F a (1\otimes 1) + \sum_\alpha (1-\cos p_\alpha)(\gamma_5 \otimes \gamma_\alpha^* \gamma_5^*) \right\} \; . \tag{9}$$

There are 4 σ_P variables per site x, which are traded for new variables $\Sigma_A(y)$, A=0,...,3 attached to the "physical" lattice built by the points x = 2y. A suitable choice of these variables finally leads to the following action:

$$S_{q,\Sigma} = \bar{q}\, \Delta_q^{-1} q \; - \; \frac{1}{2} \sum_{A,y} \Sigma_A^2 (y) \; - \; \frac{1}{4}\, g \sum_{A,y} \Sigma_A (y)\, \bar{q}\, \theta^A q \; . \qquad (10)$$

Explicit calculation gives the operators θ^A entering this equation as the sum of the unit operator $\theta^\circ = (1 \otimes 1)$, like in the continuum theory, and of (Lorentz + flavour) symmetry breaking terms involving lattice derivatives, which thus are "irrelevant". The action (10) is suitable for a perturbation expansion in $1/N$ at fixed $\lambda = g^2 N$ and defines the corresponding Feynman rules. The four point function (7) has been calculated at the first non trivial order in $1/N$. It involves the dressed Σ propagator $\langle \Sigma_A \Sigma_B \rangle_p$ as depicted in the figure,

where the fermion loop in the right hand side is given by

$$\pi_{AB}(p) = Ng^2 \int \frac{d^2 k}{(2\pi)^2} \; \mathrm{Tr} \left\{ \theta^A \Delta_q (k)\; \theta^B \Delta_q (p-k) \right\} \; . \qquad (11)$$

The trace operates on spin and flavour indices. Power counting in a at fixed *physical* values of p and m_F reveals the appearance of integrals of order $\mathrm{Ln}(a)$ at most. The $\mathrm{Ln}(a)$ contributions however are killed in all π matrix elements but π_{00}, by the explicit powers of a brought by the fore-mentioned derivatives present in all θ^A but θ°. In the $a \to 0$ limit, one is thus left with exactly the same situation as in the continuum theory regularized by a sherical cut-off[11] $\Lambda = \pi/(2a)$. Substraction at $p^2 = \mu^2$ leads to the Lorentz and flavour symmetric result of the continuum, associated (like in Eq.(8) for the mass gap) with *twice* the number N of primary fermions. This doubling is, at this order at least, the *only* effect of the discretization, its flavour interpretation with local quarks is justified, and so is the one

by momentum quarks $\tilde{\psi}(p)$ since the amplitude A_q is related to the corresponding one A_ψ by a product of w transformations (Eq.(5)) which tend to identity in the continuum limit at fixed external momenta.

In the above, the symmetry breaking contributions are of order a or $a\mathrm{Ln}a$, a consequence of the irrelevance of the breaking terms appearing in the local quark formulation of the model. In Ref.[6], Jolicoeur considers another lattice version of the Gross-Neveu model, which was discarded in Ref.[12] because the *classical* limit of the action breaks Lorentz and flavour symmetry by *relevant* terms. Specifically, the interaction is

$$S_I = \frac{1}{2} g^2 \sum_{x,i} \left[\bar{\chi}_i(x) \, \chi_i(x) \right]^2$$

$$= 2g^2 \sum_y \left[(\bar{q}q)^2 + (\bar{q}(\sigma_1 \otimes \sigma_1)q)^2 + (\bar{q}(\sigma_2 \otimes \sigma_2)q)^2 + (\bar{q}(\sigma_3 \otimes \sigma_3)q)^2 \right] .$$

Through a study of the large N limit, he obtains the following remarkable results. The field σ associated with $\bar{q}q$ acquires a non vanishing expectation value in the true vacuum as in Eq.(8), while τ_i associated with $\bar{q}(\sigma_i \otimes \sigma_i)q$ does not. The fermion loop contribution to the σ propagator contains a Lna divergence, absorbed in a wave-function renormalization of the σ field, so that the renormalized coupling constant is $g_R^\sigma = g\sqrt{z_\sigma}$, implying $g \to g^* = 0$ in the continuum limit at fixed fermion mass. No such renormalization occurs for τ_1, τ_2, so that $g_R^{\tau_1}$, $g_R^{\tau_2} = g \to 0$. It turns out that $g_R^{\tau_3} \to 0$ even faster. Hence in the renormalized theory, at least at this order, the symmetry breaking terms are (logarithmically) suppressed at large scales. Jolicoeur gives another example of a similar situation in a continuum 2-d model which breaks Lorentz invariance. These findings constitute nice illustrations of a phenomenon (the quantum theory may be more symmetric than the initial classical theory) which led[14] to the conjecture that the observed symmetry of our world at low energy emerges from a large class of less symmetric dynamics at short distances.

4. FLAVOUR SYMMETRY RESTORATION FOR A TWO-POINT FUNCTION IN QCD

Here we give a general proof of flavour symmetry restoration in QCD for a particular set of gauge invariant quark two-point correlation functions, *defined* as follows (for any dimension D):

$$\Delta_q^{(L)}(y) = \sum_{\eta, \zeta} \Gamma_\eta^{\alpha a} \Gamma_\zeta^{\dagger b \beta} \langle \chi^i(2y+\eta) \, V_{2y+\eta, \zeta}^{ij}(L) \, \bar{\chi}^j(\zeta) \rangle_s . \quad (12)$$

Here the action S contains a gauged fermionic action $S_{ks}(\chi,\bar{\chi},U)$, obtained by inserting in Eq.(1) the relevant link variables $U_{x,x+\mu}$ and $U_{x+\mu,x} = U^\dagger_{x,x+\mu}$ and adding a mass term $m\,\bar{\chi}_i(x)\,\chi_i(x)$, and a pure gauge action S_G. The i,j are the corresponding colour indices. The matrix $V_{x,x'}(L)$ is the average over a set (L) of paths ℓ going from x to x' of the product along ℓ of the U links. By (L) we denote any set such that under a lattice symmetry transformation T where $x \to x^T$, $x' \to x'^T$, then $V_{x,x'}(L) \to V_{x^T,x'^T}(L)$

when $U_{x,x+\hat{\mu}} \to U_{x^T,x+\hat{\mu}^T}$. With this choice, the quantity $\langle \chi\, V\, \bar{\chi}\rangle_S$ in Eq.(12) not only is gauge invariant, but also has the same properties as the free propagator $\langle \chi\, \bar{\chi}\rangle_{S_{free}}$ under all lattice symmetry transformations. Note that we do not rely any more on an explicit quark definition, directly constructing a correlation function by analogy with the free case explicit construction. In the detailed analysis of Ref.[5], we derive from the definition (12) a general expansion of $\Delta_q^{(L)}$ in momentum space,

$$\Delta_q^{(L)}(p) = \sum_{A,B} C_{AB}(p) \left(\Gamma_A^{\alpha\beta} \otimes \Gamma_B^{\star ab}\right) , \qquad (13)$$

where the C_{AB}'s, whose expression results from a Fierz transformation of $\Gamma_\eta \Gamma_\zeta^\dagger$, are strongly constrained by the symmetry properties of the lattice action. Our main result can be finally stated in the following way. After applying to $\Delta_q^{(L)}(p)$ the transformation w(p) defined in the free case by Eq.(5), namely *defining*

$$\Delta_\psi^{(L)}(p) \equiv w(p)\, \Delta_q^{(L)}(p)\, w^\dagger(p) , \qquad (14)$$

we show that $\Delta_\psi^{(L)}(p)$ has the remarkable decomposition

$$\Delta_\psi^{(L)}(p) = \sum_{\rho=1}^{2^D} i^{n(\rho)}\, F_\rho(p)\, (\Gamma_\rho \otimes 1) \prod_{\alpha=1}^{D} [i\sin(p_\alpha/2)]^{\rho_\alpha} . \qquad (15)$$

The D-vector ρ has components 0 or 1. The integer $n(\rho)$ is $\sum_\mu \rho_\mu(\rho_1 + \ldots + \rho_{\mu-1})$. The amplitudes $F_\rho(p)$ are real, even under

$p_\mu \to -p_\mu$, and such that the labelling of the D axis is irrelevant for the whole expression. Hence flavour is *exactly conserved* for the form $\Delta_\psi^{(L)}(p)$, and *restored in the continuum limit* for the form $\Delta_q^{(L)}(p)$ $(w(p) \to 1 \otimes 1)$. Note that the result (15) with the properties listed below it fulfills all the requirements for C, P and T invariance in the continuum limit. Finally, a study of Lorentz invariance restoration would require that of the behaviour of $F_\rho(p)$ at low p values, a problem not connected only to symmetry considerations.

5. FLAVOUR SYMMETRY RESTORATION FROM NUMERICAL STUDIES

From a numerical study of quenched SU(2) QCD with staggered fermions, Lacaze [7] has presented evidence for the existence, in the continuum limit, of degenerate mesonic multiplets associated with a U(4) flavour symmetry. (In the simulation, the singlet channel is not distinguished from the other ones: no contribution from pure glue intermediate states). An example of particular interest is given by the π case. On the lattice, only one axial generator $T_A^5 = \gamma_5 \otimes \gamma_5^*$ survives as a symmetry generator. Correspondingly, a π_0 Goldstone boson, with a vanishing mass in the chiral limit, is actually observed in Monte-Carlo simulations (see the Ref. list of [7]). The other π-like states can be excited by operators associated with axial generators $T_A^\Gamma = \gamma_5$ (or $\gamma_4\gamma_5$) $\otimes \Gamma^*$, where Γ is any U(4) generator, $\neq \gamma_5$. These are not symmetry generators, so that the corresponding masses $(am_\pi\Gamma)$ should not vanish with the quark mass m. A systematic study of all these states has been carried out by Lacaze for $\beta(4/g^2)$ = 2.3, 2.4, 2.5 at m values .3, .15, .1, .05, .025 and .0125, on a lattice of size $12^3 \times 24$. He succeeded in finding a signal/noise ratio large enough at β = 2.4 and 2.5 to extract the masses associated with *all* the $\gamma_5 \otimes \Gamma^*$ operators. The striking feature which emerges from these numbers is that the 15 non-Goldstone pions are already nearly degenerate at β = 2.4, and that they all come closer and closer to the π_0 as β increases. At the quantitative level, Lacaze compares $m_{\pi'}^2$ to m_π^2 as a function of m and β (Fig.1), π' being associated with $\Gamma = \gamma_5\gamma_i$. At moderately large m values the ratio is $m_{\pi'}^2/m_{\pi_0}^2$ is flat and decreases appreciably with increasing β. As $m \to 0$, $m_{\pi_0}^2$ behaves as sm whereas $m_{\pi'}^2$ goes to a constant, so that the ratio increases with decreasing m. But this effect, a manifestation of the

254

explicit breaking of T_A^Γ, actually becomes weaker at larger β. Similar results on the ρ-channel sixteenplet are also obtained. The overall conclusion of a careful analysis of the β-dependence is that approximate flavour symmetry and approximate scaling behaviour are simultaneously reached at β ⪢ 2.5.

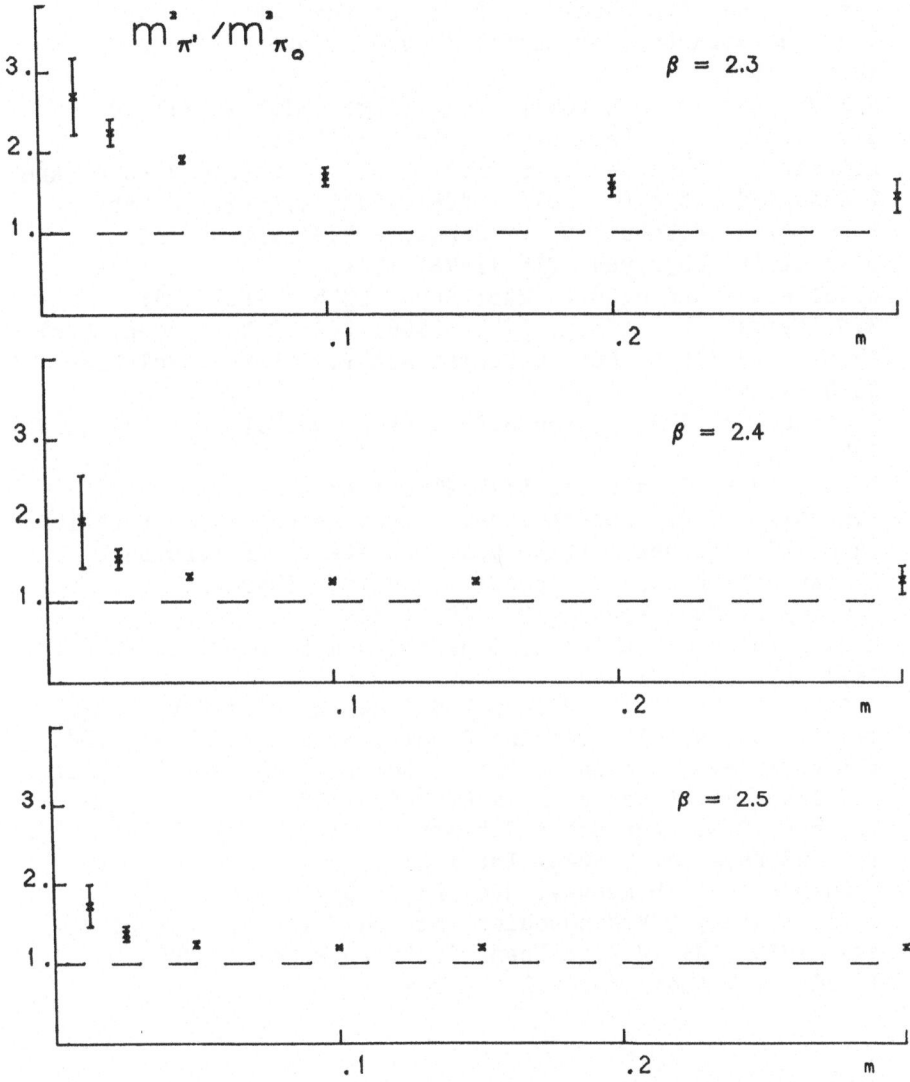

Fig.1 - The ratio $m_{\pi'}^2/m_{\pi_0}^2$ showing the approach to degeneracy as β increases[7].

It is a pleasure to thank the organizers of this Symposium for giving me the opportunity of this talk, and for the remarkable and fruitful conditions provided to the participants. I am grateful to D.Bunel for typing these notes.

REFERENCES

1. C.B.Lang and C.Rebbi, Phys.Lett. 115B (1982) 137; for recent results, see C.Michael, in Proc. of the Lattice Gauge Theories Workshop, Wuppertal (Nov.5-7,1985), Plenum Press (N.Y., 1986).
2. H.B.Nielsen and N.Ninomiya, Nucl.Phys. B185 (1981) 20; B193 (1981) 173; Phys.Lett. 105B (1981) 219.
3. L.Susskind, Phys.Rev. D16 (1977) 3031; T.Banks, J.Kogut and L.Susskind, Phys.Rev. D13 (1976) 1043; T.Banks, S.Raby, L.Susskind, J.Kogut, D.R.T.Jones, P.N.Scharbach and D.Sinclair, Phys.Rev. D15 (1976) 1111.
4. M.Golterman and J.Smit, Nucl.Phys. B255 (1985) 328; M.Golterman, Nucl.Phys. B273 (1986) 663; D.Verstegen, Nucl. Phys. B249 (1985) 685; G.Parisi and Y.C.Zhang, Nucl.Phys. B230 (1984) 97.
5. T.Jolicoeur, A.Morel and B.Petersson, Nucl.Phys. B274 (1986) 225.
6. T.Jolicoeur, Phys.Lett. B171 (1986) 431.
7. R.Lacaze, "Meson spectrum and flavour symmetry restoration in SU(2) quenched lattice QCD with staggered fermions", Saclay PhT/86-093; to appear in Nuclear Physics.
8. F.Gliozzi, Nucl.Phys. B204 (1982) 419.
9. H.Kluberg-Stern, A.Morel, O.Napoly and B.Petersson, Nucl. Phys. B220 (1983) 447.
10. H.S.Sharatchandra, H.J.Thun and P.Weisz, Nucl.Phys. B192 (1981) 205; M.Golterman and J.Smit, Nucl.Phys. B245 (1984) 61; C.Van den Doel and J.Smit, Nucl.Phys. B228 (1983) 122.
11. D.J.Gross and A.Neveu, Phys.Rev. D10 (1974) 3935.
12. Y.Cohen, S.Elitzur and E.Rabinovici, Nucl.Phys. B220 (1983) 102 and Phys.Lett. 104B (1981) 289.
13. F.Guérin and R.D.Kenway, Nucl.Phys. B176 (1980) 168.
14. J.Iliopoulos, D.V.Nanopoulos and T.N.Tomaras, Phys.Lett. B94 (1980) 141; H.B.Nielsen, Particle Physics 1980, I.Andrié, I.Dadié and N.Zovko Eds.

TOWARDS THE LIMIT OF THE QUENCHED APPROXIMATION
IN HADRON MASS CALCULATIONS

K.-H. Mütter

Physics Department, University of Wuppertal
D-5600 Wuppertal, W. Germany

in collaboration with:
Ph. De Forcrand, Cray Research, Chippewa Falls, WI
K. Schilling and R. Somer, Univ. of Wuppertal

ABSTRACT

Results are presented for a computation of hadron propagators on a $24^3 \times 48$ lattice, twice blocked, at $\beta = 6.3$. Lowering the quark mass the propagators start to fluctuate more and more. It is argued that this phenomenon is a characteristic feature of the quenched approximation and should be cured in the presence of light dynamical fermions.

INTRODUCTION

During the past two years, lattice QCD computations of hadron masses have been extended to quite large lattices within the quenched approximation [1-4]. These computations have been done at rather large quark masses, though, for an obvious reason: as one lowers the quark mass, the hadron propagators start fluctuating more and more. This prohibits a precise determination of hadron masses from these propagators.

In this paper, we intend to investigate this instability problem in the quenched approximation. For this purpose, we have computed a fairly large sample of hadron propagators on a $24^3 \times 48$ lattice at $\beta = 6.3$ and small quark masses (cf. section 3), which we subsequently analyzed in more detail than usually done (cf. section 4). Thus, by looking at the distribution of hadron propagators from 28 different background configurations, we observed for the lowest quark mass $ma = .01$, fluctuations which are quite different from normal Gaussian behaviour. This phenomenon can clearly be attributed to a 10% subsample of the configurations, which we call exceptional.

In section 4 of this paper we will argue that the appearance of these exceptional configurations is a characteristic feature of the quenched approximation. In the unquenched case, they should be suppressed by the fermion determinant. This motivates us to "clean" the sample, i.e. to compute hadron propagators just on the remaining " normal" configurations. This manipulation has two effects, which are both very welcome:

- the huge fluctuations of the hadron propagators (and of the hadron masses derived therefrom) are removed,

- the hadron masses obtained on the "cleaned" sample are in nice agreement with experimental data.

HADRON PROPAGATORS ON A $24^3 \times 48$ LATTICE AT $\beta = 6.3$

We used 28 gluonic background fields, computed by one of us (Ph.dF) in the quenched approximation at $\beta = 6.3$ on a $24^3 \times 48$ lattice. The thermalization of this large lattice was achieved as follows: first fields were produced on a $24^3 \times 6$ lattice. The final configuration was copied eight times in time direction. 510 subsequent sweeps were performed over the resulting big lattice. Finally, 30 configurations were produced, which are separated from each other by 100 sweeps. The autocorrelation time was determined from large Wilson loops and turned out to be of the order of 30 sweeps.

The computation of the quark propagators was done with the block diagonalization method [5], using two blocking steps. As explained in refs. [5], this method reduces the rank of the big fermion matrix in each blocking step by a factor 16. The resulting effective fermion matrix, however, contains more complicated couplings than the original Wilson action. Therefore, the net gain in storage is a factor 136 (not 256).

The conjugate gradient algorithm for the inversion of the blocked fermion matrix is faster than for the original problem. The gain in CPU time comes from two sources:

- a factor 12 per iteration step from the reduction in rank,

- an additional factor 2.5 in convergence rate, since the largest eigenvalue in the blocked fermion matrix is cut down by roughly a factor 5.

So far, we have computed 84 quark propagators on 28 background fields, each with 3 source points. The following values for the hopping parameter were used:

$$\kappa = .13,\ .1308,\ .1315,\ .13195,\ .13227,\ .1325$$

The critical value of κ is found to be $\kappa_C = .13285$.

Using the well-known relation between κ and the quark mass,

$$ma = \frac{1}{2}\left(\kappa^{-1} - \kappa_C^{-1}\right) \tag{1}$$

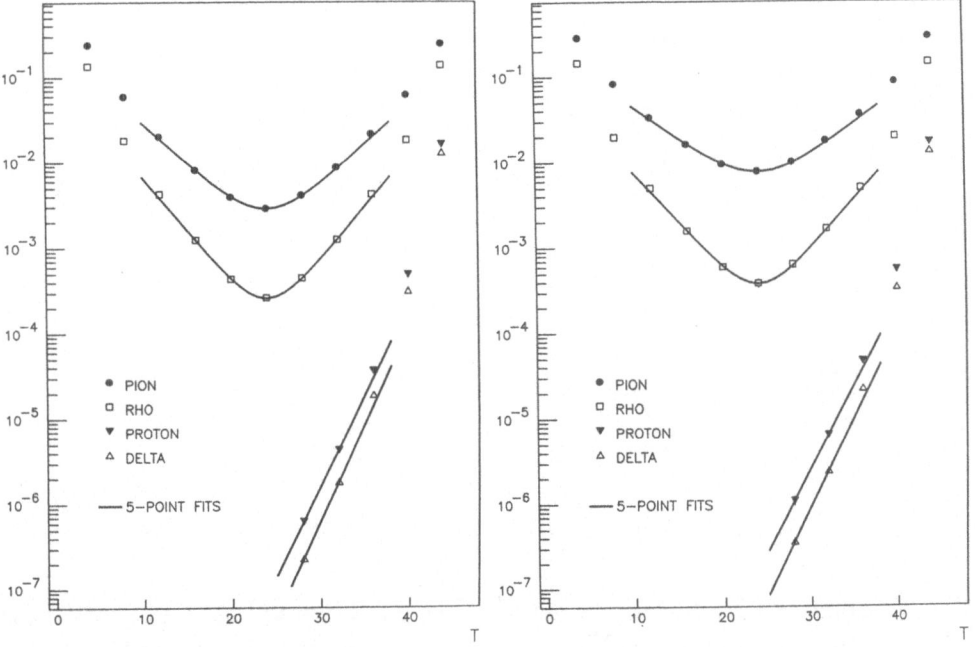

Fig. 1: Hadron propagators as functions of the time separation at quark mass $ma = .035$ ($\kappa = .1315$). The solid curves are fits to the data points at large time separations.

Fig. 2: Same as Fig. 1 at quark mass $ma = .027$ ($\kappa = .1319$).

The critical value of κ is found to be $\kappa_C = .13285$.

Using the well-known relation between κ and the quark mass,

$$ma = \frac{1}{2}\left(\kappa^{-1} - \kappa_C^{-1}\right) \qquad (1)$$

these κ-values can be translated into quark mass-values

$$ma = .083, \ .059, \ .039, \ .027, \ .017, \ .01.$$

Some preliminary results (for the larger quark masses) of this analysis were presented previously at the Wuppertal Workshop [2]. Here we show the resulting hadron propagators for the smaller quark masses (see Figs. 1-4).

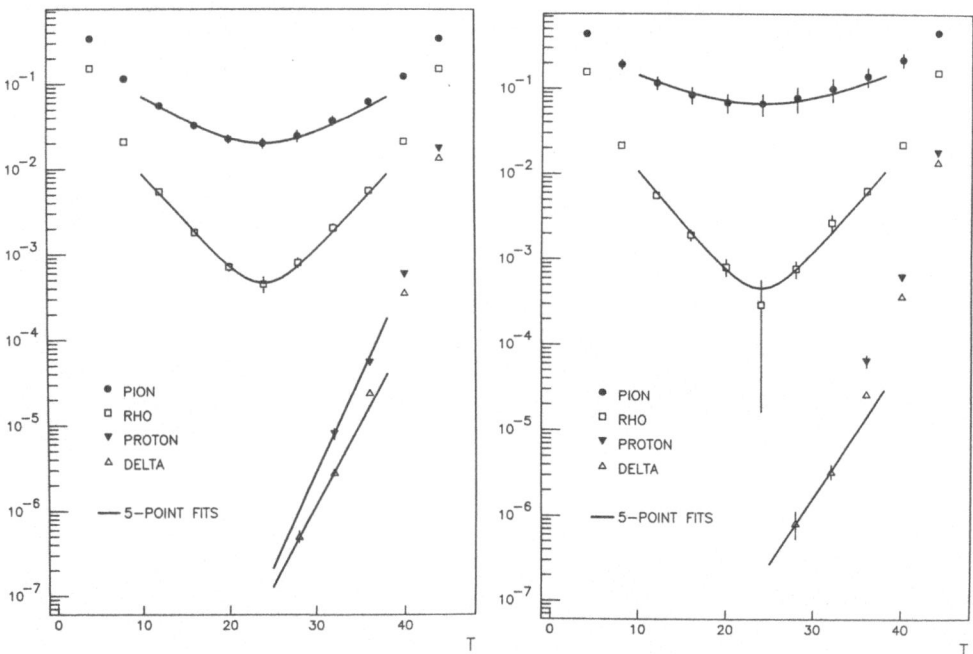

Fig. 3: Same as Fig. 1 at quark mass $ma = .017\,(\kappa = .13227)$.

Fig. 4: Same as Fig. 1 at quark mass $ma = .01\,(\kappa = .1325)$.

Looking at these plots, the reader will note the increase in errors for all hadron propagators with decreasing quark masses. Note in particular, that for the smallest quark mass, $ma = .01$, the signal of the nucleon propagator for the innermost time slices, $16 < t < 32$, is lost. We extracted hadron masses from the remaining data at large time separations, using pure cosh-fits for mesons and pure exponential ones for the baryons. The results are plotted in Fig. 5.

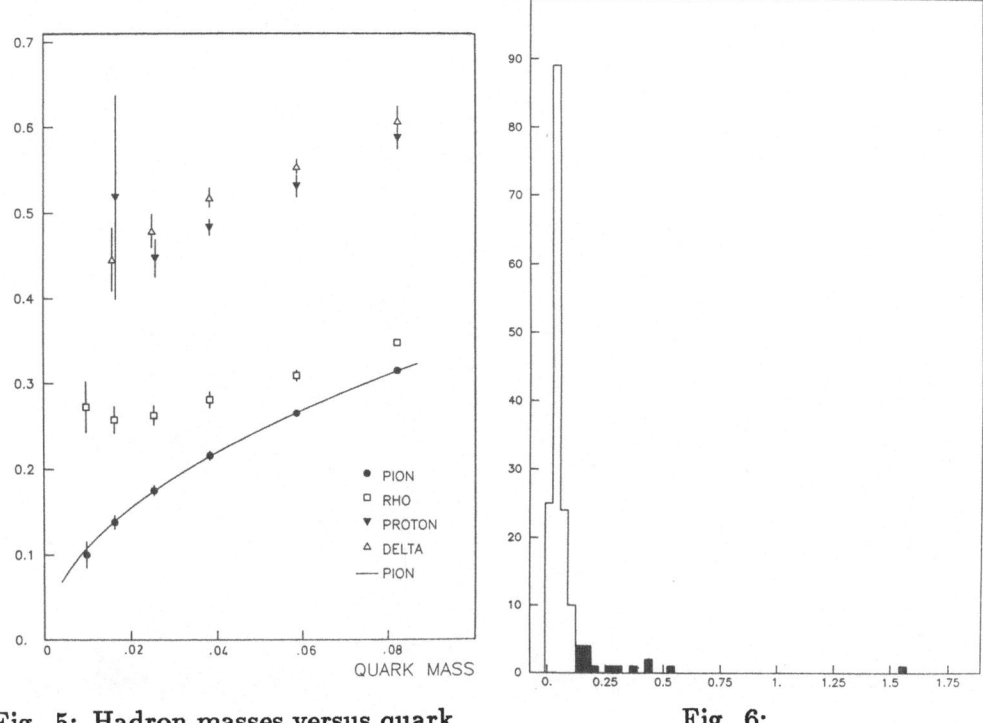

Fig. 5: Hadron masses versus quark
masses.

Fig. 6:

Fig. 6: The distribution of the measured values of the pion propagator at
the symmetric time slices $t = 20$, 28 and quark mass $m = .01$. The
histogram has 168 entries originating from 28 configurations each with
3 source points and two time slices. The largest deviations from the
mean can be traced back to three exceptional configurations. Their
entries are marked in black.

FLUCTUATIONS OF HADRON PROPAGATORS WITH THE GLUONIC FIELD

We remember, from our course in basic statistics, that the standard
procedure for the determination of statistical errors is applicable, if the central limit theorem applies and the distribution of the observables has a Gaussian form. We have tested this condition for the hadron propagators at the
smallest quark mass and for the symmetric time slices $t = 20$, 28. The data
at these large time separations are most important in the hadron mass determinations. In Figs. 6-9, we show the distributions for the various hadron
propagators, as they are measured on their time slices $t = 20$ and 28. Each
histogram shows 168 entries, due to three source points and two time slices
from 28 configurations. Glancing through these figures, it is striking that

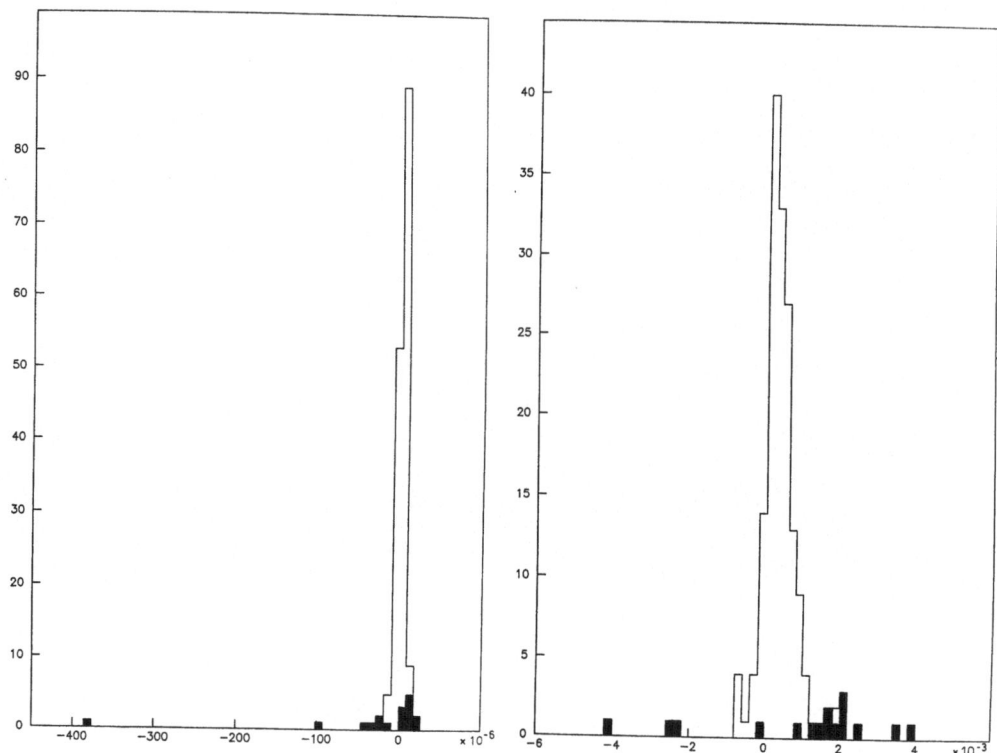

Fig. 7: Same as Fig. 6 for the nucleon propagator.

Fig. 8: Same as Fig. 6 for the rho propagator.

the mosts drastic fluctuations from the mean can be exclusively traced back to three configurations, whose entries are marked in black and which we call "exceptional". This non-Gaussian behaviour obviously forbids a naive error analysis of our data! But what is behind the exceptional configurations?

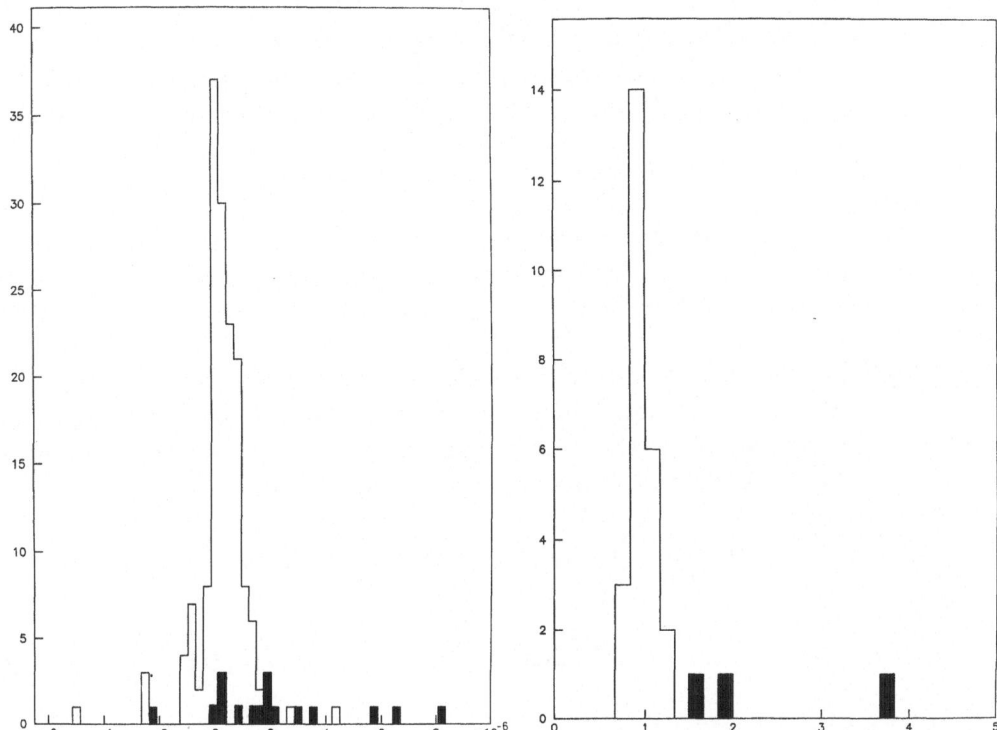

Fig. 9: Same as Fig. 6 for the delta propagator.

Fig. 10: The distribution of the measured values of the pion propagator summed over all time slices and three source points.

THE DISEASE OF THE QUENCHED APPROXIMATION AT SMALL QUARK MASSES AND ITS REMEDY

We start from the following hypothesis:

On the exceptional configurations the fermion matrix has a particularly rich spectrum of very low eigenvalues. On the basis of this hypothesis, we can draw two conclusions:

1. Exceptional configurations are very rare in an unquenched sample of background fields. They simply have a small probability to show up, due to the fermion determinant in the Boltzmann weight. Within a quenched sample, on the other hand, there is no particular reason for exceptional configurations to be suppressed.

2. The quark propagators on exceptional configurations are much larger than on "normal" ones. As a consequence, within a "mixed" sample

263

of "normal" and exceptional configurations, one will encounter wild fluctuations. This happens precisely in the quenched case, while the determinant in the unquenched situation will damp these fluctuations.

There are several ways to test our hypothesis:

a) The most straightforward and expensive method would of course be the computation of the eigenvalue spectrum of the fermion matrix in each configuration. This will be done in the future.

b) One can look for quantities which react very sensitively to the fluctuations of the small eigenmodes with the gluonic field. Such an observable is e.g. the pion propagator on a field configuration, after summation over all source points and time slices:

$$f(U) = \sum_{t,y} G_\pi(t,y) = \sum_{x,y} \Delta^{-1}(x,y)\, \Delta^{-1\dagger}(y,x)$$

$$= \mathrm{TR}\left(\Delta\Delta^\dagger\right)^{-1} = \int_0^{\lambda_0} d\lambda\, \lambda^{-1}\rho(\lambda, U) \tag{2}$$

which can be related via a trace relation to the eigenvalue distribution $\rho(\lambda, U)$ of the positive definite matrix $\Delta\Delta^\dagger$, where Δ is the fermion matrix. Only part of the sum in eq. (3) is at our disposal, since we have made computations only on three source points. But there are no cancellations in the sum since the pion propagator is positive definite. In Fig. 10, you see for our smallest quark mass the distribution of the pion propagator, summed over the time slices and three source points. In black, we have marked again the three exceptional configurations, which are clearly separated from the normal ones.

c) The appearance of small eigenvalues in the fermion matrix affects the convergence of the conjugate gradient algorithm as used for the computation of the quark propagators. Indeed, we found that on the exceptional configurations with the smallest quark mass 60% more iterations were needed to reach the required accuracy (set to a value of 10^{-7} in the rest vector).

As stated above, exceptional configurations are expected to be suppressed in an unquenched calculation. Therefore, it might be interesting to see the effect of "cleaning" the quenched sample by hand: we plot the results computed from the 25 normal configurations. This manipulation has no effect on the results with quark masses $ma > .017$. Changes are observed, however, in the pion and nucleon propagators at the smallest quark masses

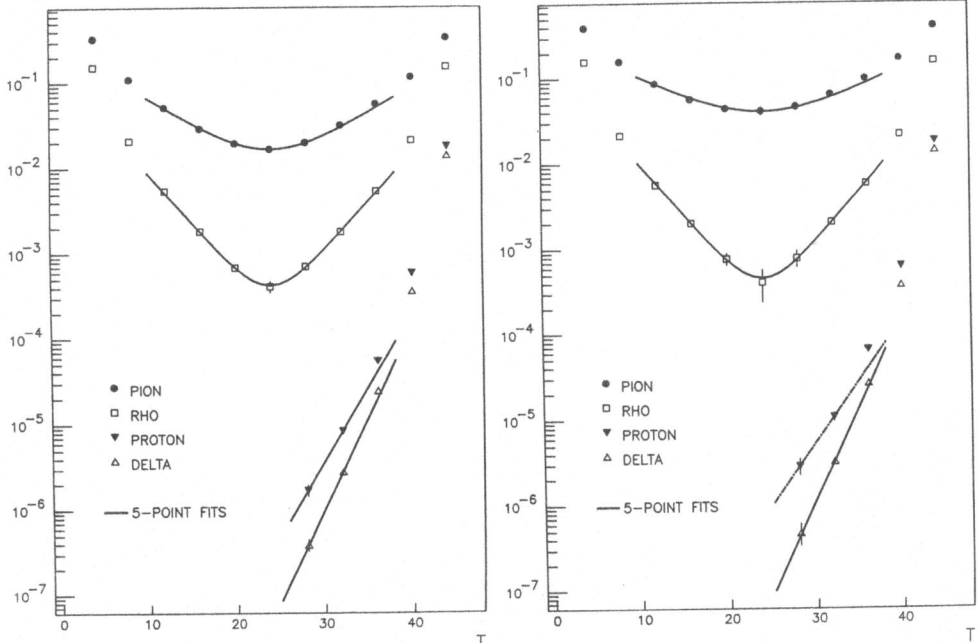

Fig. 11: Hadron propagators at quark mass $ma = .017$ computed on the cleaned sample of 25 normal configurations.

Fig. 12: Same as Fig. 11 at quark mass $ma = .01$.

$m \cdot a = .01, .017$. Compare the resulting Figs. 11 and 12 with the previous Figs. 3,4, respectively and note the following features:

- The "statistical" errors reduce, of course.

- The nucleon propagator becomes "visible" far off the source and appears to flatten,

- the pion propagator becomes slightly steeper, which amounts to a tiny shift in κ_C.

We again extracted the hadron masses from the innermost t-region, $16 \lesssim t \lesssim 32$, with the results given in Fig. 13.

With the reduced errors on the "cleaned" sample, there appears a striking new trend in the nucleon mass for the small quark mass region. For it comes out that the nucleon mass as a function of the quark mass starts to decrease significantly stronger, once the quark mass drops below $a \cdot m < .04$. It is also remarkable, that the mass differences rho-pion and delta-nucleon increase, while the rho-nucleon mass splitting decreases, as one approaches the chiral limit.

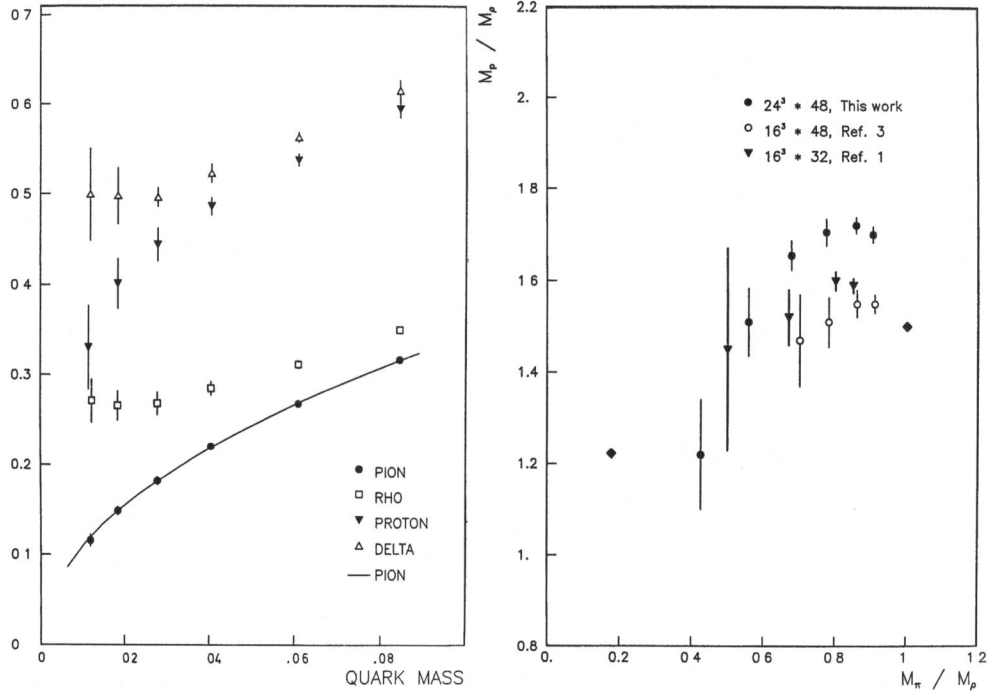

Fig. 13: Hadron masses versus quark mass computed on the cleaned sample.

Fig. 14: Mass ratio plot.

All these features bring about a substantial shift of the hadron mass ratios into the right direction, i.e. towards the experimental values. The essence of our results in comparison to previous attempts to estimate hadron masses can best be presented in the form of the "Edinburgh mass ratio plot", as shown in Fig. 14. This figure demonstrates very nicely, how close or far away we are from the physics of real hadrons! Finally, let us determine the scale $a\,(\beta = 6.3)$. We extracted it from the rho mass, which appears to have the smoothest chiral limit. We thus find:

$$a\,(\beta = 6.3) = .075\,(8)\,\mathrm{fm}$$

This is in nice agreement with the lattice spacing obtained from the string tension [6].

266

SUMMARY AND CONCLUSIONS

It remains that estsimates obtained from the "nonchiral" regime, with $m_\pi/m_\rho > .7$, appear to miss important features of QCD dynamics. Pushing lattice calculations towards the chiral limit is beyond any doubt very painful. In the quenched approximation, one encounters large fluctuations and instabilities, which - as we have argued - should be eased by the effects of dynamical fermions. This means, however, that we will have to go through the pain of hadron mass calculations with dynamical fermions at really small quark masses. In this study, we have undertaken an attempt to mimic the fluctuation damping mechanism of light dynamical fermions in a concededly crude fashion. The results look promising in the sense that we arrive at hadron masses in a ballpark which cannot be reached otherwise.

ACKNOWLEDGMENT

Work supported by Deutsche Forschungsgemeinschaft. The calculations were performed on the Cyber 205 in Karlsruhe. It is a pleasure to thank the staff of the Karlsruhe Computer Center for their support.

REFERENCES

1. D. Barkai, K.J.M. Moriarty and C. Rebbi, Phys. Lett. 156B (1985) 385.
2. Ph. deForcrand, A. König, K.-H. Mütter and K. Schilling, "Hadron Mass Calculation on a $24^3 \times 48$ Lattice", in *Proceedings of the Wuppertal Workshop 1985*, B. Bunk, K.-H. Mütter and K. Schilling (Eds.), NATO ASI Series B Vol. 140, Plenum Press, New York 1986, p. 189.
3. S. Itoh, Y. Iwasaki and T. Yoshie, "Hadron Masses in Quenched QCD", UTHEP-155, May 1986.
4. K.C. Bowler, C.B. Chalmers, R.D. Kenway, G.S. Pawley, D. Roweth, "Hadron Mass Calculations using Susskind Fermions at $\beta = 5.7$ and 6.0", Edinburgh preprint 1986.
5. K.-H. Mütter and K. Schilling, Nucl. Phys. B230[FS10](1984) 275;
   A. König, K.-H. Mütter and K. Schilling, Nucl. Phys. B259 (1985) 33.
6. K.C. Bowler et al., Phys. Lett. 163B (1985)367.

DEBYE SCREENING AND NONPERTURBATIVITY IN HOT QCD

Sudhir Nadkarni

Department of Physics and Astronomy
Rutgers University
P.O. Box 849
Piscataway, New Jersey 08854

ABSTRACT

The Debye screening effect for non-Abelian theories is investigated gauge-invariantly through static interquark potentials. Attempts to compute these potentials in the weak-coupling expansion reveal that perturbation theory is inapplicable even at the very high temperatures where it is naively expected to apply. The results suggest that the infrared divergences responsible for the perturbative breakdown might be severe enough to cause a condensation of the electrostatic potential. Whether such a condensation, which would invalidate even low-order perturbative results, does in fact occur can only be determined non-perturbatively. To this end, a lattice computation is proposed.

I. INTRODUCTION

In this talk I would like to suggest the possibility that nonperturbative methods of calculation (which usually means lattice computations) might be essential even in what has so far been regarded as a weak-coupling regime. I will discuss some perturbative calculations of the non-Abelian Debye screening mass [1,2], which raise serious questions about the applicability of perturbation theory at high temperatures. These results imply, at least to me, that one should take perturbative results in finite-temperature non-Abelian gauge theory with several large grains of salt, or, at the very least, approach any future calculations of this kind with considerable caution.

Much of the early enthusiasm for studying the properties of quantum chromodynamics at finite temperature and density ("hot QCD", for reviews see [3,4]) stemmed from the following fact: the QCD coupling constant is small at temperatures/densities well above the deconfinement/chiral symmetry restoration phase transition, making perturbative analysis possible. However, there are severe infrared divergences in hot QCD which, as was soon discovered, must cause perturbation theory to eventually break down ("the infrared problem" [5]). Thus, the $O(g^6)$ term in the thermodynamic potential receives contributions from an infinite set of Feynman graphs and so cannot be computed perturbatively; the same holds for the magnetic screening mass, the square of which starts out at $O(g^4)$. One ought, nevertheless, to be able to compute certain quantities at sufficiently low orders. One such quantity, which we shall consider here, is the Debye screening mass.

Debye screening [6] refers to the long-distance shielding of electric charge by plasma excitations, which convert the $1/R$ Coulomb potential into the $\exp(-mR)/R$ Yukawa-type potential. For an Abelian theory in its finite-temperature Euclidean formulation, the Debye screening mass m is obtained as the static infrared limit of the time-time component of the photon vacuum polarization tensor, $\Pi_{44}(k_4 = 0, \mathbf{k} \to 0)$, which is gauge invariant. The naive method of computing the non-Abelian Debye screening mass is to mimic the procedure followed for Abelian plasmas and take the same limit of the gluon vacuum polarization. At first this seems to work: at the one-loop level $\Pi_{44}^{ab}(k_4 = 0, \mathbf{k} \to 0)$ turns out to be gauge invariant and yields the commonly-accepted value for the leading-order Debye mass [3]. But this convenient situation is short-lived [7,8]: beyond the one-loop level, the gauge dependence of the non-Abelian vacuum polarization tensor manifests itself even in the static infrared limit. Thus, the naive procedure for extracting m breaks down, and one has to turn to something more refined. Since the Debye screening mass is a physical quantity which modifies the interaction between colour sources, its definition is properly given in terms of the underlying gauge-invariant interquark potentials.

In Section II we compute the potential for a heavy quark-antiquark pair with no colour correlation (each source being separately averaged over colour space), while in Section III we do the same for sources in a colour singlet state. Our results point out the failure of perturbation theory, and suggest that it may be due to a vacuum condensation of the electrostatic potential. Our conclusions are presented in Section IV. A lattice model which can be used to determine whether or not there is an electrostatic condensate is outlined in Section V.

II. THE COLOUR-AVERAGED POTENTIAL

A static quark at location $\mathbf{x}$ is represented in hot QCD by the Polyakov loop operator

$$\Omega(\mathbf{x}) \equiv P \exp\{-ig \int_0^\beta d\tau A_4(\mathbf{x}, \tau)\} \qquad (\beta \equiv 1/T).$$

The connected correlation function of Polyakov loops is then defined by

$$C_{\mathrm{PL}}(R) \equiv \langle \tilde{\mathrm{Tr}}\Omega^\dagger(\mathbf{R}) \tilde{\mathrm{Tr}}\Omega(\mathbf{0}) \rangle_c \qquad (\tilde{\mathrm{Tr}} \equiv \mathrm{Tr}/\mathrm{Tr}1),$$

and is readily evaluated at leading order to be [we use the gauge group $SU(N)$]

$$C_{\mathrm{PL}}(R) = 1 + \frac{g^4 \beta^2}{4} \frac{N^2 - 1}{2N^2} \left(\frac{e^{-m_E R}}{4\pi R} \right)^2.$$

Here m_E is defined to be the one-loop mass obtained from the static infrared limit of Π_{44}; for a pure $SU(N)$ theory with no quarks it is given by the well-known expression $m_E^2 = N g^2 T^2/3$.

A $q\bar{q}$ pair interacts via the singlet and adjoint channels and C_{PL} represents a colour average of the corresponding potentials,

$$C_{\mathrm{PL}}(R) = \frac{1}{N^2} e^{-\beta V_1(R)} + \frac{N^2 - 1}{N^2} e^{-\beta V_{\mathrm{adj}}(R)}.$$

Expanding the exponentials in the above expression to $O(\beta^2)$ and comparing the coefficients of β, β^2 with the leading order result, we find the leading expressions for the singlet and adjoint potentials:

$$V_1^{(0)} = -(N^2 - 1) V_{\mathrm{adj}}^{(0)} = -g^2 \frac{N^2 - 1}{2N} \frac{e^{-m_E R}}{4\pi R}.$$

This is precisely the Debye screened version of the usual $T = 0$ static singlet potential. The Polyakov loop correlation therefore gives satisfactory results to leading order.

At sub-leading orders, however, perturbation theory breaks down. To see this, let us define the correlation mass m and the "remainder" $f(R)$ through the general expression

$$C_{\mathrm{PL}}(R) = 1 + \frac{g^4 \beta^2}{4} \frac{N^2 - 1}{2N^2} \left(\frac{e^{-m_E R}}{4\pi R} \right)^2 + [\text{higher orders}]$$

$$\equiv 1 + \frac{g^4 \beta^2}{4} \frac{N^2 - 1}{2N^2} \left(\frac{e^{-m R}}{4\pi R} \right)^2 [1 + f(R)].$$

We determine m self-consistently in perturbation theory by writing

$$m^2 = m_E^2 + \Delta m^2$$

and choosing Δm^2 so that $f(R) \ll 1$ for $R \gg m^{-1}$.

At next-to-leading order we find for $f(R)$ (see [1] for details):

$$f(R) = \frac{R}{m}\left[\Delta m^2 + \frac{Ng^2Tm}{4\pi}(3 - \gamma - 2\ln 2)\right]$$
$$+ \frac{Ng^2T}{4\pi m}\left[-mR\ln mR + \frac{3\ln mR}{2} + \frac{6\gamma + 4\ln 2 - 1}{4} + O\left(\frac{\ln mR}{mR}\right)\right].$$

The noncancellation of the logarithmic and constant terms indicates a breakdown of perturbation theory and prevents a determination of Δm^2.

If we examine C_{PL} at higher orders, we discover a second type of problem. At the fourth order, for example, we find graphs of the following type: Each of the Polyakov loops emits and reabsorbs a timelike gluon; before being reabsorbed, these timelike gluons exchange two spacelike (magnetic) gluons. The magnetostatic mass, which at $O(g^2T)$ is smaller than the $O(gT)$ electrostatic mass at high temperatures, therefore governs the decay of the correlation at large separations. Consequently, $C_{\mathrm{PL}}(R)$ is in fact dominated by a mass gap originating in the magnetic sector and is therefore unsuitable for defining the Debye screening mass.

Thus the Polyakov loop correlation, which gives satisfactory results at leading order, reveals the following two problems at higher orders: (i) Perturbation theory breaks down at distances larger than the Debye screening length. (ii) Colour averaging washes out the Debye screening effect, since V_1 and V_{adj} have opposite signs, so that some sort of magnetic screening mass becomes dominant. The moral of the story is: avoid colour averaging if possible, or, equivalently, come up with a correlation function dominated by single-gluon exchange.

III. THE SINGLET POTENTIAL

At $T = 0$, the only gauge-invariant interquark potential available is V_1, which is obtained by computing the usual rectangular Wilson loop. At $T \neq 0$, the extra degrees of freedom created by the periodic boundary conditions enable us to define V_1 and V_{adj} separately.

Write

$$\Omega(\mathbf{x}) = \omega(\mathbf{x})\hat{\Omega}(\mathbf{x})\omega^\dagger(\mathbf{x}),$$

where ω is gauge-dependent and all gauge-invariant information is contained in $\hat{\Omega}$, the diagonal matrix of gauge-invariant eigenvalues of Ω.* Since $\text{Tr}\Omega = \text{Tr}\hat{\Omega}$, the n-point functions of $\text{Tr}\Omega$ are a subset of those of $\hat{\Omega}$. Applying $SU(N)$ projection operators to the correlation function of $\hat{\Omega}$, it can be shown that [2]

$$\langle \tilde{\text{T}}\text{r}\Omega^\dagger(\mathbf{R})\tilde{\text{T}}\text{r}\Omega(\mathbf{0})\rangle = \frac{1}{N^2}e^{-\beta V_1(R)} + \frac{N^2-1}{N^2}e^{-\beta V_{\text{adj}}(R)},$$

$$C_{\text{PLE}}(R) \equiv \langle \tilde{\text{T}}\text{r}\hat{\Omega}^\dagger(\mathbf{R})\hat{\Omega}(\mathbf{0})\rangle = e^{-\beta V_1(R)}.$$

The first equation is well known and was encountered in Section II; the second one is a new result and codifies the additional information present in $\hat{\Omega}$.

We shall now attempt a perturbative calculation of C_{PLE}, for which purpose we shall view the Polyakov loop as (the exponential of) an adjoint Higgs field. We diagonalize the Higgs by transforming to a gauge similar to the unitarity gauge familiar from the electroweak model [9]. In order to "hold down" the fields in this diagonal gauge, we introduce by hand a vacuum expectation value (VEV) for the Higgs field. (Without such a VEV, the fields would fluctuate right back out of diagonal gauge.) Through the Higgs mechanism, the VEV provides masses for the off-diagonal spacelike gluons. The magnitude of the VEV is to be determined self- consistently, order by order in perturbation theory, by setting tadpole graphs to zero.

We shall perform the calculation for $SU(2)$; the results for $SU(N)$ are similar, but tedious to obtain. Let v be the VEV of the Higgs field, so that the VEV of the Polyakov loop is given by $\hat{\Omega}_0 = \exp\{-igv\sigma^3/2\sqrt{T}\}$. Omitting the gory details of the calculation (which are given in [2]), we find to one loop that $v = 0$, which, in view of the connection between v and m_{mag} suggested above, is consistent with the well-known fact that $m_{\text{mag}} = 0$ at one loop. However, we are now stuck as far as perturbative computations go. This is because (i) m_{mag}, and therefore v, is affected by the infrared problem which creeps in at two loops and (ii) the diagonal gauge we are using is unsuitable for calculations beyond one loop. Nevertheless, we can proceed, after a certain fashion, by making the following observation.

---

* Actually, $\hat{\Omega}$ is gauge-invariant only up to permutations of the eigenvalues. Therefore, the calculations of this section have meaning only in perturbation theory, which takes place in a coordinate patch where one can trivially order the eigenvalues. It is unclear whether or not this definition can be extended to the full theory by means of some suitable global ordering scheme.

Since v is related to the magnetic mass, it must originate from the infrared divergences of hot QCD; we then have dimensional arguments [7] which tell us that to leading order v is given by

$$v = c\,g(T)\sqrt{T}$$

where c is a nonperturbative dimensionless constant. Assuming a non-zero v of the above form, we can compute the singlet potential and find that it is given by

$$V_1(R) = -g^2 K \frac{e^{-m_D R}}{4\pi R},$$

where

$$K \equiv \frac{1}{4\left[1 - \frac{11}{12\pi c}\right]}$$

$$m_D^2 \equiv \frac{m_E^2 - \frac{2c}{\pi}g^4 T^2}{1 - \frac{11}{12\pi c}}.$$

K and m_D^2 are not perturbatively calculable because of c. The nonperturbativity of m_D^2 could in turn affect low-order perturbative results, such as the thermodynamic potential which contains the $O(Tm_D^3)$ "plasmon" term.

IV. CONCLUSION

The validity of perturbative analysis in hot QCD, already weakened by the infrared problem, is further diminished by our results on Debye screening. If, as we have suggested, these results imply the generation of an electrostatic VEV, the situation becomes even worse: few perturbative results remain valid and nonperturbative techniques become indispensable even at very high temperatures. Despite its negative implications for perturbation theory, an electric condensate provides a neat explanation for the surprisingly early breakdown of perturbation theory in the electric sector and for the way in which the infrared divergences of hot QCD cure themselves.

While we have not presented any direct evidence for the condensate here, it is strongly indicated by our results in view of previous work [10] involving a variety of approaches such as analytical estimates of the effective potential and numerical simulations of hot QCD. Unfortunately, the evidence so far remains inconclusive and further work on this topic, such as a high-statistics Monte Carlo calculation to verify the occurrence of the VEV, would be most valuable.

V. OUTLOOK

We outline here a lattice model which can be used for nonperturbative studies of the infrared dynamics of QCD at very high temperatures. In the interests of computational economy, one ought to try and latticise only those aspects of the theory that are truly beyond perturbative analysis. Therefore, we shall place on the lattice not the full hot QCD lagrangian, but an effective theory which reproduces it well in the infrared. Such a theory is described by the three-dimensional action [7]

$$ S = \int d^3\mathbf{x} \left[\frac{1}{2}\mathrm{Tr}F^2(\mathbf{A}) + \mathrm{Tr}(D\phi)^2 + m_0^2\mathrm{Tr}\phi^2 + \mu(\mathrm{Tr}\phi^2)^2 \right], $$

where $\mathbf{A}$ is the magnetostatic potential, ϕ the electrostatic potential (adjoint Higgs field), m_0 the *bare* Debye mass and μ an induced quartic coupling constant. The three-dimensional gauge coupling is $g(T)\sqrt{T}$.

To economise farther, we neglect quarks (which have no significant impact on the infrared dynamics) and choose the gauge group $SU(2)$, then transform to diagonal gauge and put the resulting action on the lattice. The partition function after suitable rescaling can be written

$$ Z(\beta, \kappa, \lambda) = \int [\rho^2 e^{-\rho^2}\, d\rho][dU] e^{-S_{\mathrm{eff}}[\rho, U]}, $$

where

$$ -S_{\mathrm{eff}}[\rho, U] = \beta \sum_{\mathrm{plaqs}} \tilde{\mathrm{Tr}}U_{\mathrm{plaq}} $$

$$ + \kappa \sum_{\mathrm{links}} \rho(n)\rho(n+i)\tilde{\mathrm{Tr}}[U_i^\dagger(n)\sigma^3 U_i(n)\sigma^3] - \lambda \sum_{\mathrm{sites}} \rho^4(n). $$

The connection with the finite-temperature theory is given by the relations

$$ \kappa^{-1} = 3 + (m_0\beta_4/2T\beta)^2, $$

$$ \lambda = \mu\beta_4\kappa^2/16T\beta, $$

$$ \rho^2 = (8\beta_4/T\beta\kappa)\tilde{\mathrm{Tr}}\phi^2, $$

where the temperature T and the four-dimensional lattice coupling $\beta_4 \equiv 4/g^2(T)$ are held fixed, while the three-dimensional lattice coupling $\beta = \beta_4/aT$ is varied. In the continuum limit all masses scale as β, which itself, since the theory is superrenormalizable, scales as $1/a$.

An electrostatic condensate would manifest itself in this model as a shift in the ρ-distribution, the magnitude of which could be computed by analysing the moments of the distribution.

ACKNOWLEDGEMENTS

I thank R. Gavai for discussions pertaining to the lattice model presented in the final section. This talk is based on work supported in part by the National Science Foundation under Grant No. NSF-PHY84-15534.

REFERENCES

[1] S. Nadkarni, Phys. Rev. **D33** (1986) 3738.

[2] S. Nadkarni, Rutgers Report No. RU-86-16, 1986 (to appear in Phys. Rev. **D34**).

[3] D.J. Gross, R.D. Pisarski and L.G. Yaffe, Rev. Mod. Phys. **53** (1981) 43.

[4] J. Cleymans, R. Gavai and E. Suhonen, Phys. Rep. **130** (1986) 217;

B. Svetitsky, *ibid.* **132** (1986) 1.

[5] A. D. Linde, Phys. Lett. **96B** (1980) 289; Rep. Prog. Phys. **42** (1979) 389.

[6] P. Debye and E. Hückel, Phys. Z. **24** (1923) 185; English translation in *The Collected Papers of Peter J.W. Debye* (Interscience, New York, 1954).

[7] S. Nadkarni, Phys. Rev. **D27** (1983) 917.

[8] T. Toimela, Z. Phys. **C27** (1985) 289.

[9] S. Weinberg, Phys. Rev. **D7** (1973) 1068;

K. Fujikawa, B.W. Lee and A.I. Sanda, Phys. Rev. **D6** (1972) 2923;

E.S. Abers and B.W. Lee, Phys. Rep. **9** (1973) 1.

[10] R. Anishetty, J. Phys. **G10** (1984) 423, 439;

K.J. Dahlem, Z. Phys. **C29** (1985) 553;

J. Polonyi and H.W. Wyld, Illinois Report ILL-(TH)-85-#23 (1985);

J. Polonyi, elsewhere in these Proceedings.

ON THE COMPARISON OF STRING MODELS WITH LATTICE QCD

P. Olesen

The Niels Bohr Institute
Copenhagen, Denmark

ABSTRACT

An interesting question is how the QCD string looks. Assuming that the string picture is only asymptotic in QCD, there are several possibilities for the asymptotic potential. This has been discussed in detail in the following papers: P. Olesen, Physics Letters 160B (1985) 144; ibid. 168B (1986) 220.

NUMERICAL STUDIES OF RANDOM SURFACES

Bengt Petersson

Fakultät für Physik
Universität Bielefeld
D-4800 Bielefeld
F.R. Germany

ABSTRACT

Critical exponents of a model of discretized random surfaces are calculated by Monte Carlo simulations. The phase structure and universality properties are discussed in the light of these measurements.

INTRODUCTION

The investigation of random surfaces is of considerable interest for several topics in theoretical physics. Firstly it is a way to find a quantum theory of strings in non-critical dimensions, or to understand why it does not exist. This is of interest also for QCD, to the extent that gauge theories can be described by string variables. A non-perturbative definition of a string theory could also have an impact on present superstring theories. Another field of application is the physics of surfaces in statistical mechanics. Finally, the investigations are an exercise in the discretization of a reparametrization invariant theory. Generalizations to higher dimensions would be important for the discretization of quantum gravity.

There are also some caveats. Most of the interesting results can only be obtained by numerical methods. In fact, in this report I will discuss essentially such results. Furthermore, these results are restricted to Euclidean space. The continuation to Minkowski space may be non-trivial. Finally, the discretization of surface theories is not straightforward, as has been shown by Fröhlich and collaborators.[20] The corresponding models may have a trivial one-dimensional continuum limit, or other diseases. In our investigations,[21,22] we have used a discretization scheme proposed in ref. 4-6, where the strong coupling expansions of David[26] indicated an interesting continuum limit.

THE MODEL

The model can be seen as a natural discretization of the Polyakov string,[27]

$$Z = \int [dg(\xi)] [dx^\mu(\xi)] \exp[-\mathcal{A}(\},\S)] \tag{1}$$

where

$$\mathcal{A}(\},\S) = \int \lceil^\in \xi \sqrt{\}} \left(\}^{\alpha\beta} \partial_\alpha \S^\mu \partial_\beta \S^\mu + \beta \right). \tag{2}$$

The tensor $g_{\alpha\beta}(\xi)$ is the internal metric of the surface, and $x^\mu(\xi)$ are the d-dimensional coordinates in the imbedding space. The measure is discussed in refs. 8,9. In analogy with this formalism one may define

$$Z = \sum_S \frac{1}{C(S)} \int \prod_i q_i^\alpha \prod_{i,\mu} \frac{dx_i^\mu}{\pi} \exp[A(S,x)] \delta(x_1) \tag{3}$$

where

$$A(S,x) = \sum_{\langle ij \rangle} (x_i - x_j)^2 + \beta N_t, \quad \langle ij \rangle \text{ neighbors.} \tag{4}$$

The surfaces S are non-degenerate triangulations of given topology with symmetry factor $C(S)$. With respect to the internal metric the triangles are equilateral and of area one. This restriction is hopefully unimportant in the continuum limit. The total area is N_t. The curvature is concentrated at the vertices

$$R_i = (6 - q_i)\pi/q_i \tag{5}$$

where q_i is the number of triangles meeting at the vertex i. The action A and our choice $\alpha = d/2$ is the natural discretization of (2) if the coordinates are defined at the vertices. Green's functions $G_N(x_1, \ldots x_N)$ are defined by replacing the δ-function (which is included to remove the translational zero mode) by

$$\prod_{a=1}^N \sum_j q_j \delta(x_j - x_a). \tag{6}$$

The integral over the coordinates is Gaussian and can be performed, giving

$$Z = \sum_S \left(\prod_i q_i^\alpha \right) (\det D)^{-d/2} \exp[-\beta N_t] \tag{7}$$

where D is the connection matrix.[26] The determinant of D is equal to the number of trees of the triangulation. One may show that the entropy of the surfaces grows exponentially, giving a critical β_c. In the neighbourhood of β_c,

$$Z = \sum \exp\left[-\beta N_t\right] Z\left(N_t\right) \tag{8}$$

where

$$Z\left(N_t\right) \sim N_t^{\gamma-3} \exp\left[\beta_c N_t\right]. \tag{9}$$

γ is the susceptibility exponent, which we will measure by generating the distribution in (8). Another critical exponent is the Hausdorff dimension d_H, which we measure from

$$\bar{x}_S^2 = \frac{1}{N_t^2} \sum_{i,j=1}^{N_v} q_i \left(x_i - x_j\right)^2 q_j \tag{10}$$

and

$$\lim_{N_t \to \infty} \bar{x}_S^2 \sim \left(N_t\right)^{2/d_H}. \tag{11}$$

This exponent is measured in the canonical ensemble, N_t fixed. If

$$G_2\left(x\right) \sim \mid x \mid^{2-d-\eta} \exp\left[-m\left(\beta\right) \mid x \mid\right] \tag{12}$$

and

$$m\left(\beta\right) \sim \left(\beta - \beta_c\right)^\nu \tag{13}$$

goes to zero at the critical point, one expects[26]

$$\begin{aligned} \eta &= 2 - \gamma d_H \\ \nu &= 1/d_H \end{aligned} \tag{14}$$

so that those exponents would follow from the two which we measure. For $d = 0$ and -2, the exponent γ can be calculated analytically to be $-1/2$ and -1 respectively.[29] Some analytic arguments can be given for which surfaces dominate for $\mid d \mid$ or $\mid \alpha \mid \to \infty$.[30,31] For finite d and α, however, entropy is important, strong coupling calculations become cumbersome because of the exponentially exploding entropy, and only numerical calculations become feasible so far.

NUMERICAL METHODS

In ref. 2, we have calculated $d_H\left(d\right)$ for $\alpha = d/2$ for a series of dimensions d. We follow the proposal in ref. 13 and move in the canonical ensemble by flipping links. The acceptance is determined by a Metropolis method. One may calculate the change in $\det D$ and D^{-1} exactly, keeping

D^{-1} in memory. We found, however, this method too slow, and instead we performed the x-integration by a Monte Carlo heat-bath method.

To calculate the second exponent $\gamma(d)$, we had to find a Monte Carlo method for the grand canonical ensemble.[3] We proposed an algorithm splitting and joining vertices, and implemented a corresponding Metropolis algorithm. The choice of the weights had to be done in a somewhat clever way to get an acceptable rate of acceptance. For the detailed procedure I refer to the original publication.[3] Again the x-integration is performed by the heat-bath method. The algorithm has been checked against the analytic expression for $d = 0$. In the meantime, other algorithms have been proposed.[14,15] We believe, however, our method to be particularly efficient.

From our experience I can note the following remarks. The avoid to non-asymptotic corrections to the critical exponents, we have found it necessary to study surfaces with more than 200 triangles. In the canonical ensemble we generated surfaces up to 3100 triangles for the extraction of d_H. For the determination of γ, we used an interval with $200 \leq N_t \leq 1000$. It is also necessary to have such a large interval, because γ is a secondary term in $\log Z(N_t)$ in (9). Finally, in order to get relevant results, i.e. determine γ to within $\approx 20\%$, we found it necessary to gather a statistic corresponding to 3×10^8 accepted changes.

THE RESULTS

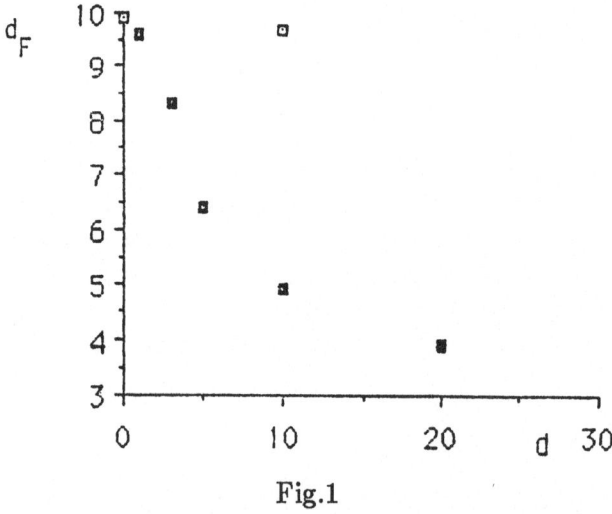

Fig.1

In figure 1, we show the Hausdorff dimension d_H as a function of d for $\alpha = d/2$. We have also included two points from an independent simulation of Billoire and David,[16] who chose $\alpha = 0$. Naively the difference should not be important in the continuum limit, because it corresponds to defining the x-variables at the corners or in the middle of the triangles. More precisely

$$\Delta S = \frac{d}{2} \sum_i \log q_i \sim \sum_i q_i \left(c_0 + c_1 R_i + c_2 R_i^2 + \ldots \right) \tag{15}$$

where the first terms are trivial, and the following dimensionally irrelevant. The results of figure 1 were in fact the first evidence that this form of universality is broken. Later, qualitative arguments were given for the existence of a phase of heavily crumpled surfaces with $\bar{x}_S^2 = \text{const}$, for $\alpha \to -\infty$.[11]

At small d we find a large, but finite Hausdorff dimension. We do not find a logarithmic behaviour

$$\bar{x}_S^2 = \log N_t \tag{16}$$

as is expected at least at sufficiently negative d, if the model relates to the saddle point expansion of the Polyakov string around $d = -\infty$ (the Liouville phase).[17] At $d = 20$, we obtain $d_H = 4$. This is the Hausdorff dimension expected if the model collapses to long tubes forming branched polymers, given a free particle continuum limit.[1] However, the simplest qualitative expectations,[11] in which there would exist only the above three phases with sharp boundaries do not seem correct. There seems to be a gradual dependence of d_H on d. These results have been confirmed by later, more detailed investigations.[18,19]

In figure 2, we show the dependence of $\langle R^2 \rangle$ on d, for $\alpha = d/2$. Clearly the tendency is in agreement with the picture above. There is a gradual flattening of the surfaces moving to higher $\alpha \sim d/2$. Note that this is not in contradiction to having branched polymers, consisting of almost everywhere flat tubes.

In figure 3, we show γ again as a function of d, for $\alpha = d/2$. For $d = 0$ we are in good agreement with the exact result $\gamma = -1/2$. Again, for large d, here at $d \geq 10$, $\gamma = 1/2$, indicating the dominance of branched polymers. Recently we calculated $\gamma = .20 \pm .06$ for $d = 10$ and $\alpha = 0$. Thus also γ depends on α. Furthermore, our data suggests that γ is not always $1/2$, when $\gamma > 0$, as would be the case if the continuum limit is a one-particle pole.

The meaning of the continuum limit for $\gamma < 0$ is under some debate,[7,12,18] because the mean number of triangles does not diverge at the critical point. Still, $d_H < \infty$ indicates $\nu = 1/d_H > 0$, i.e. a diverging

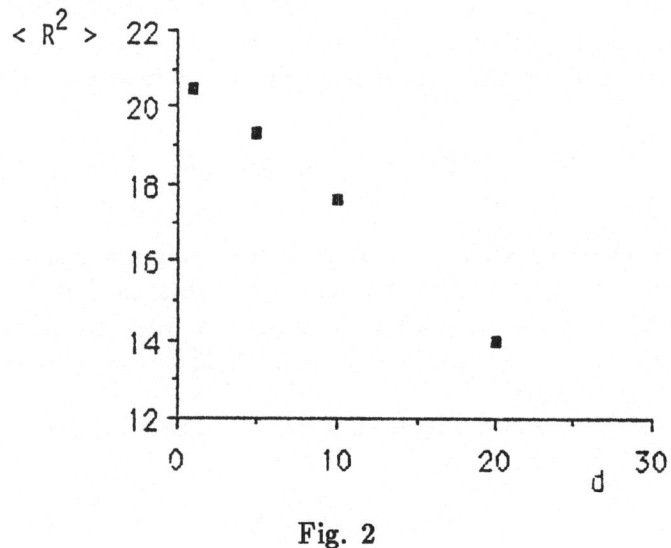

Fig. 2

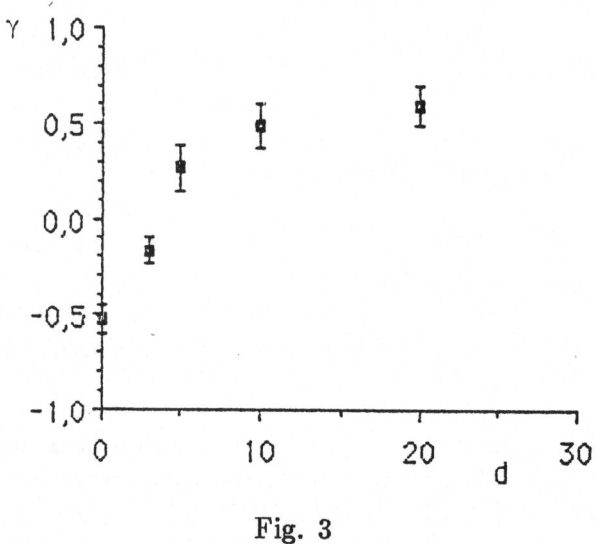

Fig. 3

correlation length for all $d > 0$. It is amusing that γ passes through 0 for $d \approx 4$, as was already indicated by strong coupling calculations. Independent numerical calculations of Ambjørn et al. using small surfaces are in general agreement with the above results on γ, although having very large errors.[14]

CONCLUSIONS

The numerical results hint to the existence of a non-trivial continuum limit of this model of random surfaces in an interesting range of dimensions. Still the connection to string theory has to be further investigated. Fröhlich has conjectured that for $\gamma > 0$, α finite, surfaces with bottlenecks will dominate, leading to a particle theory.[1] It is amusing to note that from our results $d = 4$ seems to be a "critical dimension" in this connection. However, the dependence on α of this fact has not been studied yet. In general, the breakdown of universality, found by us, should be further investigated. Other developments should concern the dependence on topology, a measurement of the string tension and the connection with Polyakov's calculation of the conformal anomaly. The first results on this last question do not give a clear answer.[19]

Our first results are encouraging enough to continue this investigation into the non-perturbative structure of surface theories.

ACKNOWLEDGMENTS

I thank J. Jurkiewicz and A. Krzywicki for a very pleasant collaboration. Furthermore, I am grateful to A. Billoire, F. David and J. Fröhlich for discussions.

REFERENCES

1. J. Fröhlich, in *Recent Developments in Quantum Field Theory*, J. Ambjørn, B. Durhuus, and J.L. Petersen, eds., North-Holland, Amsterdam (1985), and references therein.
2. J. Jurkiewicz, A. Krzywicki, and B. Petersson, Phys. Lett. 168B (1986) 273.
3. J. Jurkiewicz, A. Krzywicki, and B. Petersson, Phys. Lett. 177B (1986 89;
   see also, A. Krzywicki, Lectures, preprint LPTHE, Orsay 86/38 (1986).
4. F. David, Nucl. Phys. B257 (1985) 282.
5. A. Kazakov, Phys. Lett. 150B (1985) 282.
6. J. Fröhlich, in *Application of Field Theory to Statistical Mechanics*, Lecture Notes in Physics, 216, L. Garrido, ed., Springer-Verlag (1985).
7. F. David, Nucl. Phys. B257 (1985) 543.
8. A.M. Polyakov, Phys. Lett. 103B (1981) 207.
9. O. Alvarez, Nucl. Phys. B216 (1983) 125.
10. F. David, Phys. Lett. 159B (1985) 303, and references therein.

11. D.V. Boulatov, V.A. Kazakov, I.K. Kostov, and A.A. Migdal, Acad. of Sciences USSR, Space Research Inst. preprint $\pi\rho$-1123 (1986).
12. J. Ambjørn, B. Durhuus, J. Fröhlich, and P. Orland, Nucl. Phys. B270 (1986) 457.
13. V.A. Kazakov, I.K. Kostov, and A.A. Migdal, Phys. Lett. 157B (1985) 295.
14. J. Ambjørn, B. Durhuus, and J. Fröhlich, Nucl. Phys. B275 (1986) 161.
15. F. David, private communication.
16. A. Billoire and F. David, Phys. Lett. 168B (1986) 279.
17. J. Jurkiewicz and A. Krzywicki, Phys. Lett. 148B (1984) 148.
18. D.V. Boulatov, V.A. Kazakov, I.K. Kostov, and A.A. Migdal, Phys. Lett. 174B (1986) 87.
19. A. Billoire and F. David, preprint FSU-SCRI-86-44, Tallahassee (1986).

LATTICE GAUGE FIELDS AND TOPOLOGY

Anthony Phillips

Mathematics Dept.
SUNY
Stony Brook NY 11794

David Stone

Mathematics Dept.
Brooklyn College
Brooklyn NY 11210

Our purpose in this report is to give an informal introduction to our work on the topology of lattice gauge fields and the computation of topological charge [1] and to present some examples which have not been published, notably some naturally occurring examples of LGF's on the complex projective space CP^2.

The main problems can be stated as follows. An $SU(N)$ gauge field A_μ on a 4-dimensional space M has a topological charge O (an integer) given by $Q = (1/8\pi^2)\int_M tr F \wedge F$ where F is the su(N)-valued 2-form defined by $F_{\mu\nu} = \partial_\mu A_\nu - \partial_\nu A_\mu - [A_\mu, A_\nu]$. When we pass to a lattice gauge field defined on a (cubical, simplicial, etc.) cellularization Λ of M, then

> 1. What does Q mean?
> 2. How can it be efficiently computed?

Note first that Q is an invariant (in fact, the 2nd Chern number) of the $SU(N)$ principal fiber bundle carrying A_μ. Let us spend a couple of paragraphs on this bundle since it is an important element of a gauge field that is hidden in the usual notation.

Even though we wrote "a gauge field A_μ" just above, in general of course no single nonsingular A_μ can describe a gauge field [2]. The local descriptions $\{A^\alpha : U_\alpha \to su(N)\}$, where $\{U_\alpha\}$ is a covering of M (we will now suppress the internal index μ), are related by transition functions $v_{\alpha\beta}: U_\alpha \cap U_\beta \to SU(N)$ as follows:

$$(\ast) \qquad A^\beta = \mathrm{ad}(v_{\alpha\beta})A^\alpha + (v_{\alpha\beta})^{-1}dv_{\alpha\beta}.$$

Note that these transition functions, since they represent changes of gauge, must form a cocycle, i.e. satisfy the condition $v_{\alpha\beta}v_{\beta\gamma}v_{\gamma\alpha} = I$ at every point of $U_\alpha \cap U_\beta \cap U_\gamma$. Such a collection of functions is called a "coordinate SU(N) bundle" in Steenrod [3] and in fact determines a unique principal SU(N) bundle over M.

Since Q is a bundle invariant and since the bundle is completely determined by the $v_{\alpha\beta}$'s (in fact Q can be calculated directly from the $v_{\alpha\beta}$'s; see [4] and formulas in [5] and [6]), a reasonable Ansatz for problems 1. and 2. is to propose making a family of transition functions out of an LGF. A natural second part of this Ansatz is to take as putative trivializing neighborhoods for the bundle (the U_α's in the smooth case) the 4-dimensional cells of the dual lattice $\Lambda^\ast$. To each vertex α of Λ corresponds a 4-dimensional dual cell c_α, and two of these intersect along a 3-dimensional face if and only if the corresponding vertices are joined by a bond in Λ.

Now a LGF is supposed to represent a smooth gauge field A in the following sense: the element $u_{\alpha\beta}$ associated to the bond $\langle\alpha\beta\rangle$ describes parallel transport by A along $\langle\alpha\beta\rangle$, say from α to β. Suppose the axial gauge had been chosen in c_α and in c_β; in this gauge parallel transport by A from α to $p_{\alpha\beta} = \langle\alpha\beta\rangle \cap c_\alpha \cap c_\beta$ along the straight segment $\langle\alpha\beta\rangle \cap c_\alpha$ is the identity; and the same holds from $p_{\alpha\beta}$ to β along $\langle\alpha\beta\rangle \cap c_\beta$. So the group element describing parallel transport from α to β is just the local change-of-gauge matrix at $p_{\alpha\beta}$.

This suggests interpreting the element $u_{\alpha\beta}$ as the value of the transition function $v_{\alpha\beta}$ at the point $p_{\alpha\beta}$. The problem is then to extend this one value to a continuous function $v_{\alpha\beta}$ defined on all of $c_\alpha \cap c_\beta$, and to do this for each bond $\langle\alpha\beta\rangle$ in such a way as to satisfy the cocycle condition on triple intersections, while keeping control of the distance in the group from $v_{\alpha\beta}(x)$ to $v_{\alpha\beta}(p_{\alpha\beta}) = u_{\alpha\beta}$.

One algorithm for carrying out this extension when the group is SU(2) (it works for every LGF off a certain set of measure zero) is given in [1] and will be briefly described below. Before that it is worth expanding on the phrase "keeping control of the distance..." in the last paragraph. First it is clear that without some control over the variation of $v_{\alpha\beta}$ on c_α c_β we would lose any topological information carried by the LGF. There is no problem, for example, in arranging for each $v_{\alpha\beta}$ to be the identity on the whole boundary of $c_\alpha \cap c_\beta$, thus easily satisfying the cocycle condition. But then this cocycle could be deformed through cocycles to $\{v_{\alpha\beta} \equiv I\}$, which means that it itself defined the trivial bundle, no matter what **u** was. On the other hand we can prove that if the variation stays within certain bounds, then the bundle obtained depends only on the given LGF.

Theorem [1, Prop. 3.4] Let $\mathbf{u} = \{u_{\alpha\beta}\}$ be an SU(2)-valued LGF defined on a simplicial lattice Λ, and let $\mathbf{v} = \{v_{\alpha\beta}\}$ and $\mathbf{v}' = \{v'_{\alpha\beta}\}$ be SU(2)-valued cocycles defined on the dual lattice. If $d(v_{\alpha\beta}(x), u_{\alpha\beta}) < \pi/8$ for every α, β and x in $c_\alpha \cap c_\beta$, and if $\mathbf{v}'$ satisfies the same inequality, then $\mathbf{v}$ and $\mathbf{v}'$ determine the same bundle. Here d is distance in the unit sphere metric, where the diameter of SU(2) is π.

We do not know if this is the best available bound of this type. What we do know about the uniqueness problem can be summarized in the following table, where SU(2) results are contrasted with U(1) results, which are sharp.

Table 1. Uniqueness results. In this table, **u** is a G-valued LGF on a lattice Λ and **v** a G-valued cocycle on the dual lattice, G = U(1) or SU(2); $d(v,u) = \sup d(v_{\alpha\beta}(x), u_{\alpha\beta})$, the sup taken over all $\langle\alpha\beta\rangle$ and all x in $c_\alpha \cap c_\beta$

I 2D U(1) theory.

a) $d(v,u) < \left\{ \begin{array}{l} \pi/4 \ (\Lambda \ \text{rectangular}) \\ \pi/3 \ (\Lambda \ \text{simplicial}) \end{array} \right\} \Rightarrow$ **v** gives "nearest" bundle to **u** [7].

a') This condition is sharp (see Figure 1. below).

b) **u** admits such a **v** $\iff$ no plaquette product = -1 .

II 4D SU(2) theory.

a) $d(v,u) < \pi/8 \Rightarrow$ **v** gives "nearest" bundle to **u**.

a') ? An example [1] has **v**, **v**' giving different bundles with the same **u**, and $d(v,u), d(v',u) \cong 2\pi/3$.

b) **u** admits such a **v** $\Rightarrow$ every plaquette product is within $3\pi/8$ of I. (This follows from the inequality $d(u_{\alpha\beta}u_{\beta\gamma}u_{\gamma\alpha}, v_{\alpha\beta}(x)v_{\beta\gamma}(x)v_{\gamma\alpha}(x)) \leq d(u_{\alpha\beta}, v_{\alpha\beta}(x)) + d(u_{\beta\gamma}, v_{\beta\gamma}(x)) + d(u_{\gamma\alpha}, v_{\gamma\alpha}(x))$ applied to x in $c_\alpha \cap c_\beta \cap c_\gamma$.)

b') **u** admits such a **v** $\Leftarrow$ every plaquette product is within $\pi/24$ of I [1, Thm. A].

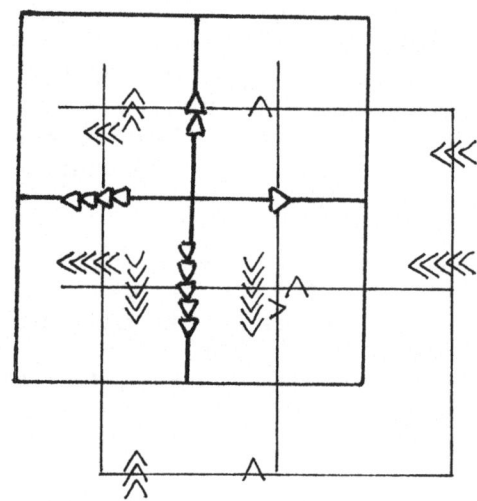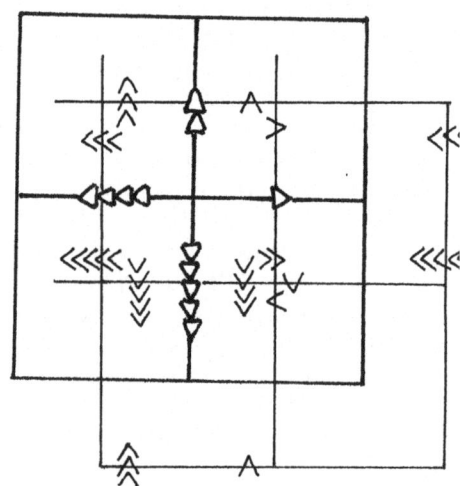

$\triangle$,$\underset{\triangle}{\triangle}$,etc.: **u**-values; $\wedge$,$\hat{\wedge}$,etc.: **v**-values; $\triangle$,$\wedge$ = exp(iπ/4)
——— 2×2 lattice on T^2 (opposite edges identified)
——— its dual (another periodic 2×2 lattice)

Fig.1. This figure shows two U(1)-valued cocycles on
the dual of the 2×2 lattice on the torus, defined from the
same LGF **u**, and both satisfying d(**v**,**u**) = π/4. The one on
the left gives a bundle with first Chern number C = 1,
while the other has C = 0. (The C's are calculated using
the formula C = (1/4π)$\iint$d(log $v_{\alpha\beta}$) from [7]). This shows
that the d < π/4 condition is sharp.

Chern number calculation for SU(2)

Once a set of transition functions has been constructed from a LGF there are two different ways (but see also
[8]) to calculate its second Chern number.

a) geometrical: reconstruct a gauge field A from the
transition functions, compute tr$F\wedge F$ and integrate. This is
essentially Lüscher's scheme [5].

b) topological: use the obstruction-theoretic definition of the Chern number (below); this was our approach.

Sections and obstructions: A principal G-bundle is
constructed from a cocycle {$v_{\alpha\beta}$: $c_\alpha \cap c_\beta \to$ G} by taking
the collection of local product bundles $c_\alpha \times$ G $\to c_\alpha$ and
using {$v_{\alpha\beta}$} to glue them together to form the total space
E of the bundle, identifying (x,g_α) with (x,$v_{\alpha\beta}$(x)g_β) for
x in $c_\alpha \cap c_\beta$. A section in a bundle π: E $\to \Lambda$ (by definition, a map Σ: $\Lambda \to$ E such that $\pi(\Sigma(x))$ = x) is given in
terms of the local product bundles by a family of maps

$x \rightarrow (x, S_\alpha(x))$ for x in c_α related by $S_\alpha(x) = v_{\alpha\beta}(x)S_\beta(x)$ on $c_\alpha \cap c_\beta$.

In the case of a SU(N)-bundle over a 4-manifold, any section can be extended cell by cell until the following configuration occurs: the section Σ has already been defined on the entire boundary ∂c of a 4-cell c. Then, writing the bundle over c as a product, the map $S: \partial c \rightarrow$ SU(N) so defined gives a nontrivial element $N(\Sigma, c)$ of the homotopy group π_3SU(N) = Z, so Σ cannot be extended over c. This homotopy element "obstructs" the section, and the obstruction-theoretic definition of the topological charge comes from the following theorem.

Theorem [3,9] Suppose Σ has been defined everywhere except on a collection $c_1, \ldots, c_k$ of 4-cells, as above. Then
$$Q = \sum_i N(\Sigma, c_i).$$

The algorithm we use for constructing the transition functions and the section is described in great detail in [1]. Here we will give a rough illustration of what happens over a typical 4-simplex $\sigma = \langle \alpha\beta\gamma\delta\epsilon \rangle$ of Λ. The vertices of Λ have been ordered, and those of σ are ordered as listed. The section is defined separately over each intersection $\sigma \cap c_\mu$ with the 4-cell dual to one of its vertices, beginning with $\sigma \cap c_\epsilon$ and working down along the ordering. Each $\sigma \cap c_\mu$ is combinatorially a 4-cube; in fact it inherits from the linear structure of σ a set of coordinates $\{s_\lambda\}$, one for each vertex $\lambda \neq \mu$, $0 \leq s_\lambda \leq 1$. These are the "modified barycentric coordinates" (m.b.c's).

Similarly the transition function $v_{\mu\nu}$ is defined separately over each $c_\mu \cap c_\nu \cap \sigma$; this intersection has m.b.c's $0 \leq s_\lambda \leq 1$, one for each vertex $\lambda \neq \mu, \nu$ of σ. The following diagram shows, at each level of the induction, the extension problem that arises. Level 0 corresponds to the case of $v_{\mu\nu}$ where μ and ν are consecutive in σ (resp. to the case of S_ϵ); level 1 to $v_{\mu\nu}$ where μ and ν have one intermediate vertex in σ (resp. to S_δ); etc. The algorithm is set up so as to use at any level only the m.b.c's coming from vertices between μ and ν (resp., for S_λ, vertices above λ). In this schematic presentation, u stands for a constant map, g for a unique shortest geodesic, h for the union of two spherical triangles and k for the union of four spherical simplexes and the cone on a doubly ruled, quadrilateral surface. At level 4 (S_α) the obstruction to extending the section over $c_\alpha \cap \sigma$ can be evaluated by elementary methods since the section on the boundary is the union of simple spherical polyhedra in SU(2) = S^3.

Table 2. Inductive construction of sections and transition functions

	Problem	Solution	"type"
level 0	•	•	u
level 1	• •u	⌣	g
level 2			h
level 3			k
level 4			

Solution obstructed by $N(\Sigma, c_\alpha \cap \sigma)$.

An example from Differential Topology: LGF's on CP^2

Complex projective 2-space is a well known 4-manifold. It can be defined as the space of complex lines through the origin in C^3 or equivalently as the quotient of $S^5 = \{(z_0, z_1, z_2): |z_0|^2 + |z_1|^2 + |z_2|^2 = 1\}$ by the relation $(z_0, z_1, z_2) \sim (\lambda z_0, \lambda z_1, \lambda z_2)$, $|\lambda| = 1$. There are two interesting principal bundles that come along with CP^2, those associated to the canonical line bundle and to the tangent bundle. The first is precisely the $U(1)$-bundle γ with total space S^5 and projection map taking (z_0, z_1, z_2) to its equivalence class $[z_0, z_1, z_2]$ in CP^2. The second is the $GL(2,C)$-bundle τ of complex frames tangent to CP^2.

The complex manifold CP^2 has canonical coordinate charts $\phi_\alpha : U_\alpha \to C^2$, $\alpha = 0,1,2$; for example $U_0 = \{[z_0,z_1,z_2] : z_0 \neq 0\}$ and $\phi_0([z_0,z_1,z_2]) = (z_1/z_0,z_2/z_0)$. The two bundles γ and τ can be represented by cocycles v and V on the covering U_0,U_1,U_2. This yields:

$$v_{01}([z_0,z_1,z_2]) = z_0/z_1 |z_1/z_0|$$

$$v_{12}([z_0,z_1,z_2]) = z_1/z_2 |z_2/z_1|$$

$$v_{20}([z_0,z_1,z_2]) = z_2/z_0 |z_0/z_2|$$

$$V_{01}([z_0,z_1,z_2]) = (1/z_0^2) \begin{bmatrix} -z_1^2 & 0 \\ -z_1 z_2 & z_0 z_1 \end{bmatrix}$$

$$V_{12}([z_0,z_1,z_2]) = (1/z_1^2) \begin{bmatrix} z_1 z_2 & -z_0 z_2 \\ 0 & -z_2^2 \end{bmatrix}$$

$$V_{20}([z_0,z_1,z_2]) = (1/z_2^2) \begin{bmatrix} 0 & -z_0^2 \\ z_0 z_2 & -z_0 z_1 \end{bmatrix}.$$

Now CP^2 has a recently discovered minimal triangulation "$CP^2{}_9$" with 9 vertices and a great deal of symmetry [10]. The cocycles v and V can be used to define LGF's on $CP^2{}_9$ as follows. First, there exists a "canonical" cellular structure on CP^2 subordinate to the canonical cover: Let $C_0 = \{|z_0|^2 \geqslant |z_1|^2 \;\&\; |z_0|^2 \geqslant |z_2|^2\}$, etc., so $C_0 \subset U_0$, etc. Now the nine vertices of $CP^2{}_9$, as given in [10], are $1 = [1,\omega,0]$, $2 = [1,\omega^2,0]$, $3 = [1,1,0]$, $4 = [0,1,\omega]$, etc., where ω is the complex cube root of 1, so $1,2,3$ lie exactly on the intersection $C_0 \cap C_1$, etc. In order to simulate general position we consider vertex i as belonging to $C_{\alpha(i)}$, $1,2,3$ to C_0, $4,5,6$ to C_1, $7,8,9$ to C_2, and then we set u_{ij} to be the value of $v_{\alpha(j)\alpha(i)}$ at the point where the bond $\langle ij \rangle$ crosses $C_{\alpha(i)} \cap C_{\alpha(j)}$, and similarly we define U from V. For example $u_{25} = v_{01}(2) = \omega^2$ and $u_{28} = v_{20}(8) = \omega$, while $U_{38} = V_{20}(8) = C$ (see below). The two LGF's are displayed in Table 3. The U-values are in the 24-element discrete subgroup of $U(2)$ containing

$$\begin{bmatrix} \omega & 0 \\ 0 & \omega^2 \end{bmatrix} \quad \begin{bmatrix} \omega^2 & 0 \\ 0 & \omega \end{bmatrix} \quad \begin{bmatrix} 0 & -\omega \\ \omega^2 & 0 \end{bmatrix} \quad \begin{bmatrix} 0 & -\omega^2 \\ \omega & 0 \end{bmatrix} \quad \begin{bmatrix} 0 & -1 \\ 1 & 0 \end{bmatrix} \quad \begin{bmatrix} -1 & 0 \\ 0 & 1 \end{bmatrix} \quad \begin{bmatrix} 1 & 0 \\ 0 & 1 \end{bmatrix}.$$

$$\;\;\;A \qquad\qquad B \qquad\qquad C \qquad\qquad D \qquad\qquad E \qquad\qquad J \qquad\qquad I$$

Since -1 does not belong to the subgroup $\{1,\omega,\omega^2\}$ of $U(1)$, the LGF u automatically satisfies condition 1b of Table 1; the bundle determined by u is in fact the canonical bundle γ. This can be checked by comparing the real cocycle on CP^2_9 given by (*) $\langle a,b,c \rangle \to (-1/2\pi i)\log(u_{ab} \times u_{bc} u_{ca})$ with the first Chern class of γ. Take a non-bounding S^2 in CP^2_9, e.g. $\partial\langle 3789 \rangle$. Table 3 and (*) yield that the value of this cocycle on $\partial\langle 3789 \rangle$ is 1.

There are problems with U. First, U is a U(2)-LGF and the Chern number part of our algorithm works only for SU(2). Note however that we would still be measuring the obstruction to a section in an S^3-bundle. Worse, the plaquette products of U are quite large, so that the bundle reconstructed from U may depend on choices made in the algorithm. Then the next step would be to find a sub-division of CP^2_9 that was equally well adapted to the canonical coordinates.

Table 3. Lattice gauge fields on CP^2_9 defined from the principal bundles γ and τ

u_{ab}	1	2	3	4	5	6	7	8	9
1	1	1	1	ω	ω	ω	ω²	ω	1
2		1	1	ω²	ω²	ω²	ω²	ω	1
3			1	1	1	1	ω²	ω	1
4				1	1	1	ω	ω	ω
5					1	1	ω²	ω²	ω²
6						1	1	1	1
7							1	1	1
8								1	1
9									1

U_{ab}	1	2	3	4	5	6	7	8	9
1	I	I	I	JA	JA	JA	D	C	E
2		I	I	JB	JB	JB	D	C	E
3			I	J	J	J	D	C	E
4				I	I	I	-JB	-JB	-JB
5					I	I	-JA	-JA	-JA
6						I	-J	-J	-J
7							I	I	I
8								I	I
9									I

[1] A. Phillips, D. Stone, Lattice gauge fields, principal bundles and the calculation of topological charge, Commun. Math. Phys. 103 (1986) 599-636
[2] C.N. Yang, Magnetic monopoles, fiber bundles and gauge fields, Ann. N. Y. Acad. Sci, 294 (1977) 86-97
[3] N. Steenrod, "Topology of Fibre Bundles", Princeton Univ. Press 1951
[4] R. Bott, H. Shulman, J. Stasheff, On the de Rham theory of certain classifying spaces, Adv. in Math. 20 (1976) 43-56
[5] M. Lüscher, Topology of lattice gauge fields, Commun. Math. Phys. 85 (1982) 39-48
[6] P. Van Baal, Some results for SU(N) gauge-fields on the hypertorus, Commun. Math. Phys. 85 (1982) 529-547
[7] A. Phillips, Characteristic numbers of U_1-valued lattice gauge fields, Ann. of Physics 161 (1985) 399-422
[8] P. Woit, Topological charge in lattice gauge theory, Phys. Rev. Lett. 51 (1983) 638-641
[9] J. Milnor, "Characteristic Classes", Annals of Math. Studies No. 76, Princeton Univ. Press 1974
[10] W. Kühnel, T.F. Banchoff, The 9-vertex complex projective plane, Math. Intelligencer 5 (1983) 11-22

CHROMOMAGNETIC MONOPOLES

J. Polonyi*

Center for Theoretical Physics
Laboratory for Nuclear Science
and Department of Physics
Massachusetts Institute of Technology
Cambridge, Massachusetts 02139

ABSTRACT

It is shown that localized configurations with chromomagnetic charge can be defined in QCD. The way how these configurations show up in lattice regularization and some of their properties are discussed. It is argued that these configurations are relevant in the confinement mechanism.

I. INTRODUCTION

The understanding of the confinement mechanism in QCD is an exciting problem of high energy physics. The usual nonperturbative analytic way to approach this question is the semiclassical expansion where one considers certain "relevant" regions in the function space of the path integral representation. However the contributions of other "random" configurations remain uncontrolled due to the lack of a small parameter to justify the expansion. The numerical methods of lattice field theory offer new directions to embark this problem. To indicate the difficulties of a possible lattice approach let us first consider the question of instanton configurations as an example.

---

* On leave of absence from CRIP, Budapest, Hungary.

There have been several geometrically motivated "pattern recognition" procedures suggested to define the winding number for lattice gauge field configurations.[1] By following the intuitive picture of the semiclassical expansion in continuum theory these constructions intend to locate the "relevant" region where a given lattice configuration belongs to. But there is no nontrivial saddle point structure in lattice gauge theory. These geometrical constructions succeed in partitioning the configuration space into non-overlapping sectors but the relation of this rearrangement to the dynamics remains unclear. Consequently properties of the winding number isolated in such a rather arbitrary way have no obvious connection with the original problem one started with. An alternative approach is to look for quantum mechanical rather than geometrical definition of the topological charge. The suggestion made in Ref. [2] is to consider the dynamics of the localized zero modes of the fermion matrix. Quantities defined in such a way prove to be useful if one can combine analytical and numerical results when using them to derive quantitative relations.

A very nice and conceptually simple mechanism for confinement in gauge theories is based on the dual Meissner effect.[3] Problems to solve along this line are: (i) What kind of localized gluon states carrying chromomagnetic charges are preferred by the dynamics? (ii) How to justify the semiclassical approximation used in the dual effective theory? We address these questions in this work. We present a definition of chromomagnetic monopoles of QCD in Section II. The way how magnetic monopoles are recognized in lattice regularization will be discussed in Section III. The field operator corresponding to magnetic monopoles is introduced in Section IV. The relevance of such states in the confinement mechanism will be demonstrated in Section V.

II.   DYNAMICAL HIGGS FIELD IN QCD

We shall show that the path integral expressions for the time evolution operator involve a space dependent field variable which transforms according to the adjoint representation of the gauge group. This field variable will be used as a matter field to construct the analogue of 't Hooft-Polyakov monopoles[4] in pure $SU(3)$ Yang-Mills theories.

Consider Yang-Mills theory in temporal gauge

$$\hat{A}_0(\vec{x}, t) = 0 \tag{1}$$

We shall use Lie algebra valued field variables $\hat{A}_\mu(\vec{x}, t) = \hat{A}_\mu^a(x, t)\lambda^a/2i$, etc., in this paper. Matrix elements of the time evolution operator $\exp\{-it\hat{H}\}$

between field diagonal states $\hat{A}_i(\vec{x})|A\rangle = A_i(\vec{x})|A\rangle$ can be written as

$$\left\langle \hat{A}^{(f)} \left| \hat{P} e^{-it\hat{H}} \right| \vec{A}^{(i)} \right\rangle = \int D\left[A_\mu(\vec{x},t)\right] e^{iS}$$

$$\vec{A}(\vec{x},0) = \vec{A}^{(i)}(\vec{x}) \tag{2}$$

$$\vec{A}(\vec{x},t) = \vec{A}^{(f)}(\vec{x})$$

Here, the operator $\hat{P}$ projects into the gauge invariant subspace to guarantee Gauss's Law and is represented by the integration over the auxiliary variable denoted by $A_0(\vec{x},t)$. Matrix elements between the states characterized by the wave functionals $\psi\left[\vec{A}(\vec{x})\right]$ and $\psi'\left(\vec{A}(\vec{x})\right)$ are of the form

$$\left\langle \psi' \left| P e^{-it\hat{H}} \right| \psi \right\rangle$$
$$= \int D\left[\vec{A}^{(i)}(\vec{x})\right] D\left[\vec{A}^{(f)}(\vec{x})\right] \psi'^*\left[\vec{A}^{(f)}(\vec{x})\right] \psi\left[\vec{A}^{(i)}(\vec{x})\right] \left\langle \vec{A}^{(f)} \left| \hat{P} e^{-it\hat{H}} \right| \vec{A}^{(i)} \right\rangle \tag{3}$$

There is no need of the projection operator $\hat{P}$ in the case of gauge invariant states $\psi\left[\vec{A}^g\right] = \psi\left[\vec{A}\right]$, $A_\mu^g = g\left(\partial_\mu + A_\mu\right) g^\dagger$. In fact, it is easy to find the time dependent gauge transformation which eliminates A_0 and leaves the matrix element (3) unchanged.

It is advantageous to label the physical states by a non gauge invariant representative field configuration $\left|\vec{A}^{\text{repr}}\right\rangle_{\text{phys}} = \hat{P}\left|\vec{A}^{\text{repr}}\right\rangle$, when using path integrals. The transition matrix element corresponding to $\left|A^{(i)}\right\rangle_{\text{phys}}$ and $\left|A^{(f)}\right\rangle_{\text{phys}}$ can be written as

$$_{\text{phys}}\left\langle \vec{A}^{(f)} \left| e^{-it\hat{H}} \right| \vec{A}^{(i)} \right\rangle_{\text{phys}} = \int D\left[h(\vec{x})\right] D\left[A_\mu(\vec{x},t)\right] e^{iS}$$

$$\vec{A}(\vec{x},0) = A^{(i)h}(\vec{x}) \tag{4}$$

$$\vec{A}(\vec{x},t) = A^{(f)h}(\vec{x})$$

We define a variable analogous to the Polyakov-line of the finite temperature formalism[5] as $\Omega(\vec{x}) = P \exp\left\{\int_0^t d\tau A_0(\vec{x},\tau)\right\}$, where P stands for path ordering. Ω transforms as $\Omega(\vec{x}) \to g(\vec{x})\Omega(\vec{x})g^\dagger(\vec{x})$ under local gauge transformations in the domain of integration defined by (4). The gauge $A_0(\vec{x},t) = 0$ is not allowed here, instead we may use the static gauge where $\partial A_0(\vec{x},t)/\partial t = 0$. Observe that the auxiliary variable $A_0(\vec{x},t)$ appearing in the path integrals has no connection with the original operator $\hat{A}_0(\vec{x})$ eliminated by the gauge choice (1).

It is instructive to consider the partition function at finite temperature

$$Z = \text{tr} \, P \, \exp\left\{-\beta \hat{H}\right\} \; .$$

The summation over different representatives in the trace operation and the integration over gauge copies in (4) amount to the integration over all possible gauge field configurations. We thus obtain the result $Z = \int D\left[A_\mu(\vec{x},t)\right] \exp\left\{-S_{\text{Eucl}}\right\}$ with the boundary condition in time $A_\mu(\vec{x},\beta) = A_\mu(x,0)$. The convention (4) leads to strict periodicity in the partition function instead of periodicity up to a gauge transformation. The periodicity for A_0 is imposed to simplify the boundary conditions. The transformation property of the Polyakov line is $\Omega(\vec{x},t) = P \, \exp\left\{\int_t^{t+\tau} d\tau \, A_0(\vec{x},\tau)\right\} \rightarrow h(\vec{x},t)\Omega(\vec{x},t)h^\dagger(\vec{x},t)$ in this case.

The boundary conditions become irrelevant when the corresponding size is large in the absence of long range interactions. So one expects A_0 to become an irrelevant quantity and to decouple from the dynamics in the limit of $t \rightarrow \infty$ or $\beta = 1/T \rightarrow \infty$ for QCD with nonvanishing gap. Nevertheless, $A_0(\vec{x})$ remains to be useful variable. Consider the expectation value of the Wilson loop $W[C]$ corresponding to the rectangular with corners at $(t,x) = (0,0)$, $(0,L)$, (t,L), $(t,0)$ computed by the path integral (4). It gives the energy of a heavy quark-antiquark pair with separation L, inserted in the states $|\vec{A}^{(i)}\rangle$ and $|\vec{A}^{(f)}\rangle$, $\exp\left\{-it\left(V(L) + \text{const.}\right)\right\} = \langle \text{tr} \, W[C]\rangle$ for $t >> L$. The correlation length of the variable $A_0(\vec{x})$ in three space is $\xi = 1/(t\sigma)$ when $V(L) = L\sigma$. The way that $A_0(\vec{x})$ decouples in the long time limit contains important information about the dynamics. This example shows that the same information about $V(L)$ comes either from the spacelike or the timelike part of the Wilson loop $W[C]$ considered in temporal or axial gauge, respectively. The Gauss's Law is satisfied by the use of proper initial and final states in temporal gauge or is introduced during the time evolution by the nonvanishing $A_0(\vec{x},t)$ in axial gauge. The rest of this paper discusses another example when the use of $A_0(\vec{x})$ leads to less obvious results.

The variable $A(\vec{x})$ raises several questions. The first is whether this field generates dynamical breakdown of the global gauge symmetry as in the case of Yang-Mills-Higgs systems.[8] Unfortunately we know no signatures of such symmetry breaking in theories like QCD without small parameter. Another question is whether the variable $A_0(\vec{x})$ can be used to construct localized configurations which are relevant in the path integral. The obvious choice is to introduce magnetic monopole configurations[4] using $A_0(\vec{x})$ in static gauge as an adjoint Higgs field.[9,10] The "unbroken" electromagnetism is characterized by the Abelian field strength tensor[4]

$$F^A_{\mu\nu}(\vec{x}, t) = \partial_\mu A^3_\nu(\vec{x}, t) - \partial_\nu A^3_\mu(\vec{x}, t)$$
$$= \frac{1}{|A_0\vec{x})|} \, \text{tr} \left\{ A_0(\vec{x}) \left(F_{\mu\nu}(\vec{x}, t) - \frac{1}{|A_0(\vec{x})|^2} [D_\mu A_0(\vec{x}), D_\nu A_0(\vec{x})] \right) \right\}$$
$$(5)$$

in $SU(2)$ Yang-Mills theory, where $A^3_\mu(\vec{x}, t) = \frac{1}{|A_0(\vec{x})|} \text{tr} \, (A_0(\vec{x})A_\mu(\vec{x}, t))$ and $|A_0| = \det A_0$. The generalization of (5) to the case of $SU(3)$ gauge group has been given in.[14] The gluon field configuration $A_\mu(\vec{x}, t)$ in (4) contains a chromomagnetic monopole at the location $\vec{x}_0, t_0$ if $F^A_{\mu\nu}$ displays a Dirac monopole [13] at x_0, t_0

$$B_i(\vec{x}, t_0) = \frac{1}{2}\epsilon_{ijk}F^A_{jk}(\vec{x}, t_0) = \frac{1}{g} \frac{\vec{x} - \vec{x}_0}{|x_i - x_{0i}|^3} \qquad (6)$$

where $x \sim x_0$. These monopole configurations are neither saddle points, nor the solutions of any physical equations.

The reason why these objects may nevertheless be relevant from the point of view of the confinement mechanism is the following: Consider QCD at finite temperature where the static confinement properties are described by the correlation functions of the Polyakov line. The three dimensional effective theory of the Polyakov lines or $A_0(\vec{x})$ has already been investigated in connection to the deconfining phase transition.[10] Enlarge this effective theory to include the component of the complete gluon vector field in the color direction of $A_0(\vec{x})$. This is the effective theory of the gauge fields which are diagonal in the static "unitary" gauge where $\partial A_0(\vec{x}, t)/\partial t = 0$ and $A_0(\vec{x})$ is diagonal. This $U(1) \times U(1)$ gauge model has the same Polyakov line as the original $SU(3)$ theory. Consequently this Abelian reduction leaves the static confinement mechanism unattached. The confinement of the compact $U(1)$ gauge theory (compactness is the result of having nonabelian gauge group to start with) is believed to be generated by the condensation of Dirac monopoles.[11,12] But the regular monopoles of the nonabelian system defined by the help of $A_0(\vec{x})$ are nothing else then Dirac monopoles [13] in this effective Abelian gauge model.

III. MAGNETIC MONOPOLES IN LATTICE REGULARIZATION

One needs nonperturbative tools to investigate the system of electric and magnetic charges. Our goal is thus to construct the operators in lattice regularization which create or annihilate the magnetic field (6) of the monopole. These operators act in the linear space corresponding to the temporal gauge Hamiltonian mentioned in the previous Section. In fact, the path integral (2) can be considered in the static gauge as an expression of amplitudes which correspond to the theory described by the canonical field variables $\hat{A}(\vec{x})$, $\hat{E}(\vec{x})$ and the Hamiltonian

$$\hat{H}_{A_0} = \int d^3x \left\{ \frac{g^2}{2} \hat{E}^2(\vec{x}) + \frac{1}{2g^2} \hat{B}^2(\vec{x}) + \frac{g^2}{2} \left(D_i A_0(\vec{x}) \right)^2 + g^2 \hat{E}_i(\vec{x}) D_i A_0(\vec{x}) \right\}$$

with $A_0(\vec{x})$ as a random external field. The monopole creation and annihilation operators shift the eigenvalues of those color components of $\hat{A}_i(\vec{x})$ which commute with $A_0(\vec{x})$. These operators are not diagonal in the basis which is used in the path integral. Thus we discuss first the form of the diagonal operator $\hat{q}(\vec{x})$ which corresponds to the magnetic charge density of these monopoles.

The regular 't Hooft-Polyakov monopoles are extended objects. $\hat{q}(\vec{x})$ defined on elementary spacelike cubes of the lattice should give the total magnetic charge of monopoles corresponding to that cube. In order to avoid double counting we use the position of the singularity in (6) to locate monopoles. Thus, the eigenvalue of $\hat{q}(\vec{x})$, $q(\vec{x})$ is supposed to be the product of the elementary magnetic charge and the number of monopoles minus anti-monopoles (which have negative magnetic charge) whose core defined by (6) lies inside the elementary cube at $\vec{x}$. There are $N-1$ different kind of regular 't Hooft-Polyakov monopoles in $SU(N)$ gauge theory. Consequently, $q(\vec{x})$ should be an $N-1$ component quantity. Moreover the magnetic charges of the lattice monopoles located by $q(\vec{x})$ should agree with the corresponding charges of the continuum monopoles.

We briefly summarize the construction of $q(\vec{x})$[8] for $SU(3)$ lattice gauge theory. The basic difficulty of such a procedure is that the magnetic field is given in terms of the gauge fields which are multi-valued functions of the link variables. We shall use different modulo functions in extracting the gauge fields and the magnetic flux from the link and plaquette variables, respectively.

The first step is to transform the configuration in the gauge where U_0 is time independent and diagonal. This gauge choice leaves the diagonal $U(1) \times U(1)$ and a discrete gauge symmetry unfixed. The latter corresponds to those nondiagonal gauge transformations which permute the eigenvectors of U_0. The construction is manifestly invariant in the continuous $U(1) \times U(1)$ part. The discrete gauge invariance is fixed by requiring that the eigenvalues of U_0 be ordered in increasing phase. This choice is close to the continuum case where this discrete gauge symmetry appears as a global one only. Gauge transformations which are non-periodical in the time direction by a global Z_3 transformation act multiplicatively on Ω so they may change the order of the eigenvalues since their phase is obtained in $\mathrm{mod}\, 2\pi$ only. Thus, the configurations are brought into the real Z_3 sector by applying such non-periodic Z_3 transformation which maximizes $\mathrm{Re} \sum_{\vec{x}} \mathrm{tr}\, \Omega(\vec{x})$.

The next step is to find the color components of the gauge fields which commute with U_0. This is achieved by writing the spatial link variables as $U_i = D_i V_i$ where D_i is diagonal and $-i \log D_{jj} = -i \log U_{jj} - \phi$, $3\phi = \sum_1^3 -i \log U_{jj} (\mathrm{mod}\, 2\pi)$. This form is the decomposition of U_i into the product of two $SU(3)$ matrices which transform a covariant way under the $U(1) \times U(1)$ gauge transformations. The spatial vectors D_i together with the diagonal $D_0 = U_0$ form the link variables of the $U(1) \times U(1)$ gauge model mentioned in the Introduction.

Finally, we have to locate Dirac monopoles in the $U(1) \times U(1)$ lattice gauge model described by D_μ. We start this process by obtaining two independent gauge fields A_μ^3 and A_μ^8 from the three diagonal elements of D_μ $A^8 = i/2 \log(D_\mu)_{33}$, $A_\mu^3 = -i \left[\log(D_\mu)_{11} + \frac{1}{2} \log(D_\mu)_{33} \right]$. The naive magnetic fields $\vec{B}_n^\alpha(\vec{x})$ are defined as $\vec{B}_n^\alpha(\vec{x}) = \vec{\nabla} \times \vec{A}^\alpha(\vec{x})$, $\alpha = 3$ or 8. These variables may jump by certain units when acting by $U(1) \times U(1)$ gauge transformations. The gauge invariant part is written as $\vec{B}^\alpha = \mathrm{mod}^\alpha (\vec{B}_n^3, \vec{B}_n^8)$. The two dimensional "modulo function" $y^3 = \mathrm{mod}^3(x^3, x^8)$ and $y^8 = \mathrm{mod}^8(x^3, x^8)$ gives that pair of numbers (y^3, y^8) which have the smallest absolute magnitude and satisfies

$$e^{ix^3 \mu^3 + ix^8 \mu^8} = e^{iy^3 \mu^3 + iy^8 \mu^8} \tag{7}$$

where $\mu^3 = \mathrm{diag}(1, -1, 0)$ and $\mu^8 = \mathrm{diag}(1, 1, -2)$. The magnetic charge density $q^\alpha(\vec{x})$ is finally given as $q^\alpha(\vec{x}) = \vec{\nabla} \vec{B}^\alpha(\vec{x})$. The div operation here involves $B^\alpha(\vec{x})$ taken on the plaquettes of the elementary cube. It is easy to verify that $q^\alpha(\vec{x})$ is an integer multiple of a basic unit. Due to (7) this basic unit satisfies the quantization conditions of the continuum theory. The double modulus function $\mathrm{mod}\,(x^3, x^8)$ removes the non-physical Dirac strings and guarantees the proper charge quantization in the same time.[8]

IV. CREATION AND ANNIHILATION OPERATORS

Consider the case of compact $U(1)$ gauge theory for simplicity. The generalization of the construction to non-Abelian models is straightforward. The expression of the magnetic charge density in the compact $U(1)$ gauge theory is $q(\vec{x}) = \vec{\nabla}\vec{B}(\vec{x})(\mathrm{mod}\, 2\pi)$, $\vec{B}(\vec{x}) = \vec{\nabla} \times \vec{A}(\vec{x})(\mathrm{mod}\, 2\pi)$, where $\vec{A}(\vec{x})$ is the phase of the link variables in spatial directions. The simple prescription[15] that the creation and annihilation operators shift the eigenvalues of the gauge field operators by a lattice version of the magnetic Coulomb field is not satisfactory. In fact, this procedure may change the magnetic charge density at any location due to the function $\mathrm{mod}\, 2\pi$ appearing in the definition of $q(\vec{x})$. A simple way to cure this defect is to define a field dependent shift $\delta A[\vec{x}, \vec{A}]$. The shift should be such that (i) it causes $q(\vec{x})$ to decrease by one unit at the chosen location if it was positive there; and, (ii) the values of $\vec{A}(\vec{x}) + \delta\vec{A}[\vec{x}, \vec{A}]$ sweep through every element of the zero monopole sector ones and only ones as $\vec{A}(\vec{x})$ goes through all fields of the one monopole sector. By the help of such a shift we define the annihilation operator for monopoles $a(x_0)$ and creation operator for anti-monopoles $b^\dagger(x_0)$ as

$$
a(\vec{x}_0) = \sum_{n(\vec{x}_0)} \sqrt{n(\vec{x}_0)+1} \sum_{n(\vec{x}),\vec{x}\neq\vec{x}_0} \prod_{\vec{x}} \hat{P}\left(n(\vec{x}),\vec{x}\right) e^{-i\sum_{\vec{x}} \vec{E}(\vec{x})\delta\vec{A}[\vec{x},\hat{\vec{A}}]} \times
$$
$$
\times \prod_{\vec{x}\neq x_0} \hat{P}\left(n(\vec{x})\right) \hat{P}\left(n(\vec{x}_0)+1,\vec{x}_0\right)
$$

$$
b^\dagger(\vec{x}_0) = \sum_{n(\vec{x}_0)} \sqrt{n(\vec{x}_0)+1}\,\hat{P}(-n(x_0)-1,x_0) \sum_{n(\vec{x}),\vec{x}\neq x_0} \times
$$
$$
\times \prod_{\vec{x}\neq \vec{x}_0} P\left(-n(\vec{x}),\vec{x}\right) e^{-i\sum_{x} \vec{E}(\vec{x})\delta\vec{A}[\vec{x},\hat{\vec{A}}]} \prod_{\vec{x}} P(-n(\vec{x}),\vec{x})
$$

$$
\tag{8}
$$

where $\vec{E}$ is the canonically conjugated operator to $\vec{A}$ and $\hat{P}(n,\vec{x})$ is the projection operator projecting into the space where $q(\vec{x})$ has eigenvalue $2n\pi$. The summations at $x = x_0$ and $x \neq x_0$ extend over non-negative and all integers, respectively. The shift $\delta\vec{A}[\vec{x}, \vec{A}]$ is chosen to decrease the magnetic charge density by one unit in the cube at $x = x_0$.

What is left to find is the actual form of the shift $\delta\vec{A}[\vec{x}, \vec{A}]$. We shall give an algorithm to obtain such shift for a wide class of configurations. Configurations where there is a cube with at least three faces which have large fluxes $\cos\left(\nabla_i A_j - \nabla_j A_i\right) < 0$ require additional considerations. But nothing we are going to say about the properties of magnetic monopoles depends on the particular choice of δA. The shift δA will be gauge invariant by definition. Its value for gauge configurations satisfying unique gauge fixing conditions is obtained in terms of the shift of the magnetic flux $B_i = \frac{1}{2}\epsilon_{ijk}\nabla_j A_k$ in the following way: First we order the elementary cubes in such a way that the first cube is the one at $x = x_0$ and every cube has common face with at least one of the preceeding ones. The shift will be computed in the order of the cubes. Since the cube at x_0 contains a monopole the change of the fluxes on the plaquettes of that cube $B \to (B + \pi) \left[1 - 2\pi / (6\pi + q(x_0))\right] - \pi$ decreases the monopole number exactly by one unit. The shift of the magnetic flux at the new, uncommon plaquettes of the next cube is $B \to (B + \pi) \left[1 + \phi_0 / (6\pi + q(x_0))\right] - \pi$ where ϕ_0 is the flux corresponding to the plaquettes common with earlier cubes. This latter procedure is to be repeated for each cubes.

It is easy to see that $q(\vec{x})/2\pi = a^\dagger(\vec{x})a(\vec{x}) - b^\dagger(\vec{x})b(\vec{x})$. The verification of the canonical commutation relation leads to a surprise: magnetic monopoles obey parastatistics. In fact, since the absolute magnitude of the magnetic charge in an elementary cube is bounded $|q(\vec{x})| \leq 24\pi$, sufficiently high power of the creation or annihilation operators vanish $e.g.$ $(a(x))^n = 0$ for $n > 24$. The actual bound on the magnetic charge density depends on the lattice used but it is finite for any other lattice geometry (The case of random lattice requires special investigations. We believe that different moments of the distribution of the upper bound remains finite in this case). We suspect that this circumstance is not a lattice artifact but rather reflects the physical properties of magnetic monopoles. What is important for this simple result is the presence of the ultraviolet cutoff (ask no question about the distribution of the magnetic charge inside of the elementary cube) and the existence of the compact phase factor (contribution of each elementary surface to the magnetic flux is bounded). It would be very illuminating to find the decomposition of a and b into the finite sum of fermionic operators.

V. CHROMOMAGNETIC MONOPOLES AT FINITE TEMPERATURE

It was mentioned in the Introduction that the elimination of non-diagonal gluons in unitary gauge leads to an Abelian model with the same static confinement mechanism. We shall now present a strong coupling picture of the condensation of magnetic monopoles and the mass generation for dual gluons in the low temperature phase. These phenomena lead to confinement via dual Meissner effect.[3] Such description of the confinement mechanism is less direct as the straightforward lattice strong coupling expansion and contains few assumptions. But we hope that those assumptions can ultimately be tested by numerical computations and the approach to the continuum will be simpler in this scenario.

First, we show that the states dominating the partition function at strong coupling have indefinite number of magnetic monopoles. Instead of computing $\langle a(\vec{x}_0)\rangle$ or $\langle b(\vec{x}_0)\rangle$, we consider a simpler quantity $\langle a'(\vec{x}_0)\rangle$,[15] $a'(\vec{x}_0) = \exp\left\{i\sum_x \hat{\vec{E}}(\vec{x})\vec{A}(\vec{x})\right\}$ where $\vec{A}(\vec{x})$ is the lattice analogue of the field of the continuum Dirac monopole located at $\vec{x} = \vec{x}_0$. All what we need is that the operator $a'(\vec{x}_0)$ changes the number of monopoles (it may changes at $x \neq x_0$ as well). It is easy to see that $\langle a'(\vec{x}_0)\rangle = 1 + c/g^2$ with finite c. In fact, strong coupling in the nonabelian theory allows local, strong coupling expansion in the effective Abelian model as well. The first correction in $1/g^2$ is finite because it is proportional to $\sum_{\vec{x}}(\vec{\nabla} \times \vec{A})^2$ and the point singularity of the magnetic Coulomb field is smoothed out by the regularization.

Numerical calculation of $\rho = \langle a^\dagger(\vec{x}_0)a(\vec{x}_0)\rangle = \langle b^\dagger(\vec{x}_0)b(\vec{x}_0)\rangle$ at finite temperature[9] shows a cusp at $T = T_{\text{dec}}$. Moreover, the ultraviolet finite, temperature dependent part of ρ is nonzero for $T \neq T_{\text{dec}}$. Assuming the absence of bulk phase transitions as $1/g^2$ leaves the realm of the strong coupling expansion we may expect that the monopole condensate is present for $T < T_{\text{dec}}$. In the analogous case of the singular Z_3 monopoles in the mixed fundamental-adjoint action gauge models[6] one finds a first order bulk phase transition at the crossover region where the condensate of these monopoles disappears.

The actual form of the effective $U(1) \times U(1)$ gauge model is not known but it should be local in the strong coupling limit. A simplest choice for the effective action would be

$$
\begin{aligned}
S &= \frac{1}{e^2}\sum \text{Re tr} P[D] \\
&= \frac{1}{e^2}\sum \left[\cos\left(F_{\mu\nu}^3 + F_{\mu\nu}^8\right) + \cos\left(F_{\mu\nu}^3 - F_{\mu\nu}^8\right) + \cos\left(2F_{\mu\nu}^8\right)\right]
\end{aligned}
\tag{9}
$$

where $P[D]$ is the plaquette formed by the diagonal matrices D_μ and $F_{\mu\nu}^3$ and $F_{\mu\nu}^8$ are the field strength tensors corresponding to A_μ^3 and A_μ^8 and e^2 is proportional to g^2. It has already been noted that the condensation of Dirac monopoles leads to area law for Wilson loops in compact $U(1)$ gauge theory.[11,12] The analytical strong coupling description of this phenomenon[11] uses auxiliary variables to define the magnetic current. To make contact with the monopole configurations defined above we replace (9) by

$$
S = \frac{1}{e^2}\sum \left[\left(G_{\mu\nu}^3\right)^2 + 3\left(G_{\mu\nu}^8\right)^2\right]
\tag{10}
$$

with $G_{\mu\nu}^\alpha = \text{mod}^\alpha(F_{\mu\nu}^3, F_{\mu\nu}^8)$. In order to simplify the notation we shall consider the case of $SU(2)$ theory where there is only one Abelian field A_μ^3. The generating functional in Minkowski space-time (which is necessary to read off masses directly) can be written as

$$Z[j] = \int_{-\infty}^{\infty} \mathcal{D}[A_\mu] e^{\frac{i}{4} \sum [F_{\mu\nu} - \frac{2\pi}{e} n(F_{\mu\nu})]^2 + i\frac{e}{2} \sum j_{\mu\nu} F_{\mu\nu}} \tag{11}$$

where $A_\mu(x) = A_\mu^3(x)/e$, $f_{\mu\nu} = \nabla_\mu A_\nu - \nabla_\nu A_\mu$, the integer valued function $n(f_{\mu\nu})$ is introduced to take care of the mod 2π prescription $\left| f - 2\pi n(f)/e \right| < \pi/e$ and $j_{\mu\nu}$ is the external current. It is advantageous to introduce the dual variables

$$\tilde{f}_{\mu\nu} = \frac{1}{2} \epsilon_{\mu\nu\rho\sigma} f_{\rho\sigma} , \quad \tilde{j}_{\mu\nu} = \frac{1}{2} \epsilon_{\mu\nu\rho\sigma} j_{\rho\sigma} \tag{12}$$

and $\tilde{A}_\mu$ which satisfies the conditions $\tilde{f}_{\mu\nu} = \nabla_\mu \tilde{A}_\nu - \nabla_\nu \tilde{A}_\mu$ and $\nabla_\mu \tilde{A}_\mu = 0$. The generator functional (11) reads as

$$Z = \int \mathcal{D}[\tilde{A}_\mu] \prod_x \delta(\nabla_\mu \tilde{A}_\mu) \times$$

$$\times e^{\frac{i}{4} \sum (\tilde{f}_{\mu\nu} + e\tilde{j}_\mu)^2 - i\frac{\pi}{e} \sum \ell_\mu \tilde{A}_\mu + i\frac{\pi^2}{e^2} \sum n^2(\tilde{f}_{\mu\nu}) - \frac{i}{4} e^2 \sum j_{\mu\nu}^2} \tag{13}$$

in terms of the dual variables and $\ell_\mu = \epsilon_{\mu\nu\rho\sigma} \nabla_\nu n(f_{\rho\sigma})$. This latter integer describes the current of the magnetic monopoles defined in the previous Sections.

We show finally in the strong coupling expansion that the dual gluon field $\tilde{A}_\mu$ acquires mass in the presence of monopole condensate. Set $j_{\mu\nu} = 0$ temporally and expand the exponential function in $1/e$. The leading order nonlocal contribution to the dual gluon self energy $\pi(\vec{x}, t)$ comes from the insertion of the second term of the exponent in (13). It contains the correlation functions of the operators $a^\dagger(\vec{x}, t) a(\vec{x}, t)$, $a^\dagger(\vec{0}, 0) a(\vec{0}, 0)$ and $b^\dagger(\vec{x}, t) b(\vec{x}, t)$, $b^\dagger(\vec{0}, 0) b(\vec{0}, 0)$, respectively. The time coordinate of the operators denotes the time slice where they shift the magnetic field. Since the creation and annihilation operators do not cluster for small $1/e^2$, arbitrarily large monopole loops are present in the path integral so $\pi(\vec{k}, \omega)$ is nonzero in the infrared limit. Static chromoelectric charges which show up through the current $\tilde{j}_{\mu\nu}$ as magnetic charges in this dual model experience linear potential due to the Meissner effect.

ACKNOWLEDGEMENT

This work was supported in part by funds provided by the U. S. Department of Energy (D.O.E.) under contract #DE-AC02-76ER03069.

REFERENCES

1. M. Luscher, *Comm. Math. Phys.* **85**, (1982) 39; J. Polonyi, *Phys. Rev.* **D29**, (1984) 716; P. Woit, *Phys. Rev. Lett.* **51**, (1983) 638; I. A. Fox, J. P. Gilchrist, M. L. Laursen and G. Schierholz, *Phys. Rev. Lett.* **54**, (1985) 749; G. Lasher, A. Phyllips and D. Stone, in *Proceedings of the LBL Meeting on "Quark Confinement and Deliberation"*, (1985); M. Teper, *Phys. Lett.* **162B**, (1985) 357; J. Hoek, *Phys. Lett.* **166B**, (1986) 199.

2. J. Smit, "Remnants of the Index Theorem on the Lattice", ITFA-86-14.

3. H. Nielsen and P. Olesen, *Nucl. Phys.* **B61**, (1973) 45; S. Mandelstam, *Phys. Rep.* **23C**,(1976) 245; G. 't Hooft, *Phys. Scr.* **25**, (1982) 133.

4. G. 't Hooft *Nucl. Phys.* **B67**, (1974) 276; S. Polyakov, *JETP Lett.* **20**, (1974) 194.

5. L. McLarren and B. Svetitsky, *Phys. Lett.* **98B**, (1981) 195; J. Kuti, J. Polonyi and K. Szlachanyi, *Phys. Lett.* **98B**, (1981) 199.

6. G. Mack and V. B. Petkova, *Ann. Phys.* **12**, (1980) 117; G. Holliday and A. Schwimmer, *Phys. Lett.* **B101**, (1981) 327, **B102**, (1981) 337; G. Bhanot and M. Creutz, *Phys. Rev.* **D24**, (1981) 3212.

7. S. Coleman and E. Weinberg, *Phys. Rev.* **D7**, (1973) 1888.

8. J. Polonyi and H. W. Wyld, in preparation.

9. J. Polonyi, in *Proceedings of the Conference "Quark Matter '86"*, (Asilomar, CA, 1986).

10. J. Polonyi and K. Szlachanyi, *Phys. Lett.* **110B** (1982) 395.

11. T. Banks, R. Meyerson and J. Kogut, *Nucl. Phys.* **B129**, (1979) 1882.

12. T. DeGrand and D. Toussaint, *Phys. Rev.* **D22**, (1980) 2478.

13. J. Arafune, P. G. O. Freund, G. Goebel, *J. Math. Phys.* **16**, (1975) 443.

14. A. Sinha, *Phys. Rev.* **D14**, (1976) 2016.

15. E. Fradkin and L. Susskind, *Phys. Rev.* **D17**, (1978) 2637.

TOWARD A PSEUDOFERMION CALCULATION OF THE HADRONIC MASS SPECTRUM

J. Potvin

Physics Department, Brookhaven National Laboratory
Upton, NY 11973

in collaboration with:

M. Campostrin, INFN, Pisa, Italy
K.J.M. Moriarty, Dalhousie University, Halifax, Canada
C. Rebbi, BNL, Upton, NY and Boston Univ., Boston, MA

ABSTRACT

We present our preliminary results of a pseudofermion simulation of QCD, on a $10^3 \cdot 32$ lattice at $\beta = 5.7$. The fermions are of the Kogut-Susskind type (3 flavors) with mass $ma = 0.05$. It is found that the hadron propagators take more than 10000 Metropolis iterations to equilibrate, a factor 5 more than what is needed in a typical quenched calculation. We present also our results on the effects of dynamical quarks on the Wilson loops.

MOTIVATIONS

The calculation of the hadron mass spectrum without including the dynamical quarks (i.e. the quenched approximation) has lead to very encouraging results. For one, the calculated hadron masses have been found consistent with the experimental values, within statistical and systematic error. Also, the breaking of chiral symmetry and the Goldstone nature of the pion have been seen.[1]

Of course, the quenched approximation is not the whole story, since QCD is a theory about gluons AND quarks. It is therefore important to verify that, upon the re-introduction of the dynamical quarks into the picture, the above results will not change much. Moreover, dynamical quark effects should be visible in the heavy quark potential, at separations beyond one Fermi where hadronization takes place.

It has been known for a long time that the unquenched simulation of QCD is much more demanding on computer resources. It is only with the recent availability of supercomputer time that algorithms with dynamical fermions have for the first time been tried out and tested on reasonably large lattices ($\geq 8^4$) and couplings (≥ 5.0), and with small masses (≤ 0.10). Examples of such algorithms are presented and discussed in this Conference Proceedings.

Most physics applications of these algorithms have been concerned with QCD at finite temperature. In contrast, only three large scale studies of the hadron mass spectrum have been carried out:

1. a Langevin calculation by Fukugita et al.[2] on a $9^3 \cdot 18$ lattice, at $\beta = 5.5$ with Wilson fermions. For more details, see Fukugita's talk in these Proceedings.

2. a pseudofermion simulation by Fucito et al.,[3] on a $12^3 \cdot 24$ lattice at $\beta = 5.4$ with three flavors of Kogut-Susskind fermions. Three mass values were tried: 0.02, 0.05, and 0.10. Their statistics was limited however, with only nine propagators per mass.

3. another pseudofermion calculation, by Laerman et al.,[4], this time with the group SU(2). The lattice size was $8^3 \cdot 16$ and $1.85 \leq \beta \leq 2.50$; four flavors of Kogut-Susskind fermions were used in the mass range $.035 \leq ma \leq 0.200$.

We have continued this exploratory effort with another pseudofermion calculation, at mass $ma = 0.05$ with three flavors of Kogut-Susskind quarks. The lattice size was $10^3 \cdot 32$, at $\beta = 5.7$. In this talk I will present our preliminary results with emphasis on the equilibrium time required for the Wilson loops ($W(1,1)$) and pion propagator (P_{tt_0}). Our conclusions will be that it is very different for both quantities: thousands more iterations (i.e. link upgrades) are needed for $W(1,1)$ than for P_{tt_0}; this is also many more than what is required in a calculation of P_{tt_0} in the quenched approximation.

Our simulation was carried out on a CDC CYBER 205 based at Rockville Md., thanks to the generosity of the Control Data Corporation.

THE ALGORITHM

According to the procedure described in Ref. 5 and Ref. 6, the pseudo-fermions ϕ_x are upgraded first with a heat bath algorithm in the background of fixed gauge fields. Here the ϕ_x are modified in the following way,

$$\phi_x{}' = \phi_x + \delta\phi_x$$
$$\delta\phi_x = \frac{1}{\sqrt{2+m^2}}V_x + \frac{1}{2+m^2}\left(D^\dagger + m\right)\left(D + m\right)\phi_x \tag{1}$$

V_x being a Gaussian random number of unit variance and $(D+m)$ the Dirac operator. Such an upgrade is done 300 times, calculating and accumulating the reaction J_x^μ every 12 sweeps. J_x^μ is needed in the calculation of the variation of the total action Ref. 6:

$$\delta S = \delta S_{\text{Gauge}} - \frac{1}{8}n_f \sum_{x,\mu}\left(Tr\langle J_x^\mu\rangle \delta U_x^\mu + \text{H.C.}\right) \tag{2}$$

$\langle J_x^\mu\rangle$ is thus evaluated from a sample of 25 values, in similarity with the calculations of Ref. 3 and Ref. 7. A distinct feature of our study is that not all ϕ_x are upgraded at each pseudofermion sweep but only a subset of approximately 1000, chosen in a uniformly random fashion (the sites being still distanced at least 3 steps away to allow for vectorization). This procedure still satisfies detailed balance, and eliminates any potential directional correlations due to upgrading. With the use of asynchronous I/O, it also shortens the pseudofermion upgrade while maintaining a reasonable vector length.

After the 300 pseudofermion sweeps, the link variables are then upgraded with a standard Metropolis procedure, at a rate of 8 upgrades per link. Here the acceptance is 81%. This coupled heat bath/Metropolis update (or "iteration") took 0.3 milliseconds per link and was run for a total of 9000 iterations (14500 after this Meeting). The Wilson loops were measured every 10 iterations and propagators every 500.

WILSON LOOPS

Figures 1 and 2 show the evolution of the 1×1, 2×2, 3×3 and 4×4 Wilson loops with respect to iteration number. Clearly, the smaller loops have equilibriated after 2500 iterations, more or less. The larger ones are of course slower and show equilibration after 4000 iterations.

After 14500 iterations, we get

$$W(1,1) = 0.5816 \pm .0002 \tag{3}$$

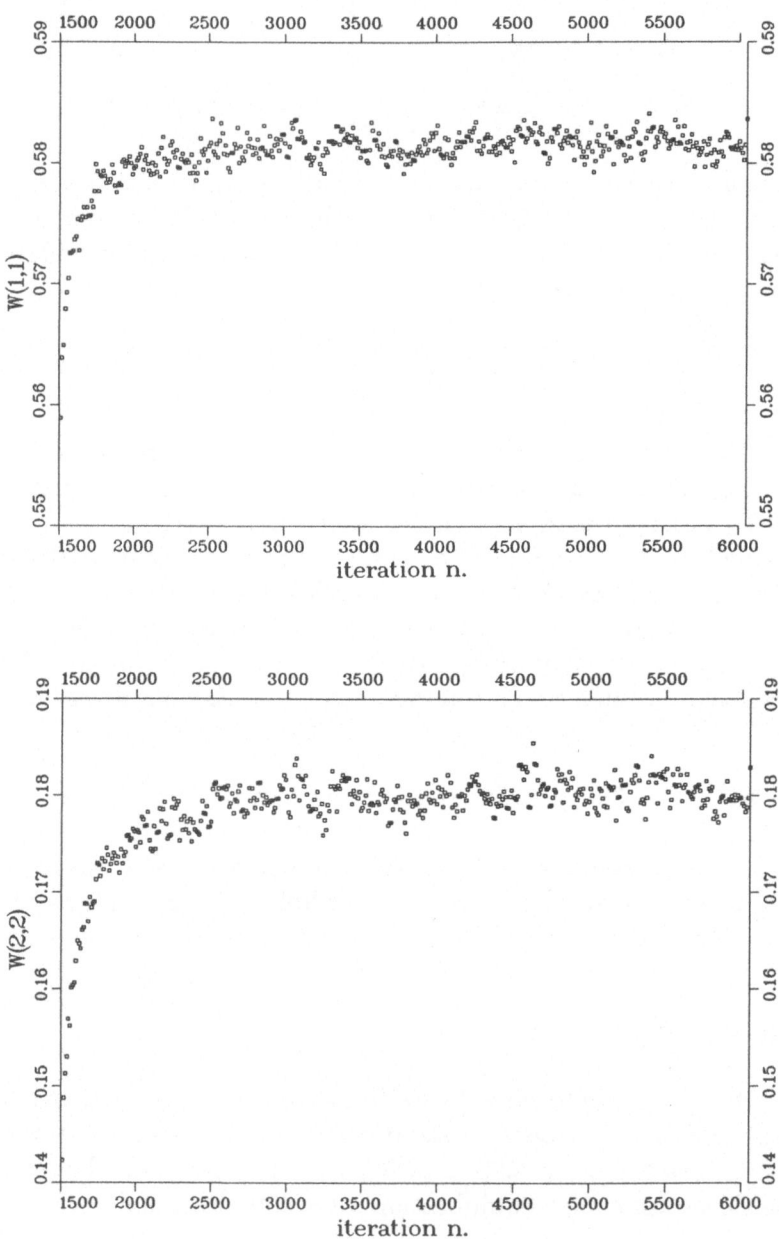

Fig. 1: 1×1 and 2×2 Wilson loops versus the iteration number.

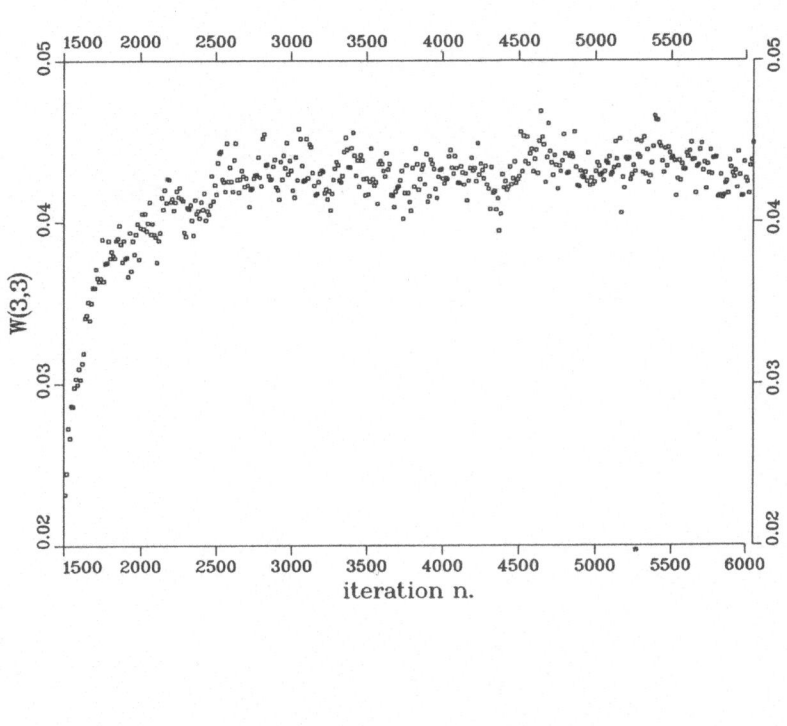

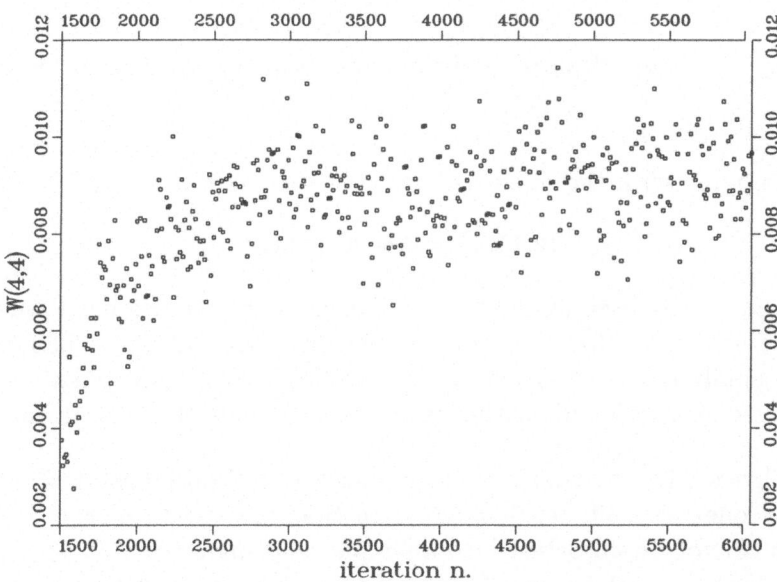

Fig. 2: 3×3 and 4×4 Wilson loops versus the iteration number.

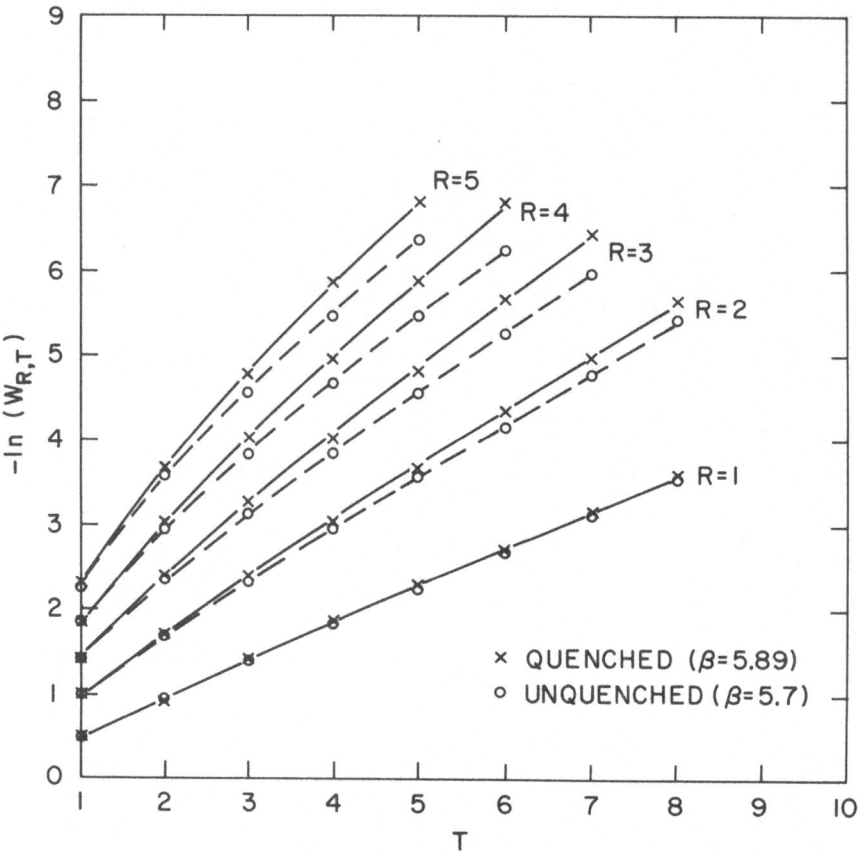

Fig. 3: Comparison of quenched and unquenched Wilson loops.

to be compared with the quenched value at the same coupling.[8]

$$W(1,1) = 0.5490 \pm .0010 \qquad (4)$$

By running a quenched simulation on the same lattice, we have found that the value shown in (3) is reproduced within errors at $\beta = 5.89$. Let us mention finally that a recent study by Gavai, Gocksch and Heller[9] suggests that the pseudofermion algorithm may underestimate $W(1,1)$ at $ma = 0.05$.

Figure 3 is a comparison of unquenched Wilson loops at $\beta = 5.7$ and quenched ones at 5.89. It demonstrates that the effects of the dynamical fermions cannot be absorbed into a simple renormalization of β, m, or n_f. This confirms early claims by Gavai and Karsch[10] ($ma = 0.10$). Let us point out however that on this issue Fukugita et al.[2] have concluded otherwise.

HADRON PROPAGATORS

The hadron propagators are defined by[3]

$$\text{meson}: P_{tt_0}^{m(k)} = \sum_x S_x^k \operatorname{tr} \Delta_{(x,t)}^{-1} \Delta_{(x,t_0)}^{-1} \tag{5}$$

$$\text{baryon}: P_{tt_0}^{b(k)} = \sum_x S_x^k \det \Delta^{-1} \tag{6}$$

$$\Delta(x, t_0) = (D + \tilde{m}a)_{x,t_0} \tag{7}$$

The matrix Δ was inverted with the conjugate gradient algorithm in the following parameter range:

$\tilde{m}a$	C-G sweeps	residue (Barkai et al.[1])
.50	50	10^{-8}
.10	100	10^{-7}
.05	200	10^{-7}
.02	350	10^{-6}
.01	500	10^{-6}

Figure 4 shows how the pion propagator P_{tt_0} evolves with the (Metropolis) iteration number, for $\tilde{m}a = .05$ and $\tilde{m}a = .01$. Each continuous line approximates the general path followed by the data as it migrates through the time slices, during a given iteration. This data is averaged over three source points ("t_0").

The figure shows the propagator near $t = 16$ to decrease by an order of magnitude over a range of at least 6000 iterations. Not surprisingly, the effect is more pronounced at $\tilde{m}a = .01$. In that case, 10000 are needed to get a good idea of how far down the minimum of P_{tt_0} has to go. Not shown is the fact that it does go up again, from iteration 10000 to 14500, without going beyond the 3000 line (this data was generated after this Lattice Meeting). In the case $\tilde{m}a = .05$, we need 6000 iterations to see the minimum go down and up again (fig. 4).

All this shows that one needs at least several thousands iterations to get a good idea of the drift of propagator values through the simulation. This is many more iterations than for $W(1, 1)$. It is also many more than the 2000 commonly used in most previous quenched studies. We have verified this last point ourselves in our quenched runs at $\beta = 5.89$ (fig. 5).

In the parameter range $ma = .05$ and $\tilde{m}a = .01$, $.05$, it appears that a pseudofermion calculation of the hadron mass spectrum with good

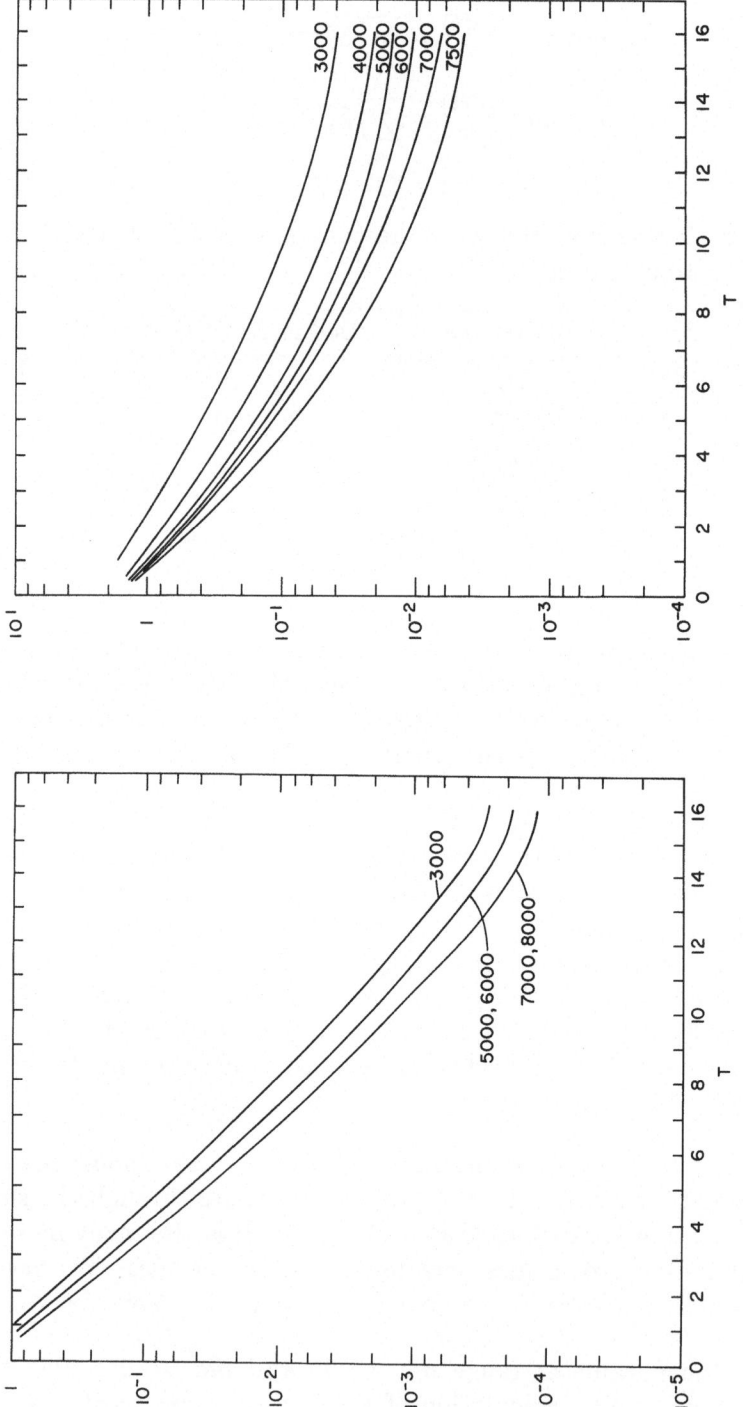

Fig. 4: The unquenched pion propagator. (a) $\tilde{m}a = .05$ and (b) $\tilde{m}a = .01$.

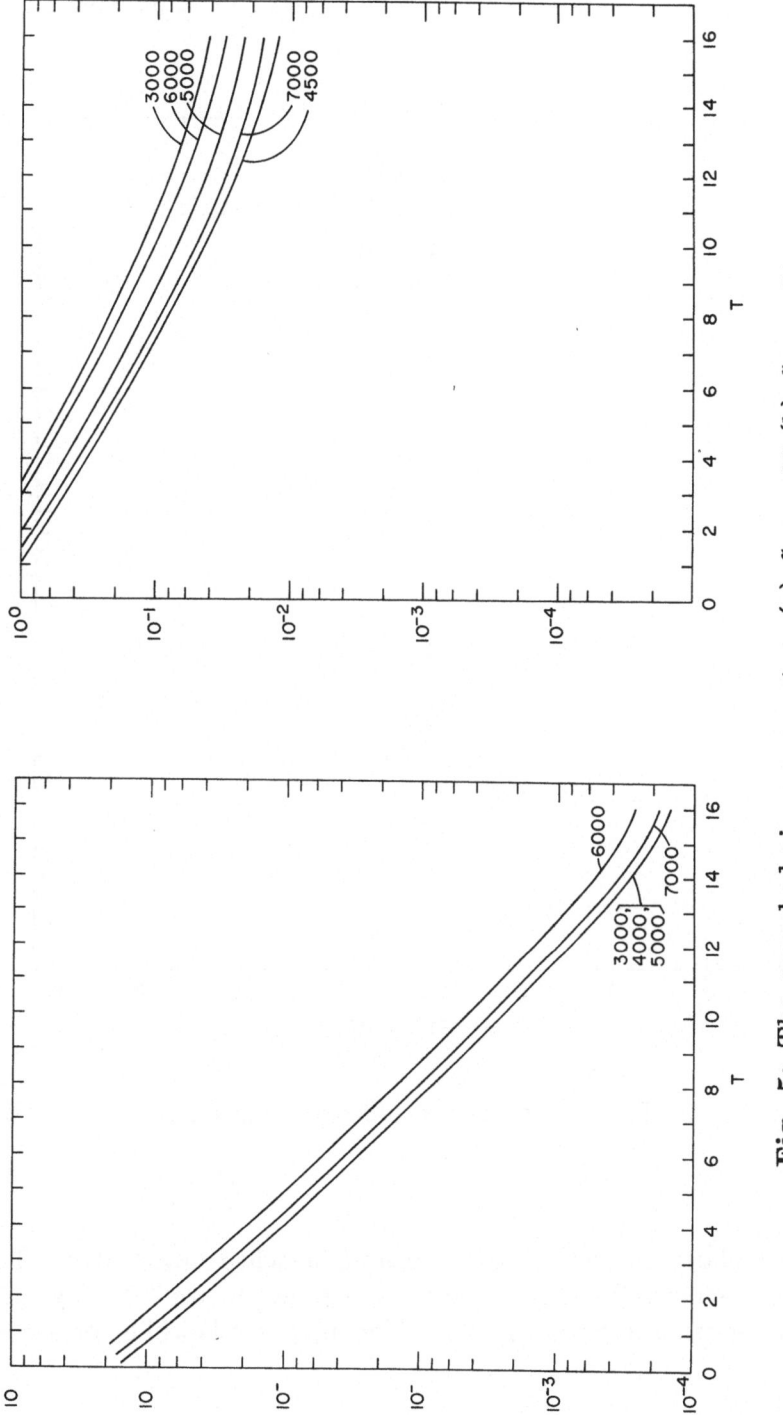

Fig. 5: The quenched pion propagator. (a) $\tilde{m}a = .05$ (b) $\tilde{m}a = .01$.

statistics requires at least 4 to 5 times more iterations than in the quenched case. This is because a pseudofermion simulation has one more $(D + ma)$-inversion,which will also be affected by critical slowdown caused by small m.

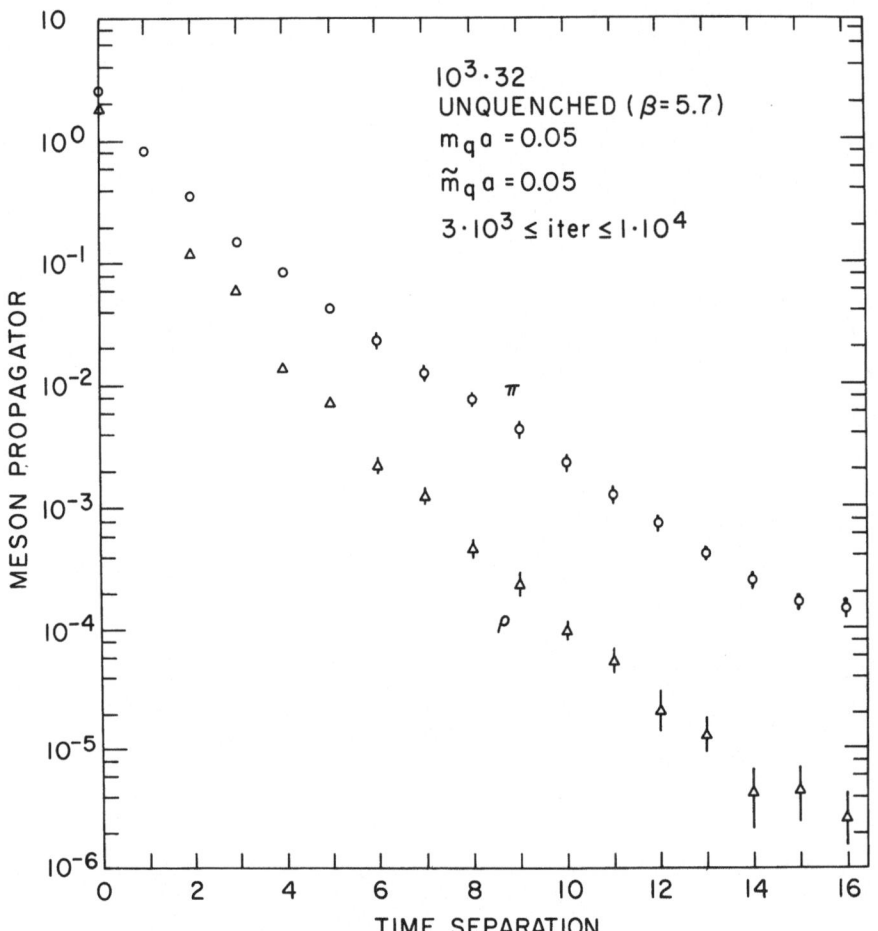

Fig. 6: The unquenched π and ρ propagators.

Figure 6 shows the resulting π and ρ unquenched propagators. The mass can be extracted according to standard methods; figure 7 displays preliminary pion masses, as function of $\sqrt{\tilde{m}a}$. The full analysis will be presented elsewhere.

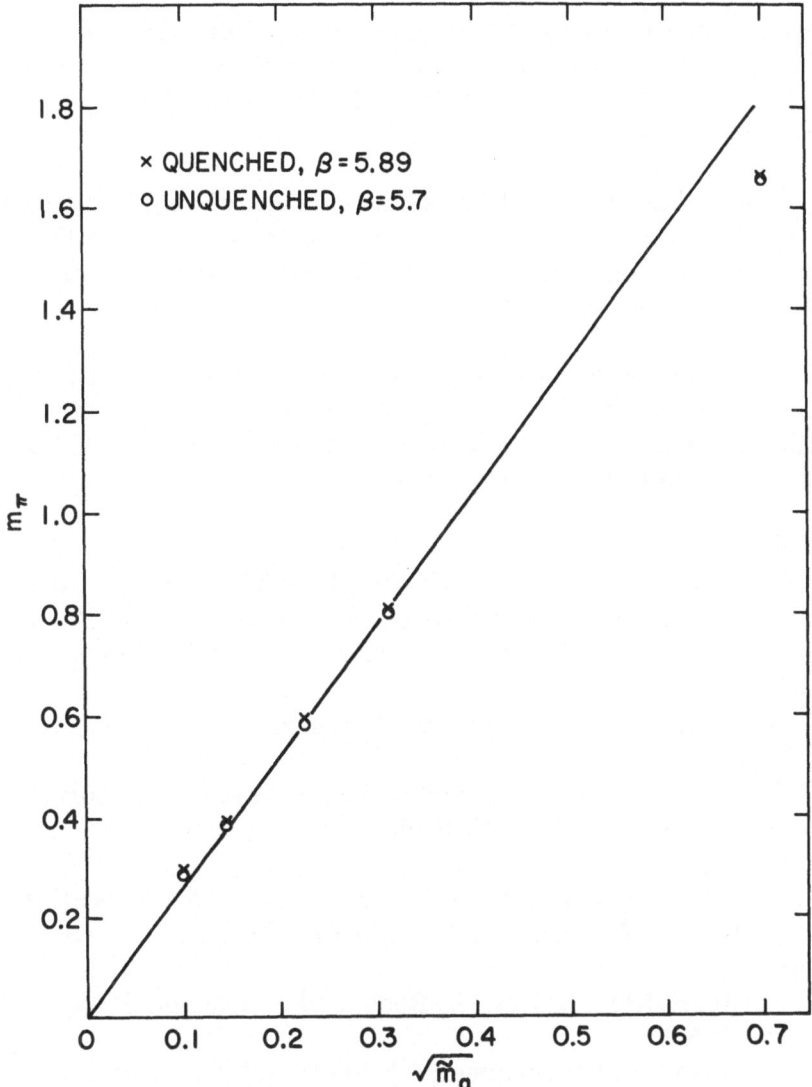

Fig. 7: Pion masses (unquenched).

CONCLUSIONS

To the best of our knowledge, this is the first demonstration of critical slow down of long wavelength structures in a background of dynamical fermions. Our propagators at $\tilde{m}a = ma = .05$ certainly need more than 8000 iteration on a $10^3 \cdot 32$ lattice in order to get reasonably good statistics. In the case of $ma = .05$ $\tilde{m}a = .01$ the outlook is worse, as at least 10000 to 14500 are needed. This indicates how hard a similar study at $ma = \tilde{m}a = .01$ will be (we anticipate a factor 2 in iteration number). Compared with a quenched calculation, this represents 4 to 5 times more iterations, a hefty 40 to 50-fold increase in computer time, if a factor of 0.1 is assumed for link upgrade in a quenched simulation.

This makes the importance of acceleration of long wavelength structures equilibration time all too obvious. Let us mention that this question tion has already been addressed by the Cornell[11] and the Kyoto-Tsukuba[12] groups in the context of the Langevin equation.

The situation looks better in the case of Wilson loops, where 3000 or so iterations are required for equilibration. We have studied the $Q\bar{Q}$-potential and found that the effects of dynamical fermions do not amount to a simple renormalization of the parameters.

References

1. For recent results see: K.C. Bowler, D.L. Chalmers, A. Kenway, R.D. Kenway, G.S. Pawley, D.J.Wallace, Nucl. Phys B240 [FS12] (1984) 213; J.P. Gilchrist, H.Schneider, G. Schierlholz and M. Teper Phys. Lett. 136B (1984) 87;
A. Billoire, E. Marinari and R. Petronzio, Nucl. Phys. B251 [FS13] (1985) 141;
D. Barkai, K.J.M. Moriarty and C. Rebbi, Phys. Lett. 156B (1985) 385;
A. Konig, K.H. Mütter, K. Schilling and J. Smit, Phys. Lett. 157B (1985) 421;
S. Itoh, Y. Iwasaki and T. Yoshie, Phys. Lett. 167B (1986) 443..

2. M. Fukugita, Y. Oyanagi and A. Ukawa, Phys. Rev. Lett. 57 (1986) 953.

3. F. Fucito, K.J.M. Moriarty, C. Rebbi and S. Solomon, Phys. Lett. 172B (1986) 235.

4. E. Laermann, F. Langhammer, I. Schmitt and P. M. Zerwas, Phys. Lett. 173B (1986) 437 and 443.

5. F. Fucito, E. Marinari, G. Parisi and C. Rebbi, Nucl. Phys. B180 (1981) 360.

6. H.W. Hamber, E. Marinari, G. Parisi and C. Rebbi, Phys. Lett. 124B (1983) 99.

7. R.V. Gavai, Nucl. Phys. B269 (1986) 530.

8. A. Hasenfratz, P. Hasenfratz, U. Heller and F. Karsch, Phys. Lett. 143B (1984) 193;
M. Creutz and K.J.M. Moriarty, Phys.Rev. D26 (1982) 2166;
A. Hasenfratz, P. Hasenfratz, U. Heller and F. Karsch, Zeitschrift f. Ph. C25 (1984) 191.

9. R.V. Gavai, A. Gocksch and U. Heller, Nucl. Phys. (in press).

10. R.V. Gavai and F. Karsch, Phys. Rev. Lett. 57 (1986) 40.
11. G.G. Batrouni, G.R. Katz, A.S. Kronfeld, G.P. Lepage, B. Svetitsky and K.G. Wilson, Phys. Rev. D32 (1985) 2736.
12. M. Fukugita, this conference proceedings.

MONTE CARLO DETERMINATION OF THE SPIN-DEPENDENT POTENTIALS

C. Rebbi

Physics Department, Brookhaven National Laboratory
Upton, NY 11973
and
Boston University, Boston, MA 02215

in collaboration with:

M. Campostrin, INFN, University of Pisa, Italy
K.J.M. Moriarty, Dalhousie University, Canada

SUMMARY

The bound states of heavy quark systems can be calculated by a Hamiltonian formulation, based on an expansion of the interaction into inverse powers of the quark mass. To second order in $1/m$ one obtains[1,2]

$$H = \frac{\vec{p}^2}{m} + V(r) + \frac{\vec{S}_+ \cdot \vec{L}_+ + \vec{S}_- \cdot \vec{L}_-}{2m^2} \frac{1}{r}\left(\frac{dV}{dr} + 2\frac{dV_1}{dr}\right)$$
$$+ \frac{\vec{S}_+ \cdot \vec{L}_- + \vec{S}_- \cdot \vec{L}_+}{m^2} \frac{1}{r}\frac{dV_2}{dr}$$
$$+ \frac{1}{m^2}\left[\left(\vec{S}_+ \cdot \hat{r}\right)\left(\vec{S}_- \cdot \hat{r}\right) - \frac{1}{3}\vec{S}_+ \cdot \vec{S}_-\right] V_3(r)$$
$$+ \frac{1}{3m^2}\vec{S}_+ \cdot \vec{S}_- V_4(r). \tag{1}$$

The potentials $V_1 - V_4$ account for the spin-orbit and spin-spin coupling between quark and antiquark and are responsible for the fine and hyperfine splittings in heavy quark spectroscopy. They can be expressed in terms of expectation values of Wilson loop factors with suitable insertions of chromomagnetic or chromoelectric fields, denoted by $\langle \ldots \rangle_W$, as in the following equation

$$\epsilon_{ijk}\hat{r}_k\frac{d\widetilde{V}_1\left(r,T\right)}{dr} = \int_0^T dt \int_0^T dt' \left(t'-t\right) \langle g^2 B_i\left(0,t\right) E_j\left(0,t'\right)\rangle_W,$$

$$\epsilon_{ijk}\hat{r}_k\frac{d\widetilde{V}_2\left(r,T\right)}{dr} = \int_0^T dt \int_0^T dt' \left(t'-t\right) \langle g^2 B_i\left(0,t\right) E_j\left(\vec{r},t'\right)\rangle_W,$$

$$\left(\hat{r}_i\hat{r}_j - \frac{1}{3}\delta_{ij}\right)\widetilde{V}_3\left(r,T\right) + \frac{1}{3}\delta_{ij}\widetilde{V}_4\left(r,T\right) = \int_0^T dt \int_0^T dt'\langle g^2 B_i\left(0,t\right) B_j\left(\vec{r},t'\right)\rangle_W,$$

$$V_i\left(r\right) = \lim_{T\to\infty} \frac{1}{T}\frac{\widetilde{V}_i\left(r,T\right)}{\langle 1\rangle_W}. \tag{2}$$

We have used a Monte Carlo simulation to evaluate the expectation values in Eq. (2) and, from them, the spin-dependent potentials. The calculation was performed on a $16^3 \times 32$ lattice, in the quenched approximation and with Wilson's action, at $\beta = 6.2$. The code has been described in detail in Ref. 3 and the results for the spin-dependent potentials have been presented in Ref. 4. Independent lattice calculations of spin-dependent potentials have also been published in Refs. 5-8.

Our results are illustrated in the figures. The x symbols represent the original Monte Carlo results, the o symbols represent the results after a correction for short distance lattice artifacts. The correction, based on the perturbative behavior of the potentials and on the actual separation of the plaquettes which enter into the formulae for $V_1 - V_4$, is not meant to be rigorous, but is introduced to estimate the size of possible short-distance lattice distortions. The values of the potentials in physical units are obtained assuming that they renormalize like the quark mass squared and using the ratio between bare and renormalized quark masses (~ 0.5), which one infers from a direct lattice calculation of hadron masses.[9] The renormalized values of the potentials are in reasonable agreement with a relation derived by Gromes.[10] One of the most important results of the Monte Carlo calculation is the clear evidence for a long range, non-perturbative component in V_1 (cfr. also Ref. 8). The calculation has now been repeated for $\beta = 6$. The new results, as well as an application to the calculation of the spin splittings, will be presented shortly.[11]

ACKNOWLEDGMENTS

We would like to thank Lloyd M. Thorndyke and Lee F. Kramer of ETA Systems, Inc. and Robert M. Price and Larry Jodsaas of Control Data Corporation for their continued interest, support and access to the CYBERNET CDC CYBER 205 at Rockville, Maryland where all our computations

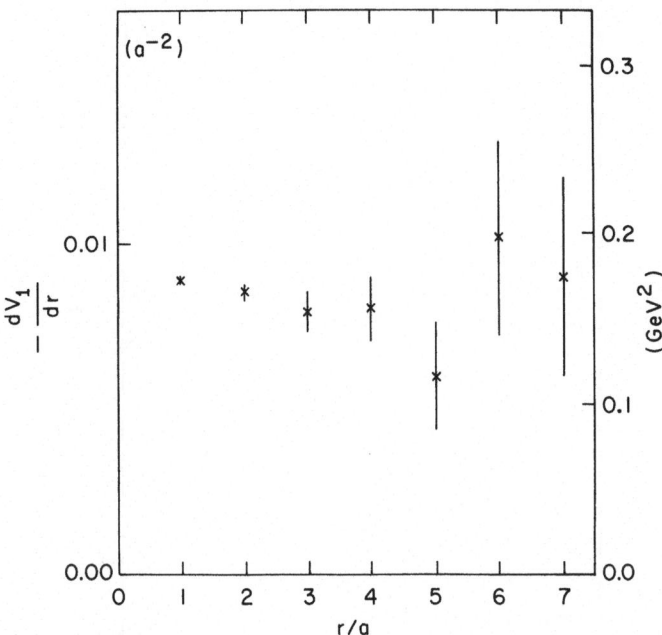

Fig. 1: Monte Carlo results for the spin-dependent potential V_1.

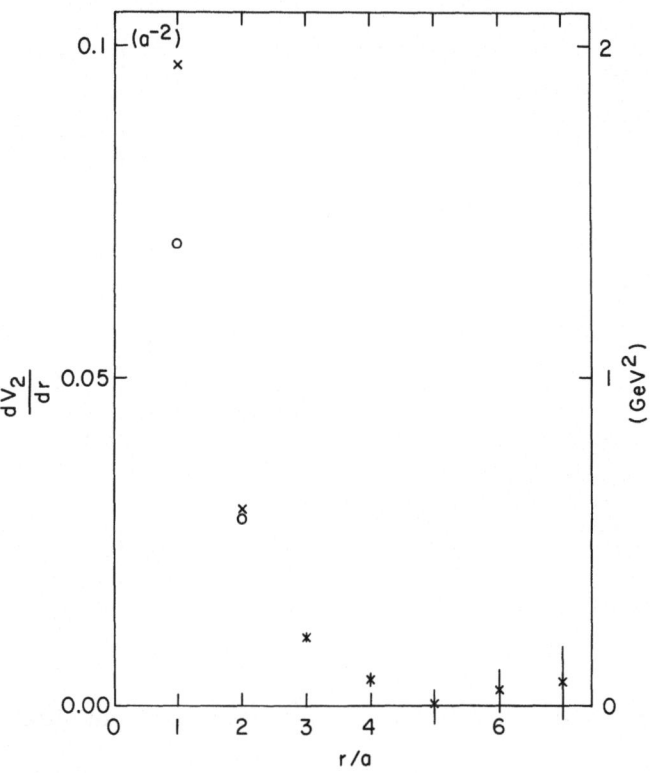

Fig. 2: Monte Carlo results for the spin-dependent potential V_2.

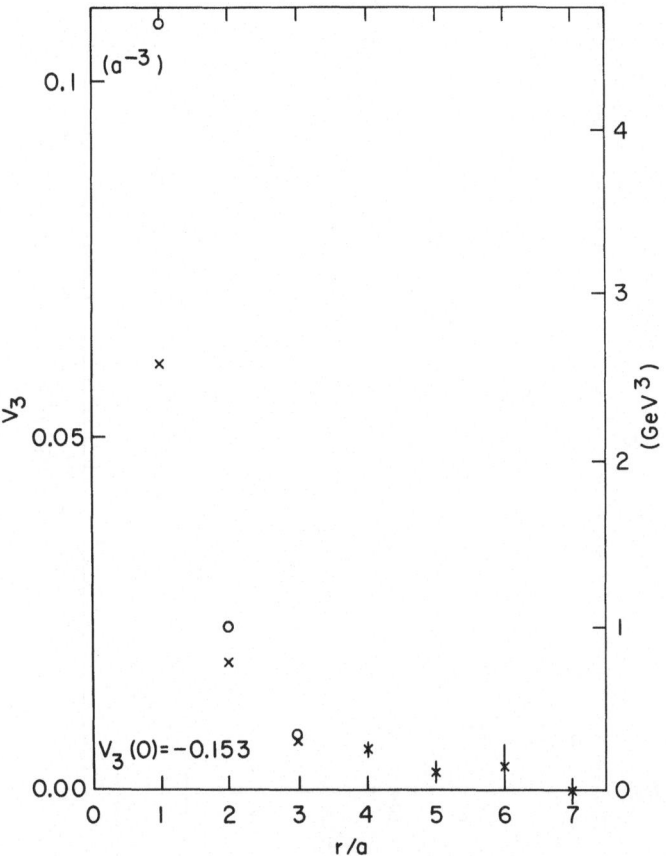

Fig. 3: Same as in Fig. 2, but for the spin-dependent potential V_3.

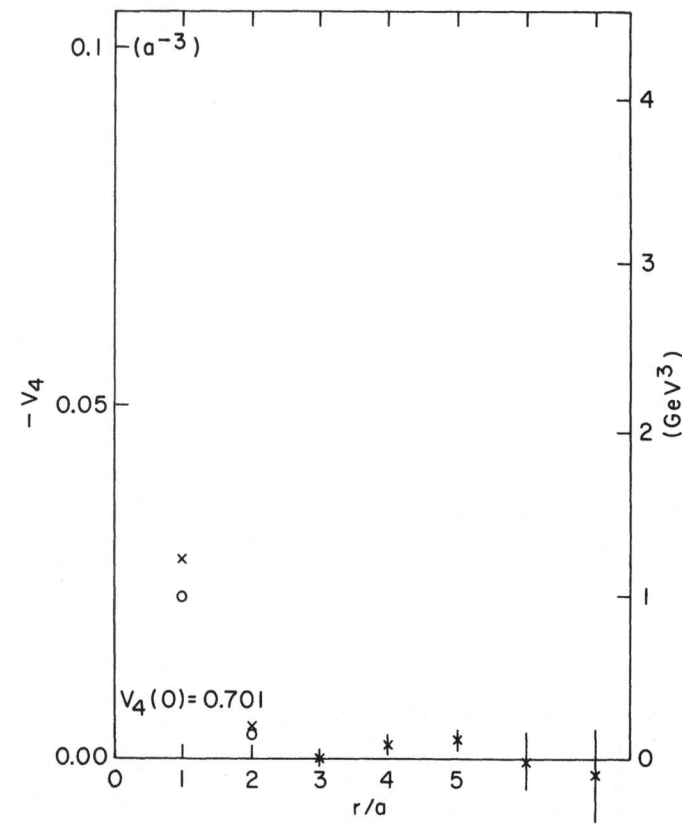

Fig. 4: Same as in Fig. 2, but for the spin-dependent potential V_4.

have been made; the Control Data Corporation PACER Fellowship grants [Grant Nos. 85PCR06 and 86PCR01] for financial support and the Natural Sciences and Engineering Research Council of Canada [Grant No. NSERC A9030] for further financial support. This research was also carried out in part under the auspices of the U.S. Department of Energy under contract No. DE-AC02-76CH00016.

REFERENCES

1. E. Eichten and F. Feinberg, Phys. Rev. Lett. <u>43</u>, 1205 (1979); Phys. Rev. <u>D23</u>, 2724 (1981).
2. M.E. Peskin, lecture in *Proc. 11th SLAC Summer Institute on Particle Physics*, (1983)SLAC Report No. 267.
3. M. Campostrini, K.J.M. Moriarty and C. Rebbi, Comp. Phys. Comm. <u>42</u>, 174 (1986).
4. M. Campostrini, K.J.M. Moriarty, and C. Rebbi, Phys. Rev. Lett. <u>57</u>, 44 (1986).
5. M. Campostrini, Nucl. Phys. <u>B256</u>, 717 (1985).
6. C. Michael and P.E.L. Rakow, Nucl. Phys. <u>B256</u>, 640 (1985).
7. P. deForcrand and J.D. Stack, Phys. Rev. Lett. <u>55</u>, 1254 (1985).
8. C. Michael, Phys. Rev. Lett. <u>56</u>, 1219 (1985).
9. D. Barkai, K.J.M. Moriarty and C. Rebbi, Phys. Rev. Lett. <u>156B</u>, 385 (1985).
10. D. Gromes, Zeit. Phys. <u>C26</u>, 401 (1984).
11. M. Campostrini, K.J.M. Moriarty and C. Rebbi, in preparation.

BARYON NUMBER CONSERVATION IN LATTICE GAUGE THEORY

Krzysztof Redlich

Fakultät für Physik, Universität Bielefeld
D-4800 Bielefeld 1, F.R. Germany
and
Institute for Theoretical Physics, University of Wroclaw
Cybulskiego, PL-50-205 Wroclaw, Poland

David E. Miller

Department of Physics
Pennsylvania State University
Hazleton, PA 18201

ABSTRACT

The exact implementation of baryon number conservation is formulated in lattice gauge theory. The canonical partition function in the model with heavy quarks in the system is derived. A mean field analysis of SU(2) gauge theory is carried out within this formulation. This analysis can contribute to an understanding of the role of the baryon number conservation in lattice QCD thermodynamics.

INTRODUCTION

It is a rather well established fact that strongly interacting matter exhibits a behaviour typical of a system with a phase transition. For sufficiently high temperature and/or baryon number density the colour is deconfined and the system undergoes a phase transition from a hadron gas to a quark-gluon plasma.[1] This critical behaviour can be observed in a pure SU(3) gauge theory on a lattice as a discontinuous jump in the energy density or in the order parameter (Wilson loop), both of which suggest a first order phase

transition.[2]   The situation for the full lattice gauge theory (QCD), including the dynamical fermions (quarks), is still not so well confirmed. There
is, however, strong evidence[1] for the deconfinement phase transition in the
full system with colour sources but the order of this transition remains unclear at this time. Some of the most recent Monte Carlo results[3] suggest a
second order phase transition in the system with dynamical quarks. There
are also difficulties in the Monte Carlo analysis of the model with non-zero
chemical potential μ. These are mainly due to the fact that in SU(N) gauge
theory for $N > 2$, the fermion contribution to the Euclidean lattice action is
complex.[4,5]   Nevertheless, a Monte Carlo investigation of lattice QCD with
$\mu \neq 0$ has recently been done,[5] where in order to avoid the problem of the
complex fermion determinant, the "partial quenching" approximation[6] has
been applied. The mean field analysis of lattice QCD with $\mu \neq 0$ has also
been studied recently in the literature.[7]

The contribution from a non-zero baryon number to the thermodynamics of lattice QCD, up to the present, has been studied strictly in the
grand canonical (GC) ensemble with respect to baryon number conservation. In relativistic statistical thermodynamics, however, we have a choice
between the GC and the canonical (C) treatments of the conservation laws.
The possible differences between GC and C descriptions of the conservation
laws together with the implications of the exact implementation of charge
conservation have been recently studied in the literature in different thermodynamical models.[8,11]   It turns out from the above discussion that in many
realistic physical situations, the application of the GC ensemble with respect
to the conservation laws can be questionable. This is especially true in the
case when we deal with a small amount of matter enclosed in a tiny volume with a fixed, but small absolute value of the quantum numbers. This
situation is found in the laboratory in the central region for heavy ion collisions with the absolute value of the baryon number $B = 0$ and also for
hadron-hadron collisions where B is small. In the above actual cases the C
description should be preferred over the usual GC treatment of the conservation laws.[8]

One of the main purposes of this paper is the formulation of the lattice
QCD in the C ensemble with respect to baryon number conservation. With
the assumption that there are only heavy quarks in the system, we shall
find the canonical partition function in SU(N) lattice gauge theory. As an
example, we use mean field (MF) analysis of SU(2) lattice QCD to show the
possible implications of the canonical formulation on the thermodynamical
behaviour of the system. We shall show in terms of the MF approximation
that in the limit of large values of baryon number and volume of the system,

but with fixed baryon number density, the GC and C ensembles are equivalent. We shall also observe that the above analysis can give some information about the validity of the "partial quenching" approximation as it has been applied in the GC ensemble.[5]

This paper is organized as follows: in the next section we briefly summarize the canonical description with respect to internal symmetries, then we formulate lattice QCD with the exact implementation of the baryon number conservation. We are then able in the fourth section to present the mean field analysis of SU(2) lattice gauge theory in the C ensemble in the strong coupling limit. Finally, we draw some conclusions about our analysis of lattice QCD in a C formulation.

CANONICAL DESCRIPTION OF INTERNAL SYMMETRY

The formulation of relativistic thermodynamics with the exact implementation of the conservation laws is carried out through a procedure based on group theoretical methods.[10] For this situation the formal structure resembles the GC description with the main difference being that we define the generating function by taking the trace over all states as follows:

$$\widetilde{Z}(g, V, \beta) = \text{Tr}\left(e^{-\beta H} U(g)\right). \tag{1}$$

where $U(g)$ is the unitary reducible representation of the symmetry group G with $g \in G$, H the Hamiltonian, V the volume and β the inverse temperature of the system. Due to the exact symmetry and decomposition of $U(g)$ into the form $\sum_\alpha \oplus U^\alpha(g)$, one can write[10]

$$\widetilde{Z}(g, V, \beta) = \sum_\alpha \frac{\chi^\alpha(g)}{d(\alpha)} Z_\alpha(\beta, V) \tag{2}$$

with $Z_\alpha(\beta, V)$ the usual canonical partition function given by $\text{Tr}_\alpha\left(e^{-\beta H}\right)$ which contains exactly that value of the quantum numbers which correspond to the α-representation of the symmetry group. $\chi^\alpha(g)$ and $d(\alpha)$ are the character and the dimension of the α-representation of the group. By using the orthogonality properties of the group character one can find[10,13]

$$Z_\alpha(\beta, V) = \int dM(\phi_1, \ldots \phi_r) \, \bar{\chi}^\alpha(\phi_1, \ldots \phi_r) \, \widetilde{Z}(\beta, V, \phi_1, \ldots \phi_r) \tag{3}$$

where dM is the Haar measure on the group and $\phi_1, \ldots \phi_r$ are the parameters of the maximal Cartan subgroup of the symmetry group G. From (2) and the definition of the GC partition function, one can also establish the following simple relation:

$$\widetilde{Z}(\beta, V, \phi_1 \ldots \phi_r) = Z^{GC}(\beta, V, \mu_1 = i\phi_1/\beta, \ldots, \mu_r = i\phi_r/\beta). \tag{4}$$

In our present analysis, we shall apply the above formalism to the simple case of the U(1) baryon symmetry group for lattice QCD in the next section. In that section we shall also develop a simple phenomenological model with the symmetry $SU(3) \times U(1)$ in order to illustrate the complex fermion contribution to the partition function.

LATTICE QCD WITH BARYON NUMBER CONSERVATION

In the lattice formulation of QCD the partition function on an isotropic Euclidean lattice with N_τ (N_σ) temporal (spatial) lattice sites and a non-zero baryon chemical potential μ can be found as follows:

$$Z(\mu, N_\tau, N_\sigma, \kappa) = \int \prod_{\text{links}} dU \exp(-S_G) \{\det Q\}^{N_f} \tag{5}$$

with

$$S_g = \frac{6}{g^2} \sum_P \left(1 - \frac{1}{3} \operatorname{Re} \operatorname{Tr} UUU^+U^+\right) \tag{6}$$

being the pure gluon part. $N_f \log \det Q$ is the quark-gluon contribution to the lattice action, obtained after the integration of the quark spinor fields. The fermion matrix in (5) has the form[16]

$$Q = 1 - \kappa \sum_{\nu=0}^{3} M_\nu \tag{7}$$

with κ the "hopping" parameter and

$$(M_\nu)_{m,n} = (1 - \gamma_\nu) U_{n,m} \delta_{n,m-\hat{\nu}} + (1 + \gamma_\nu) U^\dagger_{m,n} \delta_{n,m+\hat{\nu}} \tag{8}$$

where U is an element of the SU(N) group.

In order to bring the chemical potential into the theory, one can use the prescription of ref. 17, which is contained in the following substitution:

$$U \to U e^{\mu a}, \qquad U^\dagger \to U^\dagger e^{-\mu a}. \tag{9}$$

This is set into the $\nu = 0$ term of (8). However, this replacement implies that the fermion matrix (7) is no longer Hermitian. A direct consequence is that the fermion contribution to the Euclidean lattice action becomes complex for $N > 2$. This fact alone is the origin of the well-known difficulties in the Monte Carlo computational procedures, which require a real and positive definite measure.

The complex fermion contribution to the partition function with $\mu \neq 0$ can also be found in the model outside of lattice QCD.[12] In order to illustrate the above, let us consider a simple model which consists of a gas of quarks and gluons with exact implementation of colour charge and with the baryon number conservation in the system. Then, under the requirement that the total colour charge of the system is equal to zero, the partition function of the model can be obtained by using eqs. (1-4), in the following form:[12]

$$Z_o(T,v,\mu) = \int d\bar{\gamma} M(\bar{\gamma}) \tilde{Z}_q(T,v,\mu,\bar{\gamma}) \tilde{Z}_G(T,v,\bar{\gamma}) \tag{10}$$

with gluon $\left(\tilde{Z}_G\right)$ and quark $\left(\tilde{Z}_g\right)$ generating functions given by:

$$\ln \tilde{Z}_G = \frac{g_G}{d_G} \frac{vT^3}{2\pi^2} \sum_n \frac{1}{n^4} \chi^G(\bar{\gamma}n) \tag{11a}$$

$$\ln \tilde{Z}_q = \frac{g_q}{d_q} \frac{vTm^2}{2\pi^2} K_2(\beta m)$$
$$\times \left[(\cosh\beta\mu)\,\mathrm{Re}\,\chi_q(\bar{\gamma}) + i\,(\sinh\beta\mu)\,\mathrm{Im}\,\chi_q(\bar{\gamma})\right] \tag{11b}$$

where g_q, g_G are respectively the quark and gluon degeneracy factors and m the quark mass. The characters $\chi_q(\bar{\gamma})$, $\chi_G(\bar{\gamma})$ and dimensions d_q, d_G are those of the fundamental and adjoint representation of the SU(3) group.

From the above example, one can explicitly see that, also in terms of the considered model, the fermion contribution to the partition function is complex when $\mu \neq 0$. Furthermore, eq. (11b) has the same structure as the corresponding one obtained in lattice QCD in the leading order in the hopping parameter expansion.[12] Thus, one can conclude that the complex structure of the fermion determinant is neither directly connected with the lattice regularization scheme, nor with the way in which the chemical potential has been brought onto the lattice. Its presence has a rather general nature in gauge theories, which is related to the structure of the SU(N) symmetry group with $N > 2$.

After having established the thermodynamics of lattice QCD in the GC ensemble as given by eqs. (5-9), we now want to find a lattice partition function $Z_B(N_\tau, N_\sigma, \kappa)$ which gives an exact implementation of the baryon number conservation. However, we should note at this point that the quantity B is actually just the difference between the number of quarks and the number of antiquarks. Then, the real baryon number is simple $B/3$ for SU(3).

In the following analysis we shall restrict ourselves to the case for which there are only heavy quarks present in the system. Thus, the determinant in (5) can be evaluated using the "hopping parameter" expansion, which for very heavy quarks and $N_\tau < 4$ can be approximated by the leading term in this expansion.[14,15] With the formalism indicated in the previous section applied to the $U(1)$ internal symmetry group, the C partition function becomes

$$Z_B(N_\tau, N_\sigma, \kappa) = \int_0^{2\pi} \frac{d\phi}{\pi} \cos(B\phi) \, \widetilde{Z}(\phi, N_\tau, N_\sigma, h) \tag{12}$$

where we have used the symmetry properties of the generating function $(\phi \to -\phi)$ together with (9). In the leading order in the "hopping parameter" expansion, the generating function can be obtained as follows:

$$\widetilde{Z}(\phi, N_\tau, N_\sigma, h) = \int \prod_{\text{links}} \exp\left(-S_G - \widetilde{S}_F\right) \tag{13}$$

with S_G as in (6) and

$$\widetilde{S}_G = -h\left[(\cos\phi)\, L_R - (\sin\phi)\, L_I\right] \tag{14}$$

where $h \equiv 4N_f (2\kappa)^{N_\tau}$ is the quark parameter and

$$L_R \equiv \sum_x \text{Re } L_x$$

and

$$L_I \equiv \sum_x \text{Im } L_x.$$

L_x is the Wilson loop at the spatial site x:

$$L_x = \text{Tr} \prod_{\tau=1}^{N_\tau} U_{(\bar{x},\tau),(\bar{x},\tau+1)}. \tag{15}$$

Thus, with the generating function (13) the integration over the $U_B(1)$ group can be performed exactly so that the canonical lattice partition function becomes

$$Z_B(N_\tau, N_\sigma, h) = \int \prod_{\text{links}} dU \, e^{-S_G} I_B(h\, y)\, T_B\left(y^{-1} L_R\right) \tag{16}$$

with $y = \left(L_R^2 + L_I^2\right)^{1/2}$, $I_B(x)$ the modified Bessel function of the first kind and $T_B(\cos\phi) = \cos B\phi$ being the Chebyshev polynomial. The thermal average of any physical quantity $f(U)$ in the C ensemble can be calculated in the usual way as follows:

$$\langle f(U)\rangle_B = \int \Pi dU\, e^{-S_G} f(U)\, I_B(h\, y)\, T_B\left(y^{-1} L_R\right) \Bigg/$$
$$\int \Pi dU\, e^{-S_G} I_B(h\, y)\, T_B\left(y^{-1} L_R\right). \tag{17}$$

Now let us discuss the relation between the C and the GC partition functions as given above in (16) and (5) respectively. In the ordinary statistical thermodynamics the C partition function is the coefficient in the cluster decomposition of the GC partition function. The same relation holds for the relativistic statistical thermodynamics, where the distinction between the GC and the C ensemble are given on the level of the conservation laws.

For the case of the $U_B(1)$ symmetry this cluster decomposition has a particularly simple form

$$Z(\mu, N_\tau, N_\sigma, h) = \sum_{B=-\infty}^{\infty} e^{\mu B \beta} Z_B(N_\tau, N_\sigma, h) \tag{18}$$

with $Z_B(N_\tau, N_\sigma, h)$ given by (16). From the relation

$$\exp\left\{\frac{1}{2} Z\left(t + \frac{1}{t}\right)\right\} = \sum_{k=-\infty}^{\infty} t^k I_k(Z), \tag{19}$$

one can find from (18) that

$$Z(\mu, N_\tau, N_\sigma, h) = \int \Pi dU$$

$$\exp\left\{-S_G - h\cosh(\mu\beta) L_R\right\} \times \cos(h\sinh(\mu\beta) L_I) \tag{20}$$

which is just the GC partition function recently used in ref. 5 for the Monte Carlo evaluation of statistical QCD on the lattice with non-zero chemical potential. The last term $\cos(h\sinh(\mu\beta) L_I)$ together with the weight factor $\exp\{-S_G - S_F\}$ in (15) leads to large fluctuations and thereby produce the difficulties in Monte Carlo computation. In order to avoid this problem, the "partial quenching" contained in the substitution $L_I = 0$ in (15) has been proposed. Unfortunately, on the level of the GC ensemble one cannot test the validity of this approximation.[5]

In the C ensemble, we are in general also not free from the problem concerned with these large fluctuations in the Monte Carlo computation. For large values of the baryon number, the Chebyshev polynomial term in (16) plays a similar role to the $\cos(h\sinh(\mu\beta) L_I)$ of the GC ensemble. Nevertheless, for not too large values of the baryon number, a numerical analysis in the C ensemble may be possible. Thus, for instance, in the central region of the heavy-ion collisions where $B = 0$, as well as in hadron-hadron scattering, it could be possible to deduce the thermodynamical properties of the produced hadronic matter by using the C partition function given in (16) as a basis. Furthermore, one may well suspect that since the argument of the Chebyshev polynomial is proportional to $\left(1 + (L_I/L_R)^2\right)^{-1/2}$ and generally the ratio L_I/L_R is rather small, the numerical analysis can probably be performed for reasonably large values of the baryon number.[18]

Now we consider the "partial quenching" approximation on the level of the C ensemble. Due to the cluster expansion (18) and the relations (3) and (4), one can conclude that the approximations $L_I = 0$ in C and GC are equivalent. As we have already pointed out in the GC formulation, it is not so clear how to test the above assumption. In the C ensemble the situation is quite different. Here, it is possible to check the validity of the "partial quenching" approximation by computing first the thermal average of some quantity $f(U)$ using (17) with some given small value of the baryon number B and then comparing it with

$$\langle f(U) \rangle_B = \int \Pi dU e^{-S_G} f(U) I_B(h\, L_R) \bigg/ \int \Pi dU e^{-S_G} I_B(h\, L_R) \qquad (21)$$

In this way one can deduce the possible contribution of the imaginary part of the Wilson loop L_I in the given model.[18]

MEAN FIELD ANALYSIS

Now let us consider a mean field analysis of SU(2) gauge theory at finite temperature and in the strong coupling limit formulated on the lattice in the C ensemble. Since the previous results obtained in the mean field approximation[19,20] agree quite well with the Monte Carlo analysis, one can suspect that also in the C ensemble we can deduce some interesting features of the model in this approximation.

In the strong coupling limit the spacelike plaquettes in the Euclidean lattice partition function (5) with $\kappa = 0$ can be neglected.[19,20] Then the effective theory can be given by the partition function written in terms of the character expansion[20]

$$Z_{\text{eff}} = \int \prod_x dM(x) \prod_{x,\ell} \left(1 + \sum_\nu [Z_\nu(\beta)]\right)^{N_\beta} \chi_\nu(L_x) \chi_\nu\left(L_{x+\ell}^\dagger\right), \qquad (22)$$

where the two products run over the lattice sites and the directed links respectively. The expressions χ_ν and Z_ν are the character and the character coefficient of the ν representation of the SU(N) group, respectively.

When g^{-2} is assumed to be very small, then the leading contribution to the expansion (22) is given by the fundamental representation. Thus

$$Z_{\text{eff}} \simeq \int \prod_x dM(x) \exp\left\{\beta' \sum_\ell \text{Tr} L_x \text{Tr} L_{x+\ell}^\dagger + cc\right\} \qquad (23)$$

with $\beta' = Z_{(1,0)}^{N_\tau}$. If the quarks in the system have a very large but finite mass, and at the same time there is a non-zero net average baryon number, the effective partition function is generalized[7] as follows:

$$Z_{\text{eff}}(\mu, N_\tau, N_\sigma, h) \simeq \int \prod_x dM(x) \exp \left\{ \beta' \sum_{x,\ell} \text{Tr} L_x \text{Tr} L_{x+\ell}^\dagger + cc \right.$$

$$\left. + h \cosh(\mu\beta) L_R + i\, h \sinh(\mu\beta) L_I \right\} \qquad (24)$$

with the same notation as in (5). With this partition function and the formalism presented in the previous sections, we can establish the effective theory with exact implementation of baryon number conservation

$$Z_{\text{eff}}(B, N_\tau, N_\sigma, h) \simeq \int \prod_x dM(x)$$

$$\exp \left\{ \beta' \sum_{x,\ell} \text{Tr} L_x \text{Tr} L_{x+\ell}^\dagger + cc \right\} I_B(h\, y) T_B(y^{-1} L_R) \qquad (25)$$

following the notation of (16).

Now the above effective C partition function can be studied in terms of a mean field (MF) approximation. In the following analysis we shall restrict our consideration only to the $SU_C(2)$ gauge group for the investigation of the effective partition function in the limit where B and N_σ are large, but the ratio B/N_σ^3 defined as $\overline{B}$ is fixed (T-limit). Then from (25) the effective partition function is given by

$$Z_{\text{eff}}^c(B, N_\sigma, B/N_\sigma^3, h) \simeq \frac{2}{\pi} \int_{-1}^1 \prod_x (1 - L_x^2)^{1/2}\, dL_x$$

$$\times \exp \left\{ \beta' \sum_\ell \text{Tr} L_x \text{Tr} L_{x+\ell} + \ln I_B(B\, a_x) \right\} \qquad (26)$$

where the Bessel function $I_B(B\, a_x)$ in the T-limit can be approximated as follows:

$$\ln I_B(\overline{B}\, a_x) \simeq N_\sigma^3 \overline{B} \left[(1 + a_x^2)^{1/2} + \ln \frac{|a_x|}{1 + \sqrt{1 + a_x^2}} \right] \qquad (27)$$

with

$$a_x = h\overline{B}^{-1} \frac{1}{N_\sigma^3} \sum_x \text{Tr}\, L_x. \qquad (28)$$

In (26) the explicit expression for the Haar measure and the character formula of the fundamental representation of the $SU_C(2)$ group have been used.

The result presented in (26) indicates that in the C ensemble the quark parameter does not play the role of the external magnetic field which breaks the Z_N symmetry of the model. Because of the symmetry of the $I_B(x)$ function and due to the requirement that B must be an even number, the Z_N symmetry is not broken even for $h \neq 0$.

At this point let us consider the MF analysis of the effective partition function as it is given in (26). Assuming here the thermal Wilson loop to be constant everywhere on the lattice[21] and then using the steepest descent method for the integration over the $SU_C(2)$ group in (26), the leading contribution in the T-limit to the canonical MF free energy can be found as follows:

$$-F_{\mathrm{MF}}^C/N_\sigma^3 \simeq 3\beta' L_{\overline{B}}^2 + \overline{B}\left[\left(1 + \bar{a}^2\right)^{1/2} + \ln \frac{|\bar{a}|}{1 + \sqrt{1 + \bar{a}^2}}\right]$$
$$+\frac{1}{2}\ln\left(1 - L_{\overline{B}}^2\right) \tag{29}$$

where $\bar{a}$ is simply $h\overline{B}^{-1}L_{\overline{B}}$. Thus with this free energy the MF canonical value of the Wilson loop $L_{\overline{B}}$ can be found as the solution of the equation

$$\frac{\partial F_{MF}^C}{\partial L_{\overline{B}}} = 6\beta' L_{\overline{B}} - \frac{L_{\overline{B}}}{1 - L_{\overline{B}}^2} + \frac{\overline{B}}{L_{\overline{B}}}\sqrt{1 + a^2} = 0. \tag{30}$$

One could expect that in the T-limit the GC method for the description of the thermodynamical properties of hadronic matter is also quite adequate. Thus, in this limit, the GC and C ensembles must be equivalent. In order to show this to be the case, let us consider how the $SU_C(2)$ effective theory looks in the GC description. Starting from (24) and taking the MF approximation, which is equivalent to the one in the C ensemble, the MF free energy in the GC ensemble becomes

$$-F_{\mathrm{MF}}^{\mathrm{GC}}/N_\sigma^3 \simeq 3\beta' M^2 + \frac{1}{2}\ln\left(1 - M^2\right) + h\cosh\left(\mu N_\tau a\right) M \tag{31}$$

with a the lattice spacing. The MF value of the order parameter M in GC ensemble is given as the solution of

$$\frac{\partial F_{\mathrm{MF}}^{\mathrm{GC}}}{\partial M} = 6\beta' M - \frac{M}{1 - M^2} + h\cosh\left(\mu N_\tau a\right) = 0. \tag{32}$$

After taking the limit $\overline{B} \to 0$ in (29) one can immediately see that the canonical mean field partition function is equal to the GC one (31), with $\mu = 0$. Thus, in this limit the GC and C ensembles are equivalent. However, for arbitrary non-zero values of the baryon number density, the expressions for the MF free energy in the GC and C ensembles are different, which means that (29) and (31) cannot be directly compared. Thus, only the physical quantities should be compared. They should be the same in both descriptions when $B \to \infty$. We can see this by considering the MF value of the Wilson loop, which is obtained in both these ensembles. However, for this we still need to know the relationship between the baryon number density and the chemical potential in the GC ensemble. With the effective partition function (24) this relation has a particularly simple form, namely

$$\frac{\langle B \rangle}{N_\sigma^3} = hM \sinh (\mu N_\tau a) . \tag{33}$$

Now using the above result together with (32) we can see that if one identifies B with $\langle B \rangle$, the MF results for the Wilson loop $L_{\overline{B}}$ in the GC and those for M in the C ensemble are the same. Thus, we are able to conclude that if both B and N_σ go to very large values with B/N_σ^3 remaining constant, then the GC and C ensembles are thermodynamically equivalent.

The above result should also be valid for the model with $SU_C(N)$ as a gauge group. Nevertheless, due to the polynomial term in (12) for $N > 2$, one comes upon considerable difficulty in the evaluation of the effective partition function for the above considered T-limit for this equation. In order to find the effective theory for $N > 2$ in the T-limit, the application of some other method is required.[18]

Finally, we indicate the difference at finite baryon number density between the GC and C ensembles through numerical examples shown in fig. 1. The effective potential V_{eff} for the Wilson loop in the $SU_C(2)$ model is obtained from the MF free energy for the respective ensembles from (29) and (31). We notice in both cases of fig. 1 that there is a qualitative difference between the behaviour of V_{eff} in the GC and C ensembles. Nevertheless, the thermodynamical behaviour of any physical quantity obtained in both ensembles in the T-limit is the same. This we have already illustrated by the example of the Wilson loop. For any value of the baryon number in the C ensemble, one observes a singular structure of V_{eff} for L_{MF} approaching zero. In order to attain a finite value of V_{eff} at this point, the baryon number has to be identically zero. The structure of V_{eff} indicates the spontaneous breaking of the Z_2 symmetry of this model. In both ensembles there is a general quantitative sensitivity of the value of V_{eff} to changes in the parameters β' and $\overline{B}$. However, in the GC ensemble we note a qualitative change in the structure of V_{eff} for the different values of these parameters as illustrated in figs. 1a and 1b.

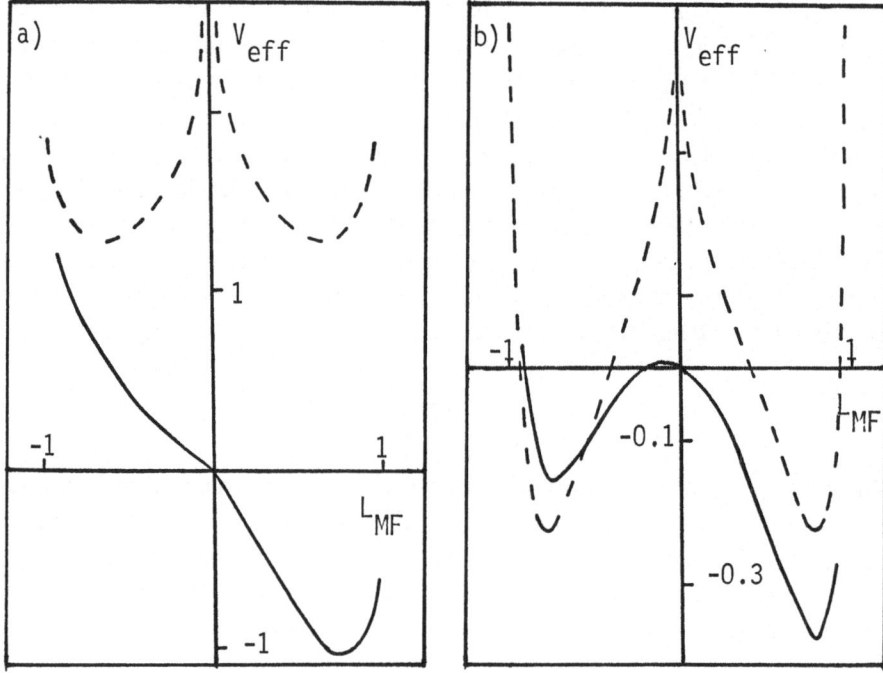

Fig. 1: The effective potential V_{eff} in $SU_C(2)$ model for the canonical ensemble (broken line) and the grand canonical ensemble (solid line) as a function of the mean field values of the Wilson loop L_{MF} for different value of the parameters: (a) $\beta' = 0.1$, $\bar{B} = 0.5$, $h = 0.06$; (b) $\beta' = 0.4$, $\bar{B} = 0.1$, $h = 0.06$.

CONCLUSIONS

Having in mind lattice QCD as the realistic theory which can possibly give some useful information about the properties of hot hadronic matter as it might be produced from hadronic collisions, we have formulated lattice QCD in the C ensemble respecting baryon number conservation. The obtained C partition function $Z_B(N_\tau, N_\sigma, h)$ in (16) can be presumably considered to be the starting point for the more detailed Monte Carlo analysis of the model.[18] It can also give some information about the validity of the "partial quenching" approximation proposed in the GC ensemble.[5]

We have also indicated that there can be significant differences between the GC and C treatment of the charge conservation law. In particular, the quark contribution to the partition function in the C ensemble does not play the role of an external magnetic field which breaks the Z_N symmetry in our SU(2) model. However for any finite value of the baryon number, the Z_N

symmetry is spontaneously broken. In the limit of large B and N_σ with fixed B/N_σ^3, the GC and C ensembles are thermodynamically equivalent. This has been shown by the example of the MF analysis of the $SU_C(2)$ model.

ACKNOWLEDGMENTS

We would especially like to thank H. Satz for many essential discussions on this subject. We are grateful to F. Karsch for important discussions and helpful remarks. One of us (D.E.M) would like to recognize the partial support of the Faculty Scholarship Support Fund at Pennsylvania State University. (K.R) would like to express his appreciation for many discussions with R. Hagedorn in related fields. (K.R) also acknowledges support by the Alexander von Humboldt Stiftung and Bundesministerium für Forschung und Technologie (BMFT). We would also like to acknowledge discussion with J. Engels and B. Petersson.

REFERENCES

1. For recent review, see e.g.:
   J. Cleymans, R. Gavai and E. Suhonen, Phys. Rep. 130 (1986) 218;
   H. Satz, Ann. Rev. Nucl. Part. Sci. 35 (1985) 245;
   B. Svetitsky, Phys. Rep. 132 (1986) 1.
2. T. Çelik, J. Engels and H. Satz, Phys. Lett. 129B (1983) 323;
   J. Kogut, M. Stone, H.W. Wyld, W.R. Gibbs, J. Shigemitsu, S.H. Shenker and D.K. Sinclair, Phys. Rev. Lett. 50 (1983) 353.
3. P. Hasenfratz, F. Karsch and I.O. Stamatescu, Phys. Lett. 133B (1983) 221;
   R.V. Gavai, M. Lev and B. Petersson, Phys. Lett. 140B (1984) 397; 149B (1984) 492;
   F. Fucito, C. Rebbi and S. Solomon, Caltech Report CALT-68-1127 (1984) unpublished;
   J. Polonyi et al., Phys. Rev. Lett. 53 (1984) 644;
   T. Çelik, J. Engels and H. Satz, Nucl. Phys. B256 (1985) 670;
   K. Redlich and H. Satz, Phys. Rev. D33 (1986) 3747;
   R.V. Gavai and F. Karsch, Nucl. Phys. B261 (1985) 273;
   S. Duane and J.B. Kogut, Phys. Rev. Lett. 55 (1985) 2774.
4. J. Kogut, H. Matsuoka, M. Stone, H.W. Wyld, S. Shenker, J. Shigemitsu and D.K. Sinclair, Nucl. Phys. B225 (1983) 93;
   I. Barbour et al., Illinois preprint, ILL-TH-86-23, (1986) unpublished.
5. J. Engels and H. Satz, Phys. Lett. 1515B (1985) 151;
   B. Berg, J. Engels, E. Kehl, B. Waltl and H. Satz, Z. Phys. C31 (1986) 167.
6. For more detailed arguments for the particle quenching approximation see ref. 5.

7. E. Dagotto, F. Karsch and A. Moreo, Phys. Lett. 169B (1986) 421;
   T. Çelik, T. Firat, Y. Gündüc and M. Önder, Ankara preprint (1986);
   T. Çelik, in preparation.

8. R. Hagedorn and K. Redlich, Z. Phys. C27 (1985) 541;
   K. Redlich in *Local Equilibrium in Strong Interaction Physics*, eds. D.K.
   Scott and R.M. Weiner, World Scientific Publ. (1985).

9. See e.g.:
   H. Th. Elze, W. Greiner and J. Rafelski, Phys. Lett. 124B (1983) 515;
   Z. Phys. C24 (1984) 361;
   A.T.M. Aertz, T.H. Hanson and B. Skagerstam, Phys. Lett. 145B
   (1984) 123; 150B (1985) 447;
   M.J. Gorenstein, S.I. Lipshikh, V.K. Petrov and G.M. Zinovjev, Phys.
   Lett. 123B (1985) 437;
   B. Müller and J. Rafelski, Phys. Lett. 116B (1982) 274;
   K. Redlich, Z. Phys. C21 (1983) 69;
   L.D. McLerran and A. Sen, Phys. Rev. D32 (1985) 279;
   H. Th. Elze and W. Greiner, Phys. Rev. A33 (1986) 1979;
   LBL preprint, LBL-21924 (1986).

10. K. Redlich and L. Turko, Z. Phys. C5 (1980) 201;
    L. Turko, Phys. Lett. 104B (1981) 153;
    Similar results can be obtained based on coherent states method;
    see ref. 13.

11. P. A. Amundsen and B.S. Skagerstam, Phys. Lett. B168 (1985) 375
    and references therein.

12. H. Th. Elze, D.E. Miller and K. Redlich, Bielefeld preprint BI-TP 86/15,
    to appear in Phys. Rev. D.

13. B.S. Skagerstam, Z. Phys. C24 (1984) 97; J. Phys. A18 (1985) 1; Phys.
    Lett. 133B (1985) 419.

14. J. Engels, F. Karsch and H. Satz, Nucl. Phys. B205[FS5](1982) 239.

15. See e.g.:
    H. Satz, Nucl. Phys. A400 (1983) 541c.

16. K. Wilson, Phys. Rev. D10 (2974) 245, and in *New Phenomena in
    Subnuclear Physics*, ed. A. Zichichi, Plenum, New York 1977.

17. P. Hasenfratz and F. Karsch, Phys. Lett. 125B (1983) 308;
    H. Matsuoka and M. Stone, Phys. Lett. 136B (1984) 204.

18. Under consideration.

19. J.B. Kogut, M. Snow and M. Stone, Nucl. Phys. B200[FS4](1982) 211.

20. F. Green, Nucl. Phys. B215[FS7](1983) 83;
    F. Green and F. Karsch, Nucl. Phys. B238 (1984) 295.

21. J. Polonyi and K. Szlachanyi, Phys. Lett. 110B (1982) 395.

EXPERIMENTING WITH LANGEVIN LATTICE QCD

S. Sanielevici

Physics Department
Brookhaven National Laboratory
Upton, NY 11973

in collaboration with:
R.V. Gavai and J. Potvin
Brookhaven National Laboratory

ABSTRACT

We report on the status of our investigations of the effects of systematic errors upon the practical merits of Langevin updating in full lattice QCD. We formulate some rules for the safe use of this updating procedure and some observations on problems which may be common to all approximate fermion algorithms.

1. INTRODUCTION AND SUMMARY

Understanding the origins and effects of systematic errors in the algorithms used for the numerical simulation of lattice QCD is increasingly recognized as one of the major problems in the field [1]. The updating procedure based on the discretized version of the Langevin equation [2-4] is well suited for such studies, because the origin of its systematic errors is well understood [3,4] and because the leading errors are under analytic control [2-5]. Since Langevin updating can be connected to microcanonical updating via "hybrid" algorithms [6] and to pseudo-fermion updating [7] via the procedure suggested in Ref. [8], any findings on Langevin systematic errors may also turn out to be relevant to these other procedures. This is especially likely for the specific effects of the fermion determinant.

Our numerical study of the systematic errors in Langevin algorithms for QCD with dynamical quarks [9,10] is restricted to basic (non-hybrid, real-action, non-Fourier-accelerated) updating. It takes place on a 4^4 lattice having periodic boundary conditions in 3 directions and antiperiodic ones in the fourth. The coupling is fixed to $\beta = 4.8$; we simulate four flavors of Kogut- Susskind quarks of masses $ma = 0.1$ and $ma = 0.05$ using the Cornell group's bilinear noise approach [2].

We measure planar Wilson loops at various values of the discrete step size in Langevin time. The systematic error originates in the finite value of this step size and becomes worse as the step size increases. It is interesting to assess the effect of the bias upon planar Wilson loops, because these are the fundamental building blocks of most lattice observables. Therefore, the errors on the fundamental Wilson loops will in general propagate non-trivially to plague the eventual physical quantities one wants to extract.

As a benchmark to assess the effects of the error, we use the measurements of the same loops performed with an algorithm which computes the fermion determinant directly [11]. We also compare the Langevin results to those obtained with the pseudo-fermion method at various acceptances [11]. Since the number of iterations necessary for equilibration and the size of time correlations in equilibrium are important in determining the total cost of a numerical experiment which aims for a given statistical error, we also monitor these quantities by various methods. Note that these quantities decrease as the time step increases. What we require is a procedure which allows us to run at the largest possible step size with an acceptable systematic error.

The main result we want to present here is a set of "safety rules" for Langevin updating:

1. In order to obtain Wilson loop measurements in reasonable agreement with the benchmark for quark mass $ma = 0.1$, one may choose out of three methods:

    A. One run with the first-order (Euler) algorithm using a time step $\epsilon \lesssim 0.001$. Since time correlations are rather large with this method, the run must be sufficiently long.

    B. Use the Euler algorithm, but do 2 or 3 runs in the step size interval $\epsilon \in [0.005, 0.01]$. A linear extrapolation to $\epsilon = 0$ gives a good central value for your loop.

    C. One can do one run with the second-order (Runge-Kutta) algorithm with $\lesssim 0.001$. This will give results similar to the ones obtained by method A; time correlations will be smaller but the overall cost is slightly larger because this algorithm requires two conjugate gradient inversions instead of one.

2. For the values of the step size mentioned above, the non-integrable contribution to the leading systematic error is numerically unimportant.

3. There are physically important problems of a qualitative nature, such as the presence or absence of a first-order phase transition, which can be investigated without worrying about the systematic error. One can study these using either first order or second order algorithms, with a step size (say, $\epsilon = 0.01$) which should still be small enough that the absolute values of the monitored quantities (plaquette,$\bar{\chi}\chi$, Polyakov line) are reasonable. When using the Euler algorithm, one should shift the coupling, number of flavors and mass according to the formulae given in Refs. [2,5,10].

4. The effect of the systematic error becomes more severe when the quark mass is decreased to $ma = 0.05$. In this case, runs at $\epsilon = 0.001$ are no longer sufficient: one should go to still smaller step sizes.

2. ALGORITHMS

We shall begin by briefly reviewing the basic notations and formalism of discrete Langevin diffusion processes in SU(3) group space, the origin of systematic errors in Langevin updating and the algorithms we have been testing. Any SU(3) lattice link variable $U_\mu(x, \tau)$ (x runs over lattice sites and μ runs over Euclidean directions) is updated from Langevin time τ_N to τ_{N+1} by the formula

$$U_\mu(x, \tau_{N+1}) = U_\mu(x, \tau_N) \exp(-if_a T_a) \qquad (1)$$

where T_a are the generators of SU(3) ($a = 1, 2, \ldots 8$; $tr T_a T_b = \delta_{ab}/2$) and $f_a = f_a[U, \eta_a]$ is called the driving force. The driving force is built out of link matrices and out of 8 SU(3)-matrices η_a which generalize the white noise of stochastic quantization [2] ($a = 1, 2, \ldots 8$; $\langle \eta_a \rangle = 0$, $\langle \eta_a \eta_b \rangle = 2\delta_{ab}$). It contains the discrete time step of the Langevin simulation, $\epsilon \equiv \tau_{N+1} - \tau_N$.

Any given choice of the driving force defines a Langevin updating algorithm for the theory. One can then use the Fokker-Planck equation associated to (1) to determine the equilibrium action $\bar{S}$ of the Langevin diffusion as a function of the QCD action S. The result is of the form [3-5]

$$\bar{S} = S + o(\epsilon^n), \qquad (n = 1, 2, \ldots) \qquad (2)$$

This difference between $\bar{S}$ and S is the origin of the systematic error in Langevin updating. The explicit form of the relationship (2) can be used to engineer f_a such that the error be minimal (n in Eq. (2) maximal). Unfortunately, (2) can be worked out analytically (so far) only to first order in ϵ, so that algorithms can only be improved to yield $n = 2$ [2-5]. Therefore, numerical experiments remain the only way to assess the effects of the systematic bias upon lattice observables.

We report here on the performance of the following Langevin algorithms for full QCD:

1. First-order (Euler) QCD with fermionic noise. The driving force is [3]

$$f_a = \epsilon\left(\partial_a S_g\left[U\right] - \frac{1}{4}\xi^\dagger A_a\left[U\right]\xi\right) + \sqrt{\epsilon}\eta_a \tag{3}$$

where S_g is the usual Wilson action,

$$A_a\left[U\right] = M^{-1}\left[U\right]\partial_a\left(M^\dagger M\right)M^{-1}\left[U\right], \tag{4}$$

$M = \not{D} + ma$ such that $S = S_g - (1/2)\,Tr\,\ln M^\dagger M$ for 4 flavors of staggered quarks and ξ is a bilinear noise, normalized like the η_a's. This leaves $\bar{S} - S$ of order of order ϵ [3,5].

2. "Naive" second-order (Runge-Kutta) QCD with fermionic noise. This is defined by the driving force [5]

$$f_a = \frac{\epsilon}{2}(1 + \frac{C_2\epsilon}{12})(\partial_a S_g[U] + \partial_a S_g[\tilde{U}]$$
$$- \frac{1}{4}\xi^\dagger A_a[U]\xi - \frac{1}{4}\chi^\dagger A_a[\tilde{U}]\chi) + \sqrt{\epsilon}\eta_a \tag{5}$$

where $\tilde{U} \equiv U\left(x, \tau_{N+1/2}\right)$ represents an intermediate update obtained by Eqs. (1) and (3) and χ is a second bilinear noise. The difference between $\bar{S}$ and S is now of order ϵ^2, up to a so-called "non-integrable term" which arises from averaging over the bilinear noise and which is of order $\epsilon^{3/2}$ [5].

3. A "true" second-order algorithm, which leaves only errors of order 2 and higher, can be obtained by cancelling the non-integrable term explicitly [5]. This is done by replacing η_a in Eq. (5) by

$$\bar{\eta}_a = [\delta_{ab} - \frac{\epsilon}{128}\text{Re}\{\xi^+ A_a\chi\chi^+ A_b\xi\}]\eta_b \tag{6}$$

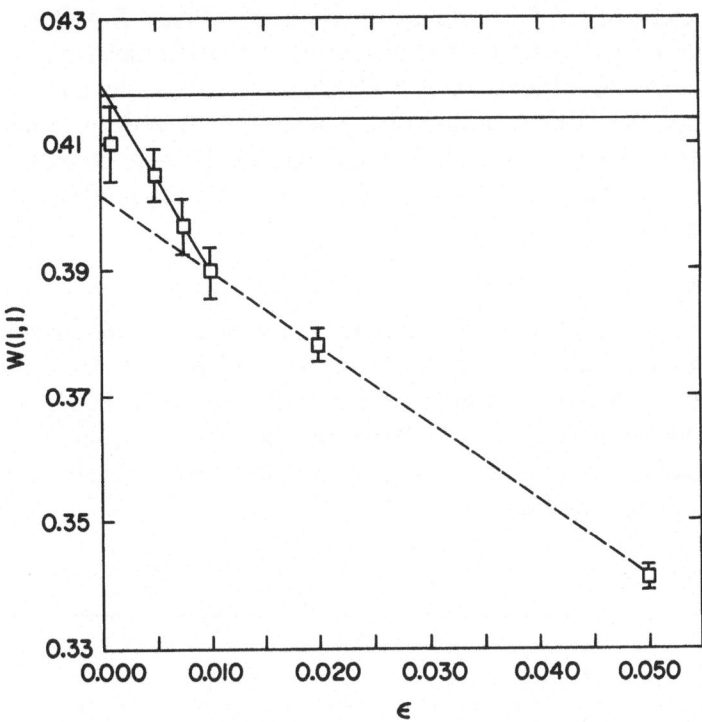

Fig. 1: Performance of the Euler algorithm for Langevin QCD. The full corridor indicates the benchmark result [11]. Error bars are corrected for time correlations in equilibrium [12]. Straight lines represent possible extrapolations.

3. THE EULER ALGORITHM

Figure 1 presents the dependence of the 1×1 Wilson loop on the step size. It has been obtained by using the Euler algorithm under the following conditions: $\beta = 6/g^2 = 4.8$, the quark mass in lattice units $ma = 0.1$, four flavors of Kogut-Susskind quarks. The inversion of M was done by conjugate gradient. We imposed the stopping condition $r = \left(||M\vec{x} - \vec{\xi}||^2 \right)^{1/2} < 0.05$. Note that our definition of the residue differs from other popular definitions by not dividing out the length of the vector $\vec{\xi}$ (our $r = 0.05$ is of the order 10^{-5} in the other normalization). In the mean, the required accuracy was reached after 65 conjugate gradient iterations. The updating time per link resulted to be about 0.88 milliseconds on the CRAY X-MP 22 at NMFECC (as compared to about 0.77 milliseconds for a similarly optimized pseudo-fermion code). Equilibration was checked in all cases by comparing averages over successive batches of 3000 iterations, for all Wilson loops up to 3 × 3. At $\epsilon = 0.01$ we also checked that hot and cold starts converge to the same average plaquette. The final averages were then obtained over 3000 iterations at equilibrium.

The corridor in Fig. 1 represents the result of the recent direct computation of the fermion determinant [11], which was done under exactly the same conditions as the present study. We see that we can come close to the benchmark either by running at sufficiently low ϵ ($\epsilon \lesssim 0.001$) or by doing two runs in the interval $\epsilon \in [0.005, 0.01]$ and extrapolating linearly to $\epsilon = 0$. An extrapolation based on runs below $\epsilon = 0.01$ would underestimate the value of the plaquette by far.

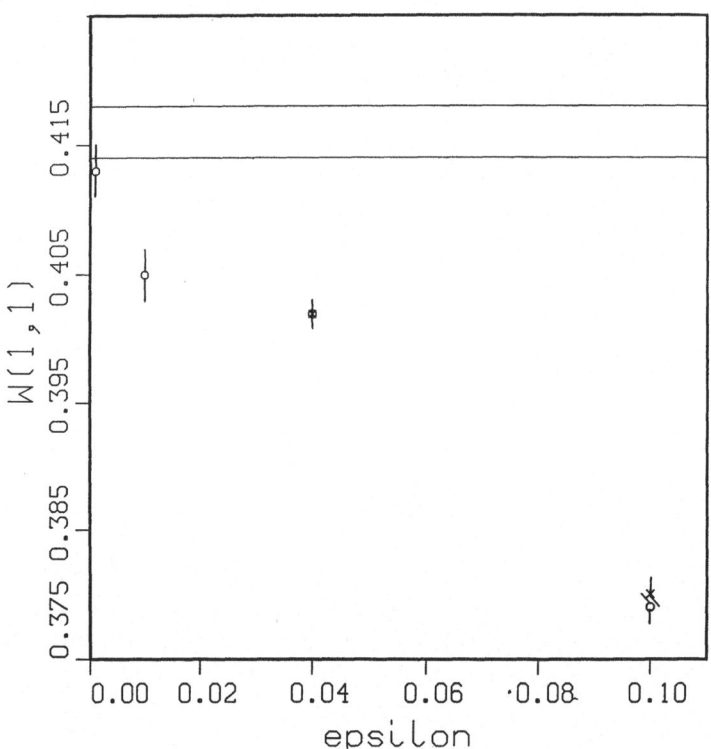

Fig. 2: Performance of Runge-Kutta algorithms for Langevin QCD. Open circles represent the results of the algorithm of Eq. (5) and crosses represent the results of Eq. (6). The full corridor shows the result of the exact computation [11].

350

TABLE I: Relative performance of Langevin and pseudo-fermion algorithms $(ma = 0.1)$. Algorithm D is the direct computation [11], P reproduces the relevant pseudo-fermion results of [11] and L is the Runge-Kutta algorithm of (5).

Alg.		Acc.	W(1,1)	W(1,2)	W(1,3)	W(2,2)	W(2,3)	W(3,3)
D			.042 (.003)	.039 (.003)	.023 (.003)	.016 (.003)	.005 (.002)	.001 (.002)
P		.86	.042 (.002)	.039 (.002)	.024 (.001)	.017 (.001)	.005 (.001)	.001 (.001)
L	.001		.039 (.002)	.038 (.002)	.024 (.002)	.018 (.002)	.004 (.002)	.003 (.002)
P		.74	.030 (.002)	.027 (.002)	.016 (.001)	.010 (.001)	.003 (.001)	.001 (.001)
L	.01		.031 (.002)	.029 (.002)	.018 (.001)	.011 (.001)	.002 (.001)	.002 (.001)

4. RUNGE-KUTTA ALGORITHMS

Figure 2 contains the same information as Figure 1 for the case of the Runge-Kutta algorithms (5) and (6). Table I assesses the quality of various planar Wilson loops measured using Eq. (5) with respect to the benchmark set in Ref. [11] and with respect to the pseudo-fermion algorithm at various acceptances [11]. Using Eq. (5), one must do 2 conjugate gradient inversions per Langevin step; the updating time per link becomes about 1.65 milliseconds and the memory requirement becomes about 5/4 that for the Euler or pseudo-fermion schemes. Eq. (6) requires 3 inversions per time step and takes about 2.37 milliseconds per link update.

We see that the higher-order errors are still substantial unless $\epsilon \lesssim 0.001$, in which case both first-and higher-order errors are rather small. The improvement gained by cancelling the non-integrable term is seen to be insignificant. There appears to be a correpondence between the Langevin and the pseudo-fermion algorithms: Runge-Kutta at $\epsilon = 0.001$ is comparable to pseudo-fermions at 86% acceptance and Runge-Kutta at $\epsilon = 0.01$ to pseudo-fermions at 74% acceptance. Time correlations and hence the corrected statistical errors [12] are systematically smaller for the Runge-Kutta than for the Euler scheme. However, the total run time required to achieve a given accuracy would still be higher for the second-order scheme because of the additional matrix inversion.

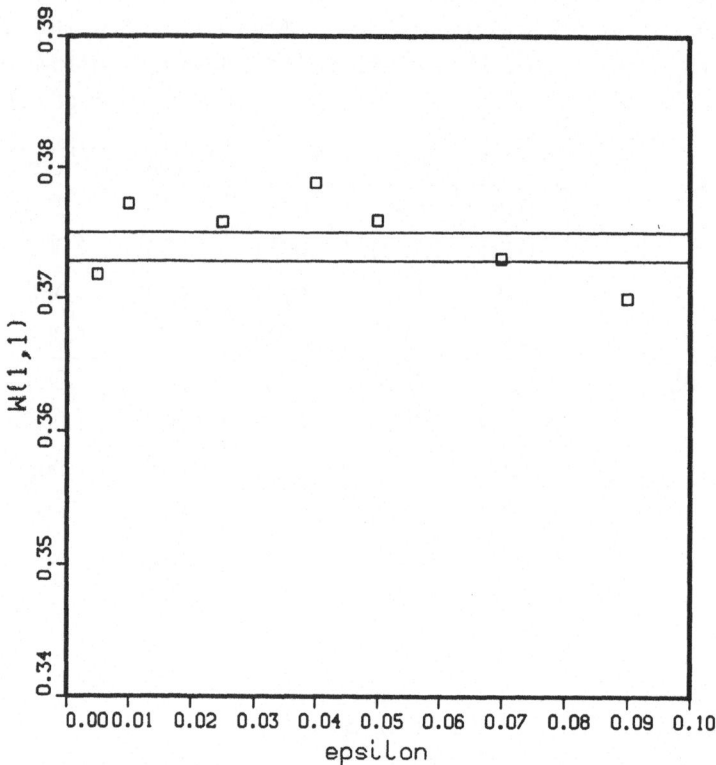

Fig. 3: The Runge-Kutta algorithm for Langevin pure SU(3). All the runs
are on a 4^4 lattice with $\beta = 4.8$. The benchmark is based on runs
with the standard Metropolis et al. method.

5. LOWERING THE QUARK MASS

Ref. [11] has observed that the pseudo-fermion algorithm performs
less well with respect to the benchmark at quark mass $ma = 0.05$ than at
$ma = 0.1$. Would this also be the case for Langevin updating? We have
run the Runge-Kutta algorithm based on Eq. (5) at $ma = 0.05$ in order to
answer this question. Using $\epsilon = 0.001$ and a stopping residue of 0.2 (as in
[11]) which is reached after 121 conjugate gradient iterations on the average,
we spend about 2.13 milliseconds per link update and find

$$\langle W(1,1) \rangle = 0.404 \pm 0.003$$

as compared to [11]

$$\langle W(1,1) \rangle = 0.417 \pm 0.002 \qquad \text{(DIRECT ALGORITHM)}$$

$$\langle W(1,1) \rangle = 0.410 \pm 0.006 \quad \text{(PSEUDO} - \text{FERMIONS 86\%)}$$

352

We tried lowering the stopping residue to $r = 0.01$ (reached after about 189 conjugate gradient iterations: 3.32 milliseconds per link update) but all the Wilson loops remained the same within errors (for instance, $\langle W(1,1) \rangle = 0.405 \pm 0.003$).

It would seem that Langevin at $\epsilon = 0.001$ is still comparable to pseudo- fermions at 86% acceptance and that the effective systematic error increases for both algorithms as the quark mass is decreased. That this is indeed an effect of the non-linear fermion determinant can be seen by comparing Fig. 2 to Fig. 3, which shows the ϵ-dependence of $W(1,1)$ in pure SU(3) [10]. The range of ϵ where Langevin algorithms give good results appears to shrink progressively as the quark mass decreases from infinity. This tends to confirm the observation [13] that the residual systematic error in Eq. (2), for fermionic schemes, is actually of the order ϵ^2/λ^4, where λ is some average over field configurations of the lowest eigenvalue of the lattice Dirac operator. It is known that this average decreases with the quark mass. It is tempting to speculate that the same effect is also seen in the pseudo-fermion scheme.

6. CONCLUSIONS

Our numerical experiments have shown that the analytically intractable terms of order ϵ^2 and higher in Eq. (2) have substantial effects upon planar Wilson loops in full QCD. Based on the known first-order correction terms, one had hoped that all systematic errors could either be eliminated by improving the naive Euler algorithm or they would be irrelevant in the continuum limit. Even though the quantitative insignificance of the leading-order non- integrable term is good news in this respect, the large effects of higher- order terms mean that such an optimistic conclusion cannot be taken for granted in the general case. One should therefore respect the "safety rules" we recommended above.

A study such as the present one obviously depends upon the existence of reliable and accurate benchmarks. It is therefore very important to pursue the effort initiated in [11], to make sure the direct algorithm has no significant hidden bias of its own and to improve the statistics of these benchmark runs.

Acknowledgment

This work was supported by the U.S. Department of Energy under contract number DE-AC02-76CH00016. The present author wishes to thank the Natural Sciences and Engineering Research Council of Canada for financial support.

REFERENCES

1. D. Weingarten, Nucl. Phys. B257 [FS 14](1981) 629.
2. G. Parisi and Y.S. Wu, Sci. Sin. 24 (1981) 483.
3. G.G. Batrouni, G.R. Katz, A.S. Kronfeld, G.P. Lepage, B. Svetitsky, and K.G. Wilson, Phys. Rev. D32 (1985) 2736.
4. A. Ukawa and M. Fukugita, Phys. Rev. Lett. 55 (1985) 1854.
5. G.G. Batrouni, Phys. Rev. D33 (1986) 1815;
   A.S. Kronfeld, Phys. Lett. B172 (1986) 93.
6. See e.g. J.B. Kogut, Nucl. Phys. B270 [FS16](1986) 169.
7. F. Fucito, E. Marinari, G. Parisi, and C. Rebbi, Nucl. Phys. B180 [FS 2](1981) 369.
8. M. Creutz and R.V. Gavai, BNL-38204 (1986), to appear in Nucl. Phys. B.
9. R.V. Gavai, J. Potvin, and S. Sanielevici, BNL-38526 (1986), to appear in Phys. Lett. B.
10. R.V. Gavai, J. Potvin, and S. Sanielevici, in preparation.
11. R.V. Gavai and A. Gocksch, Phys. Rev. Lett. 56 (1986) 2659;
    R.V. Gavai, A. Gocksch, and U.M. Heller, BNL-38449 and NSF-ITP-86-89 (1986); to appear in Nucl. Phys. B.
12. K. Binder, in "Phase Transitions and Critical Phenomena", C. Domb and M.S. Green eds., vol. 5B (Academic, New York, 1976).
13. M. Fukugita, these Proceedings.

KAON DECAY AMPLITUDES USING STAGGERED FERMIONS

Stephen R. Sharpe

Stanford Linear Accelerator Center, Bin 81
P. O. Box 4349, Stanford, CA 94305

ABSTRACT

A status report is given of an attempt, using staggered fermions, to calculate the real and imaginary parts of the amplitudes for $K \to \pi\pi$. Semi-quantitative results are found for the imaginary parts, and these suggest that ϵ' might be smaller than previously expected in the standard model.

INTRODUCTION

This talk describes a calculation of weak interaction matrix elements done in collaboration with Rajan Gupta, Gerry Guralnik, Greg Kilcup and Apoorva Patel (the Los Alamos Advanced Computing Group). Theoretical details can be found in reference [1]; detailed numerical results will appear elsewhere [2].

Present lattice measurements incorporate physics from the range of scales $\pi/L \approx .5\text{GeV} \leq \mu \leq 1/a \approx 2\text{GeV}$. Here a is the lattice spacing, and $L = N_s a$ is the physical size of the spatial box. At the ultraviolet end of this range we hope to match onto perturbative calculations: for weak interaction calculations we use the Renormalization Group machinery to scale down from M_W to $1/a$. This is reliable for small enough a, roughly $1/a > 2\text{GeV}$, corresponding to $g < 1$ on the Wilson axis.

The lower limit to μ is the infrared cut-off provided by the physical size of the lattice. Clearly we cannot simulate processes involving real pions until the smallest non-zero momentum is less than m_π. Further, as stressed here by Ken Wilson, we cannot look in detail at the wavefunctions of hadrons until the smallest momentum is less than the typical transverse momentum of quarks in these particles, i.e. $\simeq 200\text{MeV}$. However, we can overcome the first of these problems using the chiral Lagrangian to extrapolate from the lattice world with $m_\pi \simeq \pi/L$ to the real world with light pions. To do this we have to match our lattice results onto the forms expected for small m_π.

Combining these two matchings, we are almost in a position to evaluate those matrix elements of the weak interaction Hamiltonian which are relevant to Kaon decays, though only in the quenched approximation. This talk will explain what "almost" means for staggered fermions. I will first discuss the state of the theory, then present our results, and close with some conclusions.

THEORY

We transcribe fermions onto the lattice using the staggered formulation rather than that of Wilson. The pros and cons of staggered, relative to Wilson are:

PRO staggered	CON staggered
$U(1)_A$ symmetry for $m \to 0$	4 staggered species per continuum flavor
$\Rightarrow$ Ward Identities	$\Rightarrow$ Continuum theory has
$\Rightarrow$ Restricted Operator mixing	$\qquad U(4N_f)_V \times SU(4N_f)_A$ symmetry
	$\Rightarrow$ Extra factors of $N_f = 4$
	Operators with up to 4 links
	$\Rightarrow$ Noisier results
	$\Rightarrow$ Possibly large $O(g^2)$ corrections (?)

One other possible CON — the inability to project onto states of definite parity — should not be a problem for lattice pions (though see below). We have chosen to live with the CONS in order to make use of the PROS; this talk will show how this choice has worked out so far. The only part I will not comment on below is the possible CON of large $O(g^2)$ corrections; these have not been calculated yet for staggered fermions, although simpler calculations give some cause for worry [3]. The absence of these calculations also means that the short distance matching cannot be done in detail, and so only qualitative results can be given.

I will concentrate on the matrix elements (ME) $\langle K | \mathcal{H}_W | \pi\pi \rangle$. Our aim is to calculate their real and imaginary parts, for both charged and neutral kaons. Experimentally, the real part of the K^0 amplitude is 20 times larger than that of the $K^\pm$; this is the long-standing puzzle of the $\Delta I = 1/2$ rule. As for the imaginary parts, it is the relative phase between $I = 1/2$ and $I = 3/2$ amplitudes that determines the magnitude of ϵ'. In the standard model both amplitudes get phases from penguin diagrams (strong and electromagnetic) with t and b quarks in the loops.

The direct measurement of $K \to \pi\pi$ amplitudes is beyond present lattice technology. Instead, the standard trick [4] is to use the chiral Lagrangian to relate the amplitudes to those of $\langle K | \mathcal{H}_W{}^{subtracted} | \pi \rangle$. Aside from the fact that this is an approximation, to which I will return later, this trick brings with it a nasty problem – the fact that subtractions have to be done. This is the worst problem to be overcome in order to extract numbers. It might be thought that, given this problem, it is worth putting a lot of effort into a direct calculation of the $K \to \pi\pi$ amplitude, for which no subtractions are needed. This is far from clear. To avoid subtractions, one must calculate on-shell matrix elements. This is hard on the lattice because of the discrete momenta. Furthermore, one has to understand final state interactions; these take a complicated form in Euclidean space since one cannot have a phase.

So we proceed by calculating $\langle K | \mathcal{H}_W{}^{subtracted} | \pi \rangle$. In the continuum $\mathcal{H}_W$ contains a slew of operators multiplied by coefficients. When transcribed to the lattice, a lot more operators are needed. For staggered fermions this is because the flavor and spin degrees of freedom are spread out over 2^4 points [1]. The operators thus contain varying numbers of gauge links, up to four in each bilinear. We know how this works at $O(g^0)$, but not yet at $O(g^2)$. However, we do know the general features of the operators to all orders. As in the continuum, there is a natural division of the operators into four types: (1) $I = 3/2$, LL operators; (2) $(8_L, 1_R)$, $I = 1/2$, LL operators; (3) $(8_L, 1_R)$, $I = 1/2$ LR operators; (4) $(8_L, 8_R)$ LR operators. As the scale is changed these operators mix; at m_W one has only operators of the first two types, but strong interaction "penguin" diagrams (with t and b loops) produce type (3) operators as the scale is reduced, and electromagnetic penguins produce operators of type (4). At a scale corresponding to about m_c, the real parts of the coefficients are largest for the LL operators, and thus these dominate the K decay rates. Conversely, the imaginary parts are larger for the LR operators, and these probably dominate the contributions to ϵ'.

Before proceeding I want to make a comment about the scale of the lattice calculation. We will be working at a scale $1/a \approx m_c$, for which the charm quark can be ignored, to first approximation. By this I mean that the coefficients are calculated by running down to m_c, but then the charm quark is dropped from the operators. This implies that the usual penguin solution of the $\Delta I = 1/2$ puzzle cannot be tested directly. In this solution, it is the RG scaling below m_c that induces the real part of the coefficient of operators of type (3), and then the enhanced matrix elements of these operators give the $\Delta I = 1/2$ rule. In fact, this idea can never really be tested, because if one runs much below m_c one is outside the range of perturbation theory.

The four types of operator yield four classes of contraction. The first is the eight contraction of the LL operators – known below as LL8. Type (1) operators have only these contractions, and so these are the only contractions contributing to $K^\pm$ decays. They also give the main contribution to the imaginary part of $K^0 \leftrightarrow \overline{K^0}$ mixing, i.e. to ϵ. They are straightforward to calculate, partly because they are eights rather than eyes, and partly because they do not require subtractions. However, because of their simplicity, one can make reasonable estimates of them using various continuum approximations, e.g. vacuum saturation. For the lattice to improve upon these estimates, it must give a result accurate to better than a factor of two.

The second class is the eye contractions of the LL operators of type (2), which I call LLI. These are purely $I = 1/2$. At the charm quark scale these are the only possible source of the $\Delta I = 1/2$ rule within the standard model. To the extent that it makes sense to discuss lower scales, these contractions may also be dominant, a view emphasized by Donoghue [5]. These contractions are harder to evaluate for two reasons. First, they require the use of source techniques, which reduces the statistics. Second, they require subtractions. These contractions are also harder to estimate in the continuum: e.g. they vanish in vacuum saturation.

The third contractions are those of the LR octet operators. Unlike the LL operators, these do not retain their spinor structure upon Fierz transformation. For the most important such operator, one has:

$$\bar{s}_a \gamma^\mu (1 + \gamma_5) d_b \ \bar{q}_b \gamma^\mu (1 - \gamma_5) q_a \ = \ -2 \, \bar{s}_a (1 - \gamma_5) q_a \, \bar{q}_b (1 + \gamma_5) d_b$$

Here, a and b are color indices, and q is summed over u, d and s. These operators have both eight and eye contractions, so I refer to them as LR8I. Source methods are again needed, as are subtractions.

The appearance of densities, rather than currents, in the Fierzed form leads one to expect an enhancement over LL operators by $m_K^4/(m_s^2 \Lambda^2)$, where $\Lambda \approx 1 \text{GeV}$ is the cut-off in the chiral Lagrangian [6]. The factors can be explicitly worked out in the large N_c limit [7][8]. It is this enhancement which has led to all the speculation about the role of penguins in the $\Delta I = 1/2$ rule. However, here I am interested in the penguins as the source of ϵ'.

The final contraction is that of the LR singlet operators. The dominant contribution comes from eight contractions, so I refer to them as LR8. Compared to LL operators, one expects an enhancement of m_K^2/m_s^2 due to the LR structure. They are straightforward to calculate, needing no sources, and no subtractions (at least for the dominant part). They are also easy to estimate in the continuum.

The great advantage of staggered fermions is that these 4 types of contractions separately satisfy exact lattice Ward Identities (WI) [9]. This is true separately for each of the many operators that appear, and is true configuration by configuration. These WI are precise lattice analogues of the continuum WI of PCAC. They constrain the behavior of the ME as one varies m_q. For a general operator $\mathcal{O}$ in $\mathcal{H}_W$:

$$\langle K | \mathcal{O} | \pi \rangle = f^2 (\alpha f^2 + \beta m_\pi^2 + \gamma m_K^2 + \delta m_\pi m_K) + O(m_q^2)$$

where f is the value of f_π extrapolated to $m_\pi = 0$, and α, β, γ and δ are dimensionless. The WI imply that for the LL8 $\alpha = \beta = \gamma = 0$, for the LLI and LR8I that $\alpha = \beta = 0$, but no relations for the LR8. For the LR8I (which are made up of contractions like the LR8 together with eyes) it is the addition of eights to eyes that cancels the α and β terms.

One can show [4][9] that, to $O(m_q)$, the subtraction needed for the LLI and LR8I will remove the γ term. After subtraction, then, the LLI and LR8I have the same form as the LL8 contractions. One can further show that, to the same order, one can measure γ using

$$\langle K | \mathcal{O} | 0 \rangle = \gamma \sqrt{2 N_f} f^3 m_K^2 (m_d - m_s)/(m_d + m_s)$$

where $N_f = 1$ in the continuum, but $N_f = 4$ for staggered fermions. This shows how the subtraction removes the effect of $s \leftrightarrow d$ mixing.

This can all be phrased equivalently in the language of operator mixing [10][9]. All the operators in $\mathcal{H}_W$ mix with other operators of d=6, constrained by the lattice symmetries. In addition, operators of types (2) and (3) mix with the d=4 operator

$$S = (m_d + m_s)\bar{s}d \; + \; (m_d - m_s)\bar{s}\gamma_5 d$$

which is also a $SU(3)_L$ octet, $I = 1/2$ operator. It is exactly this operator which gives the γ terms in the above equations. This suggests the following method to remove the γ terms. Choose ρ such that

$$\langle K|\mathcal{O}^{subtracted}|0\rangle \equiv \langle K|\mathcal{O} - \rho S|0\rangle = 0$$

and then

$$\langle K|\mathcal{O}^{subtracted}|\pi\rangle = \delta f^2 m_\pi m_K + O(m_q^2)$$

This method allows a time by time subtraction, and we use it below.

As stressed by the CERN/Rome group [10], the coefficients of the mixing with S are non-perturbative, i.e. of $O(1/a^2)$. Thus one should use a non-perturbative method of calculating γ, such as that outlined above. But for Wilson fermions this method is not available [10]. One can proceed by performing a perturbative evaluation of γ, as suggested by the UCLA group [11]. However, this is suspect, because non-leading terms will be of $O(g^n/a^2)$, and thus diverge as $a \to 0$. Nevertheless, they claim that for $g \approx 1$ their method might be viable. It seems to me to be important to check first on simpler quantities such as $\langle \bar{\psi}\psi \rangle$.

Whatever method of subtraction one uses, the entire procedure rests upon an expansion in meson masses. This has two consequences. First, as emphasized by Martinelli, the output is only the $O(m^0)$ term in the K decay amplitude; higher order terms cannot be obtained. Second, it is essential that one finds the advertised chiral behavior. Without this, the low energy matching cannot be done. One can also check this in other ways, e.g. by looking at the variation of f_π with m_π.

RESULTS

After a trial run on an $8^3 \times 16$ lattice [1], we have now completed an analysis on a $12^3 \times 30$ lattice [2]. This is long enough to unambiguously expose the lightest states. We use an improved action, that of ref [12] with $K_F = 10.5$. This corresponds to $\beta = 5.96$ on the Wilson axis, and thus is nearly in the scaling region. We have used 25 lattices, and have attempted to address the issue of low energy matching by using small quark masses. To do this, we have calculated with $m_q = .040$ and .005. The larger mass corresponds quite closely to the physical strange quark mass, so I refer to it S. The lighter mass is as close as we can get to a realistic u or d quark, and I call it the U quark. This allows us to consider three psuedo-Goldstone bosons: SS ($m \approx 700\text{MeV}$), US ($m \approx 500\text{MeV}$), and UU ($m \approx 300\text{MeV}$). Using these we have measured three ME: $\langle K(SS)|\mathcal{H}_W|\pi(SS)\rangle$, labelled SS; $\langle K(US)|\mathcal{H}_W|\pi(UU)\rangle$, called US; and $\langle K(UU)|\mathcal{H}_W|\pi(UU)\rangle$ named UU. All our propagators have been calculated with antiperiodic boundary conditions (APBC).

First I comment on the chiral behavior of quantities derived from two point correlators. For UU, US and SS respectively, we have $m_\pi = .180, .358, .469$, $f_\pi = .056, .072, .088$, and $Z_\pi = .264, .325, .43$. The Z factor is defined through the two point psuedoscalar correlator C(t)

$$C(t) = e^{-|t|m_\pi} Z_\pi/N_\pi; \quad N_\pi \equiv 2 \sinh m_\pi$$

One expects f_π and Z_π to have the form $a + b\overline{m}$ ($\overline{m}$ is the average quark mass), while $m_\pi^2 = c\overline{m}$. If we are to be in the region where $O(m_q)$ expansions are valid, as we must be to do the low energy matching, the b terms must be small for $\overline{m} = m_s$. We are clearly at the limit of this region with the SS states. Extrapolating our numbers to $m_q = 0$, we find for the physical π and K that $f_K/f_\pi \approx 1.35$, compared to the experimental 1.25. Thus the $O(m)$ terms are in rough agreement with those in the continuum.

I now turn to the correlators from which we extract the ME. I denote these by $C(t_\pi, t_K)$. Here the operator is at $t = 0$ (or $t = 0, 1$ for two timeslice operators), and the π and K are respectively at t_π, t_K. All the figures have $t_\pi = 7$, a distance large enough to remove heavier states, yet small enough to retain a reasonable signal. For eight contractions we have data for all t_π, but for the eyes, what you see is all we've got. The region for $t_K > 15$ has the π and K on opposite sides of $\mathcal{H}_W$, and so corresponds to the ME we want to measure. For $t_K < 15$ we are measuring the off-shell ME $\langle 0|\mathcal{H}_W|K\pi\rangle$, together with final state interaction effects. If one ignores such effects, then the chiral behavior we desire corresponds to the correlator being antisymmetric about the midpoint.

The correlator is related to the ME $\mathcal{M}$, for $t_K > 15$, by

$$C(t_\pi, t_K) = \mathcal{M} \, N_f (e^{-t_\pi m_\pi} Z_\pi/N_\pi)(e^{(t_K - 30)m_K} Z_K/N_K)$$

If the ME has the correct chiral behavior $\mathcal{M} \propto m_\pi m_K$, and we ignore variations in Z_π, etc., then the coefficient of the exponentials should be constant. In the figures this means that the extrapolation of the exponential decay to $t = 30$ (30.5 for two timeslice operators) should not depend on m_π, m_K. But this is not a fair test for our range of m_q. A better approach is to compare the lattice correlators to those calculated in the vacuum insertion approximation (VIA) directly on the lattice. The VIA correlators automatically have the correct chiral behavior, apart from the variation of Z_π etc. with m_q. Comparison of the data with VIA removes this spurious variation. It also makes for simpler comparison between different calculations.

Finally I come to the pictures. In all of them, the y-axis is logarithmic, but I do not show the scale, as it is not relevant here. I should also stress that I am using the $O(g^0)$ transcription of continuum operators onto the lattice. Figure 1 shows the results for the lattice equivalent of the eight contraction in the ME:

$$\langle \pi^+|O_1|K^+\rangle \equiv \langle \pi^+|\bar{s}_a\gamma_4\gamma_5 u_a\ \bar{u}_b\gamma_4\gamma_5 d_b|K^+\rangle$$

This LL8 contraction is shown for masses UU and SS, along with the VIA results. The bump evident for $t_\pi \approx t_K$ is mainly due to wrap-around effects allowed by the APBC, to which I will return. I note the following: 1. the exponential decays are clear; 2. the correlators are roughly antisymmetric; 3. VIA works well for SS; 4. VIA works poorly for UU. Since the VIA results correspond to the correct chiral behavior, the data appear to be growing too fast as $m_q \to 0$.

In Figure 2 the LL8 results for the operator $O_2 = \bar{s}_a\gamma_4\gamma_5 u_b\ \bar{u}_b\gamma_4\gamma_5 d_a$ are compared to those of O_1. In continuum VIA the ME of O_2 are 3 times smaller. For SS this is roughly true, but for UU it is clearly false. The growth of the O_2 ME at small m_q is completely inconsistent with the required chiral behavior. In fact, it is consistent with M independent of m_π, m_K.

Though I don't have space to show it here, all other LL8 channels show similar violations of VIA and, consequently, the wrong chiral behaviour. A typical example is $\bar{s}_a\gamma_4 u_a\ \bar{u}_b\gamma_4 d_b$. This operator has three links in each bilinear, and is zero in VIA, yet yields a clear signal. This signal is small for SS, but for UU it is $\approx 1/5$ of that for O_1. This violation of chiral behavior

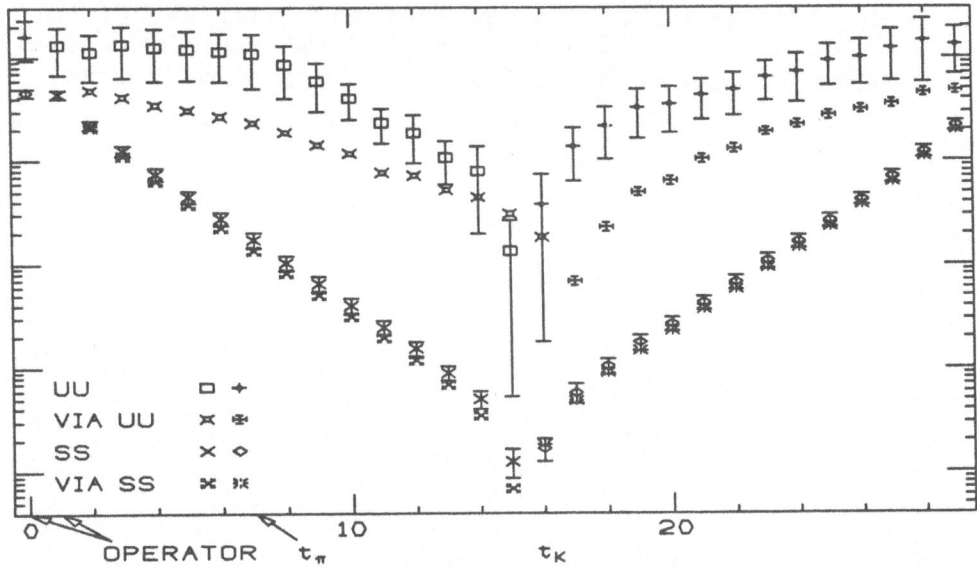

Fig 1. LL8 data for $\langle K|O_1|\pi\rangle$. The errors in the VIA data have been removed for clarity. They are comparable to those on the data they approximate. Of the two symbols, the first is for positive data, the second for negative.

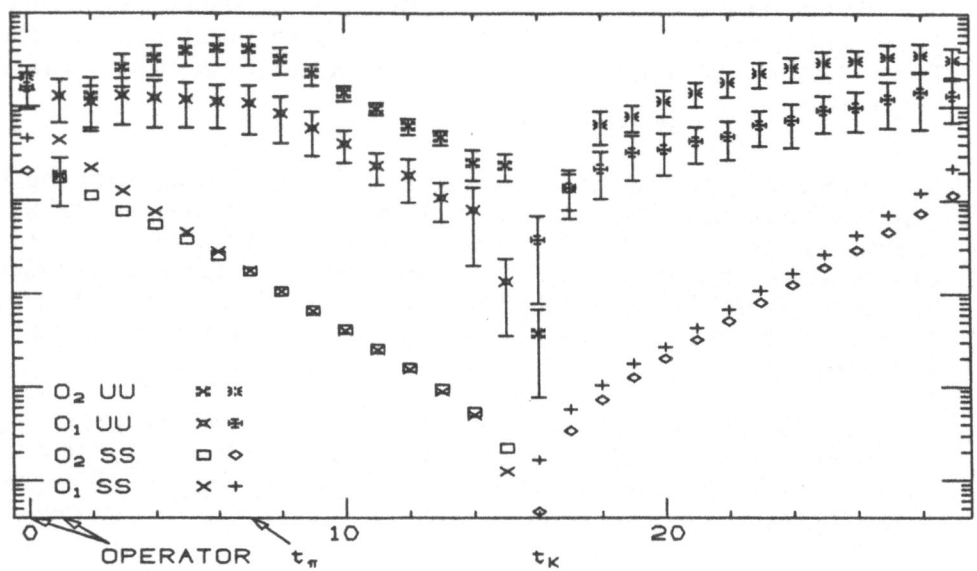

Fig 2. Comparison of O_1 and O_2 LL8 data. Error bars for SS data are about the size of the symbols.

and VIA is both good and bad. It is good because VIA gives a poor description of Kaon decays [7]. It is bad since the wrong chiral behavior can mean only two things: (a) we do not yet have small enough m_q; (b) the APBC effects are dominant. These wrap-around effects do not violate the WI, but do affect the argument leading from the WI to the chiral behavior.

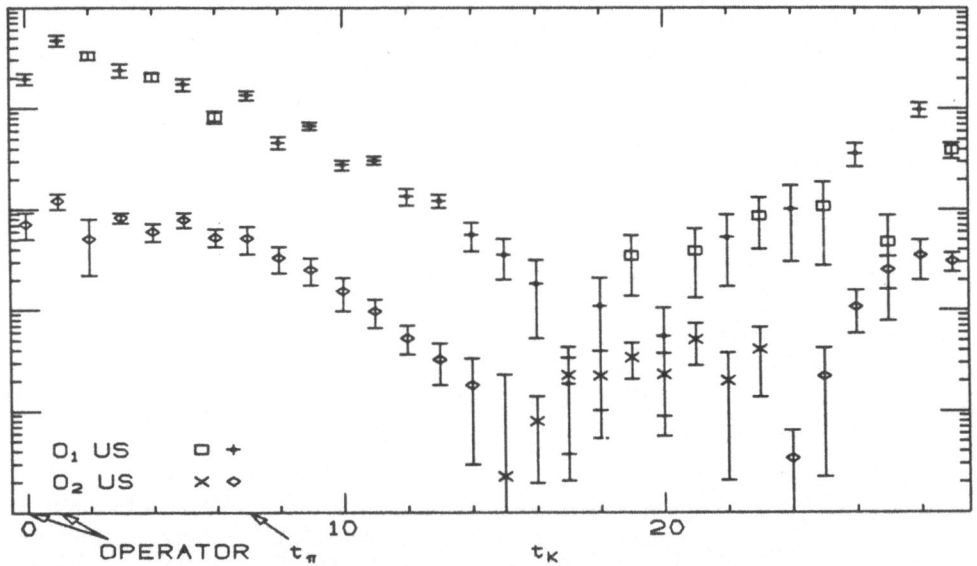

Fig 3. LLI data for O_1 and O_2 after subtraction.

Next I turn to the LLI contractions. Here we must make the subtraction, and this can only be done for the masses US. Figure 3 show the results for the eye contractions of operators O_1 and O_2, after subtraction. We want the correlator to be antisymmetric, and there are signs of this. However, the O_2 data is too poor to extract a number, and the O_1 data shows a dominant oscillatory behavior. This, we think, is due to wrap-around effects. So one of the staggered fermion CONs has really come home to roost. For what it is worth, the typical magnitude of the O_2 ME *is* large enough to yield the $\Delta I = 1/2$ rule.

So much for the bad news. For the LR8I contractions we *can* extract some useful conclusions. Here we can do the subtraction for all m_q. We show results for the dominant part of the operators:

$$O_5 = \bar{s}_a(1+\gamma_5)d_a \; \bar{q}_b(1-\gamma_5)q_b \; ; \; O_6 = \bar{s}_a(1+\gamma_5)d_b \; \bar{q}_b(1-\gamma_5)q_a$$

Figure 4 shows the SS results for $\langle K|O|0\rangle$ with operators O_5, O_6, S, and the VIA to O_6. It is apparent that (a) VIA does very well; (b) the continuum VIA expectation that $O_6 = 3O_5$ works extremely well; and (c) the determination of the subtraction coefficient ρ can be done easily.

Figure 5 shows the SS results for $\langle K|O_6|\pi\rangle$, $\langle K|O_6^{sub}|\pi\rangle$ and their VIA values. This shows how the subtraction removes a large symmetric part to expose the antisymmetric residue. The data, however, agrees extremely well with VIA, and, though not shown, it remains true that the $O_6 = 3O_5$.

For the UU LL8I results VIA does less well. It falls above the data for $\langle K|O_6|0\rangle$ by $10-20\%$. The $\langle K|O_6|\pi\rangle$ data are shown in Figure 6. Here the VIA result is much cleaner than the actual data, the latter showing signs

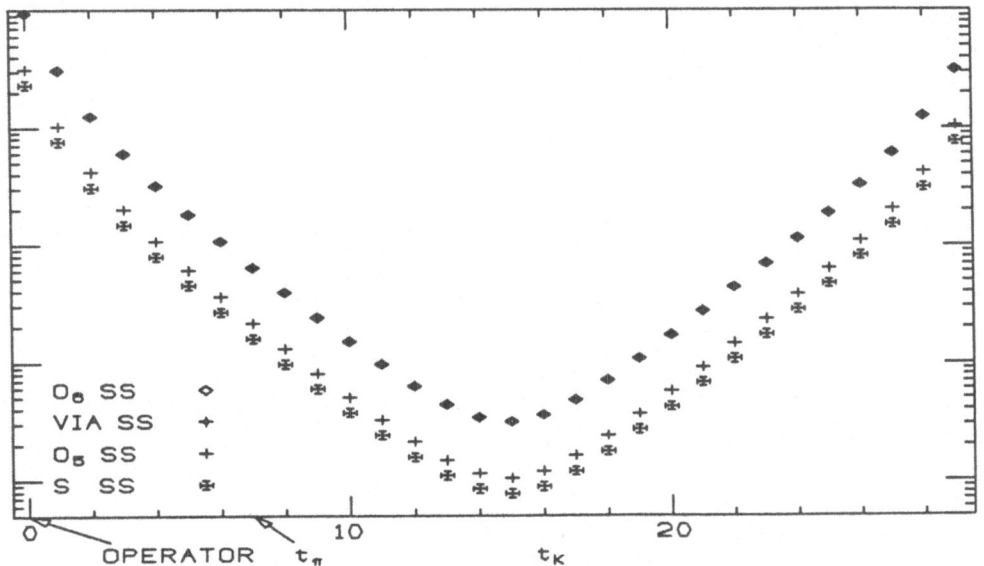

Fig 4. $\langle K|O|0\rangle$ data for LR8I contractions.

363

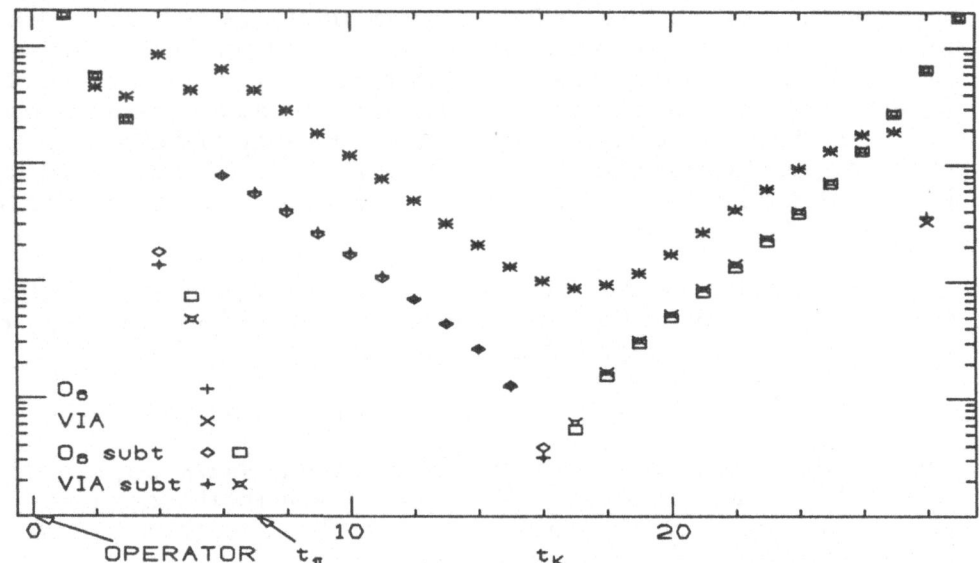

Fig 5. $\langle K|\mathcal{O}|\pi\rangle$ data for LR8I, with and without subtractions.

of oscillations again. Nevertheless, the data are much better than the LLI, with the antisymmetry being clear. Because of the oscillations, it is hard to extract quantitative conclusions, but it is clear that the data gives a ME substantially smaller than that in VIA. This is in striking contrast to the LL8 (and the LR8) data. Furthermore, the chiral behavior is fine, if anything a little to soft.

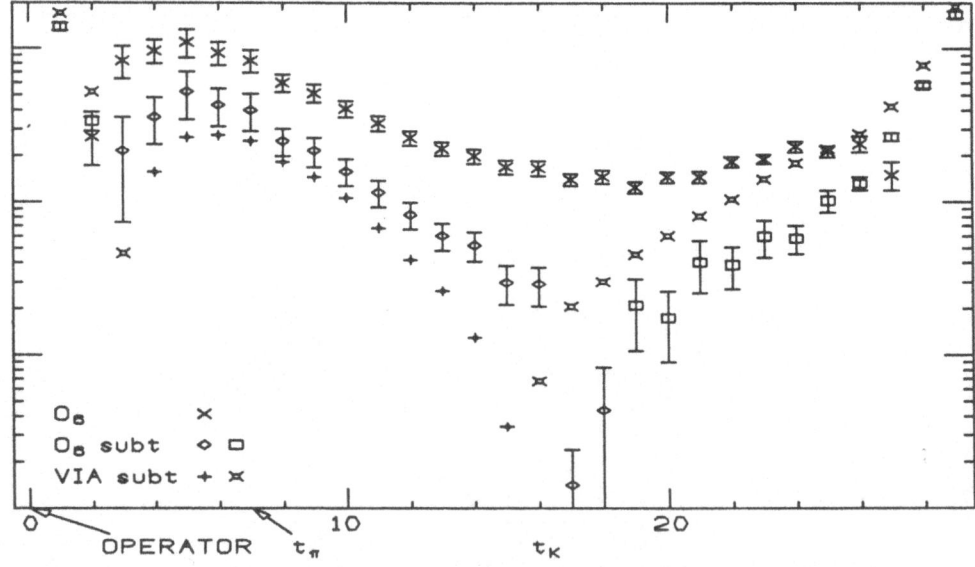

Fig 6. Same as Fig 5, but for UU.

364

The final result I want to discuss is for the LR8 contractions, though I have no space for pictures. These are dominated by a symmetric part corresponding to M being independent of the meson masses. We find that the relation $3O_5 = O_6$ works well for all masses. VIA works very well for SS, but lies significantly below the UU data.

CONCLUSIONS

The two major problems with our study, in purely numerical terms, are poor statistics and wrap-around contributions. The latter occur for all types of lattice fermion, but are exacerbated by the use of staggered fermions. These two problems conspire to make an extraction of even a qualitative result on the $\Delta I = 1/2$ rule impossible. Together with the lack of a perturbative operator mixing calculation, they also do not allow even a semi-quantitative result for the B parameter of $K^0 - \overline{K^0}$ mixing. This last point is true for both LL and LR operators.

We can, however, make some general comments. For both LL8 and LR8, the data agree with VIA at large m_q, but exceed VIA for small m_q. Clearly, fluctuations are damped at large m_q (recall that VIA is exact on a single configuration). An optimistic interpretation is that there is region for small m_q where VIA is violated, but in such a way that the data has the correct chiral behavior. We need more low mass data to check this. A more pessimistic possibility is that the bad chiral behavior is intrinsically related to our use of APBC. The wrap-around contributions cannot invalidate the WI, but can remove the connection between the WI and the chiral behaviour of the ME. Even assuming the optimistic scenario, one should not forget the caveat raised here by Mütter. Some of the fluctuations at small m_q are artifacts of the quenched approximation. They will be damped out by the fermion determinant in the full theory.

I have placed much stress on the utility of a comparison with VIA. Of course, VIA cannot work for all values of the lattice spacing, because the anomalous dimensions of the true operators and their approximants differ. Nevertheless, at the present stage, when the calculations are not quantitative, this is a small effect. The usefulness of VIA is most clear for the LR8 contractions. One expects these to be enhanced by factors $\propto 1/m_s^2$ relative to LL8. But what is m_s? In this calculation, and in other calculations in the quenched approximation, one finds $m_s \approx 50\text{MeV}$. This is at least a factor of 3 smaller than the continuum $m_s(\mu = 1\text{GeV})$. This discrepancy *cannot* be explained by saying that the appropriate scale is not $\mu = 1/a$ but $\mu = (\Lambda_{mom}/\Lambda_{lat})(1/a)$ [13]. The logarithmic scaling of masses is too slow. The only reasonable explanation is that the small m_s is due to the quenched approximation. In essence, one fixes the combination $m_s\langle\overline{\psi}\psi\rangle$ to be correct. Since $\langle\overline{\psi}\psi\rangle$ is too large in the quenched approximation, m_s must be too small. Returning to the LR8I, this means that naively extracted ME are ≈ 10 times too large. The correct approach is to compare the calculation to the VIA, which can then be evaluated in the continuum with the correct m_s. The same comments apply to the LR8.

The continuum VIA to the LR8I has been worked out in references [7] and [8]. Using $m_s(\mu \approx m_c) = 125\text{MeV}$, one finds roughly that $\epsilon'/\epsilon = .002\,(\tilde{c}_6/.1)$. Here, $\tilde{c}_6$ is the imaginary part of the Wilson coefficient evaluated at $\mu \approx m_c$. Its value is controversial, but is not likely to be much larger than .1, though it could be smaller. The electromagnetic penguins also contribute to ϵ', and in VIA reduce it by about 20%. There are also isospin violating effects which combine to reduce ϵ' by another $30-40\%$ [14]. These numbers should be compared to a present experimental limit of $\approx .005$, and a future sensitivity of .001. Our results suggest that a bad situation may be worse still. If the penguin contribution is smaller than VIA, yet the electromagnetic penguin is larger than VIA, ϵ' will be further reduced. A more detailed discussion will be given in [2].

In summary, progress has been made towards the calculation of ϵ'. For this the use of staggered fermions is essential. To improve this calculation, and to extract values for the decay rates of Kaons, ϵ and the B parameter, we need to do the following. (1) Replace APBC with fixed BC; (2) Increase the number of small quark masses used; (3) Increase the statistical sample; (4) Check asymptotic scaling; (5) Increase the size of the infrared cut-off; (6) Include dynamical fermions; and, last but not least, (7) Do the perturbative operator mixing calculation.

This work was supported by the Department of Energy, under contract DE-AC03-76SF00515, and by a grant of time at the MFE computing center. I thank Greg Kilcup for reading the manuscript.

REFERENCES

[1] S. R. Sharpe, A. Patel, R. Gupta, G. Guralnik, and G. Kilcup, UW/40048-11 P6 (August 1986), to be published in *Nucl. Phys.* **B**.
[2] S. R. Sharpe *et al.*, in preparation.
[3] M. F. L. Golterman and J. Smit, *Nucl. Phys.* **B245** (1984) 61.
[4] C. Bernard, T. Draper, A. Soni, H. D. Politzer and M. Wise, *Phys. Rev.* **D32** (1985) 2343.
[5] J. F. Donoghue, Talk presented at the XXIII International Conference on High Energy Physics, Berkeley, CA, July 1986
[6] M. Shifman, A. Vainshtein and V. Zakharov, *Nucl. Phys.* **B120** (1977) 316.
[7] R. S. Chivukula, J. Flynn and H. Georgi, *Phys. Lett.* **171B** (1982) 453.
[8] W. Bardeen, A. Buras and J. Gérard, *Phys. Lett.* **180B** (1986) 133.
[9] G. W. Kilcup and S. R. Sharpe, HUTP-86/A048 (June 1986), to be published in *Nucl. Phys.* **B**.
[10] L. Maini, G. Martinelli, G. C. Rossi and M. Testa, *Phys. Lett.* **176B** (1986) 445; CERN-TH.4517/86 (August 1986).
[11] C. Bernard, these proceedings.
[12] A. Patel, R. Gupta, G. Guralnik, G. Kilcup and S. R. Sharpe, *Phys. Rev. Lett.* **57** (1986) 1288.
[13] C. Bernard, T. Draper, G. Hockney, A. M. Rushton and A. Soni, *Phys. Rev. Lett.* **55** (1985) 2770.
[14] J. F. Donoghue, E. Golowich, B. R. Holstein and J. Trampetic, UMHEP-262 (August 1986).

MICROCANONICAL AND HYBRID SIMULATIONS OF LATTICE QUANTUM CHROMODYNAMICS WITH DYNAMICAL FERMIONS

D. K. Sinclair

High Energy Physics Division
Argonne National Laboratory
9700 South Cass Avenue
Argonne, Illinois, 60439

ABSTRACT

Lattice QCD is simulated using Microcanonical and Hybrid (Microcanonical/Langevin) methods to facilitate the inclusion of dynamical fermions (quarks). We report on simulations with 4 flavours of light dynamical quarks on a $10^3 \times 6$ lattice to study the finite temperature deconfinement/chiral transition which should be observable in relativistic heavy ion collisions, as a function of quark mass. A first order transition is observed at large mass, weakens at intermediate mass and strengthens for very small quark mass.

INTRODUCTION

The grassman nature of fermion fields precludes their direct simulation by Monte Carlo methods. All methods which are practical except on the smallest lattices are based on the observation that one can replace the fermi fields by bose fields if we replace the Dirac operator by its inverse. The main problem then is calculating this inverse as efficiently as possible, and avoiding having to calculate it more than once per sweep of the lattice. This second restriction requires that the changes in the fields during each sweep be small. The most promising methods to date appear to be the Microcanonical and Langevin schemes and more recently a hybrid of the two which preserves the best features of each. The principal advantage of these schemes is that the calculation of the inversion of the Dirac operator is replaced by the solution of a set of linear equations which is faster to implement. The small change requirement is met by the fact that the fields evolve continuously, described by a differential equation. The microcanonical and hybrid algorithms have been used extensively by the Argonne/University of Illinois group to study Lattice QCD.

I will only report on recent and ongoing work studying thermodynamics on a $10^3 \times 6$ lattice to study the finite temperature deconfinement and chiral transitions which could be observed in relativistic heavy ion collisions (e.g. at RHIC). These simulations were done in collaboration with E.V.E. Kovacs (Argonne) and J. B. Kogut (University of Illinois).

Section 2 gives a brief description of the microcanonical method. Section 3 describes the hybrid scheme. In section 4 the simulations on the $10^3 \times 6$ lattice are described. Finally section 5 contains the summary and conclusions.

MICROCANONICAL METHOD

In the microcanonical approach [1] of Callaway-Rahman, adapted to include dynamical fermions by Polonyi-Wyld, one notes that the classical Lagrangian

$$L = \frac{1}{2} \sum_{\substack{\mu \\ \text{links}\,\ell}} \dot{U}_\mu^\dagger(\ell) \hat{P} \dot{U}_\mu(\ell) + \sum_{\substack{\text{even} \\ \text{sites}\,mn}} \dot{\psi}_m^\dagger \left[(-\not{D} + m)(\not{D} + m) \right]_{mn} \dot{\psi}_n$$
$$- \omega^2 \sum_{\substack{\text{even} \\ \text{sites}\,n}} \psi_n^\dagger \psi_n - \beta \sum_\square \text{tr}\left[UUUU + \text{h.c.} \right],$$

where $\hat{P} = \text{diag}(1,1,0)$ in colour space and "·" indicates derivative with respect to a new, fictitious time, has canonical partition function

$$Z = \int_{\text{phase space}} e^{-H},$$

where the fields have been rescaled to make the classical temperature $kT = 1$. Integrating out the (pseudo) fermion fields and both gauge and fermi momenta yields

$$Z = \text{constant} \int dU \, \det(\not{D} + m) e^{-\beta \sum_\square \text{tr}[UUUU + \text{h.c.}]}$$

368

Hence we obtain the physics of 4-flavour (staggered fermions) Lattice QCD.

One now uses the known equivalence between canonical and micro-canonical statistical mechanics to transform to the microcanonical formulation. Here the same physics is obtained from solving the classical equations of motion derived from L or H, and calculating QCD observables as long time averages, assuming ergodicity.

Simulations in this approach involve the numerical integration (Verlet) of these equations of motion (molecular dynamics). This in turn requires the numerical solution of linear equations of the form

$$(-\not{D} + m)(\not{D} + m)\psi = \phi.$$

We currently use the conjugate gradient algorithm to perform this inversion.

HYBRID ALGORITHM

The hybrid algorithm of Duane [2] involves periodic replacement of the "momenta" in the microcanonical Hamiltonian with random "momenta" having a Maxwell-Boltzmann distribution at a fixed "temperature" (usually taken to be 1), during the simulation. Clearly, once equilibrium obtains, such a replacement does not disturb this equilibrium, since all that is required for different parts of the system to be in thermodynamic equilibrium is that they have the same temperature. One can consider that the "momenta" have been allowed to thermalize with a heat bath at the chosen temperature.

This guarantees ergodicity, and tuning the "refreshment"" frequency gives more rapid coverage of phase space than either the microcanonical (no refreshment) or Langevin (refresh every sweep) methods. Because of its stochastic nature it is not quite as sensitive to metastability as the microcanonical scheme.

THERMODYNAMICS ON A $10^3 \times 6$ LATTICE

We have used both methods (microcanonical and hybrid) to study the mass dependence of the finite temperature deconfinement/chiral transition on a $10^3 \times 6$ lattice. This augments previous simulations on $8^3 \times 4$ lattices [3] where the transition occurs on the strong coupling side of the cross over region (Here it occurs in the cross over region, nearer the continuum physics.) We use staggered fermions, yielding 4 light flavours of quarks. Studies at fermion masses of .05 to .25 (lattice units) have already been completed [4]. Extension down to mass of .025 is nearing completion [5] and we include preliminary data here.

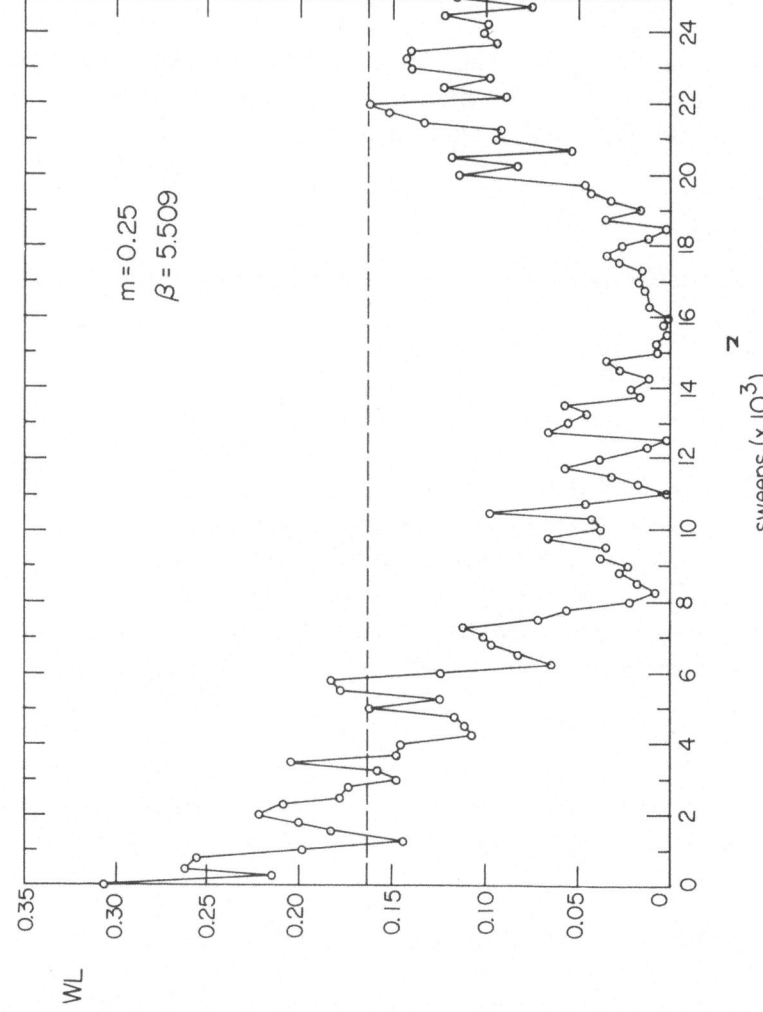

Fig. 1. Wilson Line vs. sweep ($dt = .02$) for $\beta = 5.509$, $m = .25$ from an ordered start.

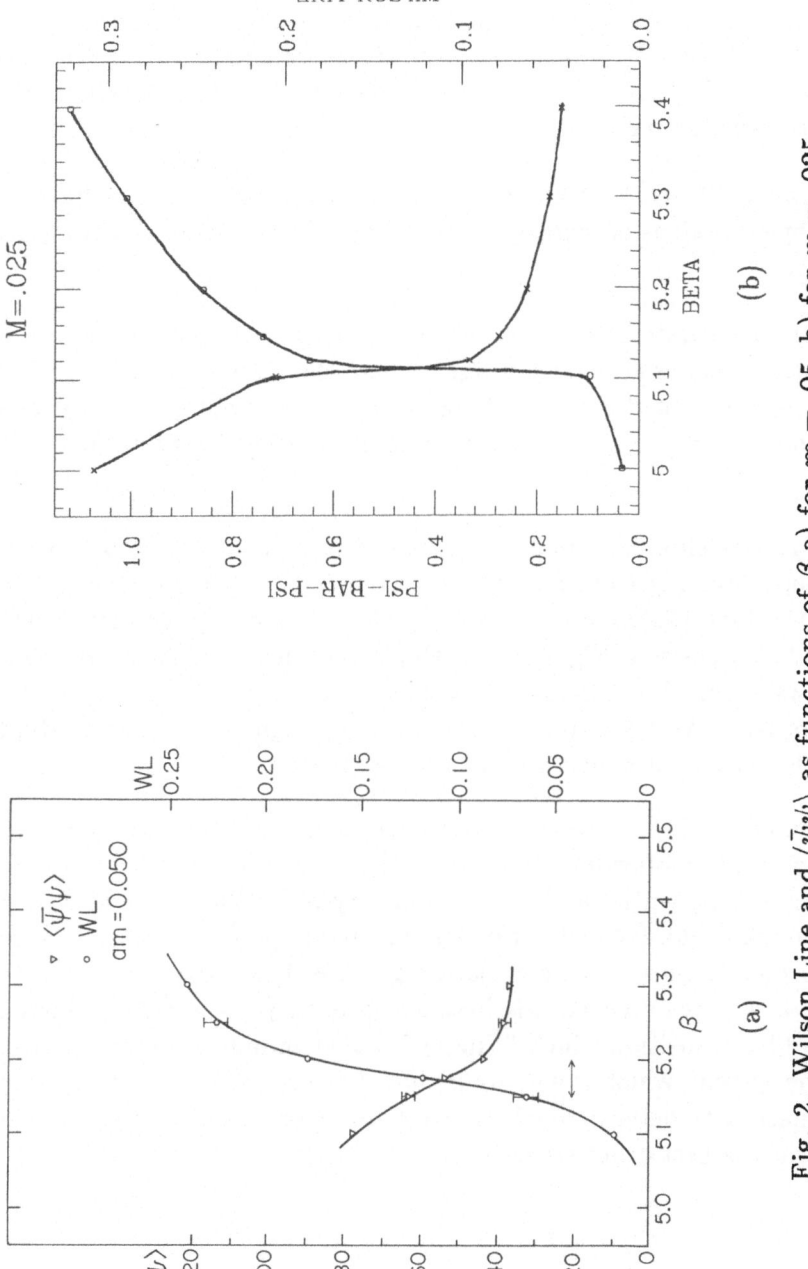

Fig. 2. Wilson Line and $\langle\bar{\psi}\psi\rangle$ as functions of β a) for $m = .05$, b) for $m = .025$.

So far these simulations have involved over 4500 ST-100 hours (equivalent to 3000 CRAY X-MP hours) and over 500 CRAY X-MP hours of computer time.

At $m = .25$ we still observe the first order deconfinement transition of the pure gauge theory [6], which drives the chiral transition for the quenched fermion case [7]. This is indicated by the metastability in the Wilson Line close to the transition (Fig.1).

For $m = .05$ to .1, although there is a rapid crossover, we have not observed any clear signs of metastability or any other clear signals of a phase transition [4].

Preliminary data at $m = .025$ show a much more rapid crossover than at intermediate masses as shown in fig. 2 where the wilson line and $\langle \bar{\psi}\psi \rangle$ are plotted as functions of β for $m = .05$ and $m = .025$. (Note, the absence of error bars on the graph at $m = .025$ indicates the preliminary nature of the data).

In fig. 3 we show the time evolution of the Wilson Line at $\beta = 5.1$ for 50,000 sweeps, starting from an ordered start. Fig. 4a shows an expanded version of the first 10,000 sweeps showing the rapid crossover between the 4000th and 7000th sweep. Fig. 4b is a histogram of these 10,000 sweeps. This clearly shows a bimodal distribution as might be expected if the transition were first order. We have not, as yet seen any sign of the metastability normally associated with such a first order transition.

Calculation of the time correlation lengths from this preliminary data is summarized in the following table. Since the time autocorrelation function is not a simple exponential and the data is very noisy these should only be taken as a rough guide. (In fact, a further 40,000 sweeps at $\beta = 5.1$ performed since this presentation showed correlation times ~ 1, an indication that the long time correlations seen at this beta are probably the system "ringing" after the sudden transition. Such "ringing" is also an indication of long time modes in the system, which is why we include this result here.) Clearly more analysis is needed to determine whether we are seeing critical slowing down, or the return to a first order transition.

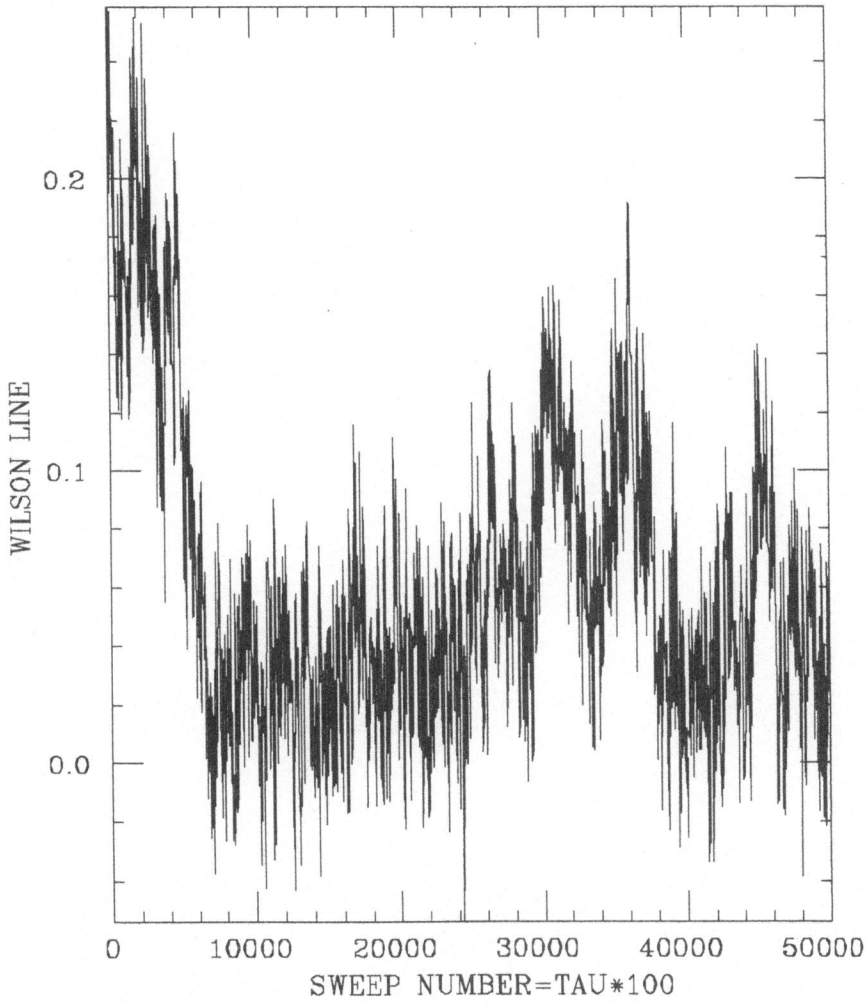

Fig. 3. Wilson Line time evolution $(dt = .01)$, $m = .025$ 50,000 sweeps.

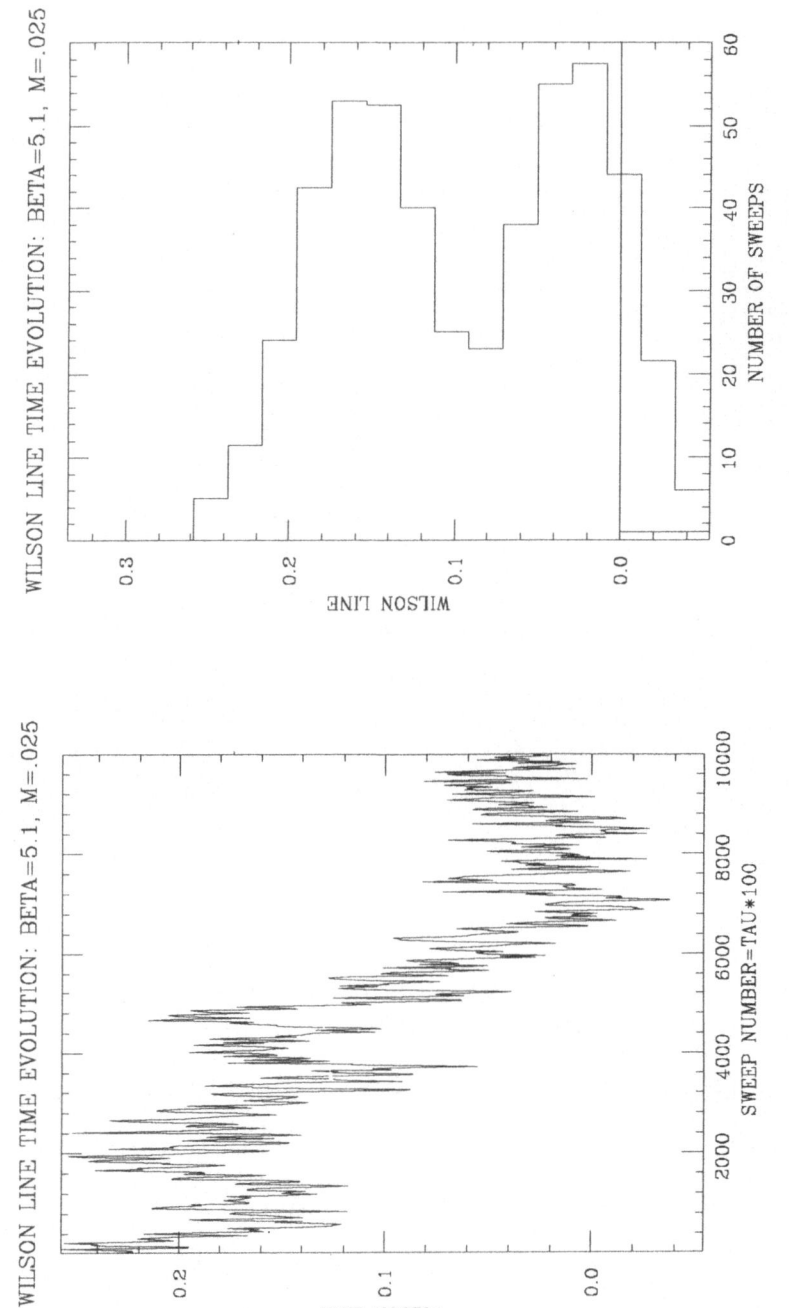

Fig. 4. a) Enlargement of first 10,000 sweeps of fig. 3,
b) Distribution of values of the Wilson Line for these 10,000 sweeps.

TABLE: Time correlation lengths at $m = .025$

β	$\tau_{\langle\bar{\psi}\psi\rangle}$	$\tau[\text{Wilson Line}]$
5.4	0.7	2.1
5.3	0.9	1.2
5.2	1.5	1.7
5.15	5.1	2.4
5.12	7.2	4.7
5.1	19.*	13.*
5.0	1.0	0.3

*—see note in text

CONCLUSIONS

The behaviour of the deconfinement/chiral transition in lattice QCD as a function of the fermion mass, as observed on a $10^3 \times 6$ lattice with 4 degenerate fermion flavours can be divided into 3 regions.

1. LARGE FERMION MASS: In this range of fermion mass $(> .1 - .25)$ we observe the first order deconfinement transition of the pure gauge theory.

2. INTERMEDIATE FERMION MASS: Here the transition softens and possibly disappears.

3. FERMION MASS$\rightarrow 0$ (THE CHIRAL LIMIT) The chiral symmetry restoration transition appears and drives the phase transition. It is not clear whether this phase transition appears for small but finite mass or only in the chiral limit. The order of this transition has yet to be resolved. Further simulations at this mass are in progress.

The critical value of beta at $m = 0$ is determined to be $5.01 \pm .03$ by linear extrapolation, yielding a chiral transition temperature

$$T_c = (2.14 \pm .10)\Lambda_{\overline{ms}}$$

Clearly we need to go to larger lattices in order to determine the continuum behaviour of the finite temperature phase transition. Algorithm improvements are needed to reduce the CPU time needed for such simulations, and those to determine other hadron properties such as masses and matrix elements.

ACKNOWLEDGEMENTS

We thank the HEP division at Argonne and Dr. R. T. Hagstrom for access to their ST-100 array processor, and the N.C.S.A. at the University of Illinois for access to their CRAY X-MP/24.

REFERENCES

1. D. Callaway and A. Rahman, Phys. Rev. Lett. 49, 613 (1982); J. Polonyi and H. W. Wyld, Phys. Rev. Lett. 51, 2257 (1983).

2. S. Duane, Nucl. Phys. B257[FS14], 652 (1985); S. Duane and J. B. Kogut, Phys. Rev. Lett. 55, 2774 (1985); S. Duane and J. B. Kogut, Illinois preprint ILL-TH-86-15 (1986).

3. J. B. Kogut, J. Polonyi, J. Shigemitsu, D. K. Sinclair and H. W. Wyld, Phys. Rev. Lett. 53, 644 (1984); F. Fucito and S. Solomon, Phys. Rev. Lett. 55, 2641 (1985); M. Fukugita and A. Ukawa, Phys. Rev. Lett. 57, 503 (1986).

4. J. B. Kogut and D. K. Sinclair, Argonne/Illinois preprint ANL-HEP-PR-86-61/ILL-TH-86-46 (1986).

5. E.V.E. Kovacs, D. K. Sinclair and J. B. Kogut (in preparation).

6. The following are recent publications on the pure gauge theory deconfinement transition. References to earlier work are to be found in these. S. A. Gottlieb, J. Kuti, D. Toussaint, A. D. Kennedy, S. Meyer, B. J. Pendleton, and R. L. Sugar, Phys. Rev. Lett. 55, 1958 (1985); N. H. Christ and A. E. Terrano, Phys. Rev. Lett. 56, 111 (1986).

7. J. B. Kogut, H. Matsuoka, S. H. Shenker, J. Shigemitsu, D. K. Sinclair, M. Stone and H. W. Wyld, Phys. Rev. Lett. 51, 869 (1983). J. B. Kogut, J. Polonyi, J. Shigemitsu, D. K. Sinclair and H. W. Wyld, Nuclear Physics B251[FS13], 311 (1985).

PSEUDOSCALAR MASSES AND INDEX THEOREM

Jan Smit and Jeroen C. Vink

Institute of Theoretical Physics
Valckenierstraat 65
1018 XE Amsterdam, The Netherlands

ABSTRACT

A lattice derivation for the neutral pseudoscalar masses in lattice QCD leads to a formula for the topological susceptibility through the fermion fields, in which the gauge field topological charge is recovered via the Atiyah-Singer index theorem. A numerical study in two dimensional QED gives insight into corrections due to a non-zero lattice distance.

In QCD, the masses of the neutral pseudoscalars π^0, η and η' can be explained in terms of the topological susceptibility χ [1,2],

$$m_{\eta'}^2 - \tfrac{1}{2}(m_\eta^2 + m_{\pi^0}^2) = \frac{6}{f_\pi^2} \chi \quad , \tag{1}$$

$$\chi = \int dx\ \langle q(x)q(0) \rangle = \langle Q^2 \rangle/(\text{space-time volume}) \quad . \tag{2}$$

It is desirable to have a lattice derivation of (1). The resulting formula should give a prescription for the topological charge Q or charge density q(x) through the quark fields. We have given such a derivation in [3] for Wilson fermions and it will be presented in [4] for staggered fermions. The derivation uses the Ward-Takahashi identity method and a convenient way of expressing the results is that a quantity $\bar{Q}$ is to be used as Q in formula (2):

$$\bar{Q} = \kappa_P \frac{m}{N_f} \ \text{Tr} \ \Gamma_5 \ G \qquad . \tag{3}$$

Here N_f is the number of (light) flavors, G is the fermion propagator in a given gauge field, m is the quark mass parameter, Γ_5 represents γ_5, and κ_P is a renormalization factor. For Wilson fermions, Γ_5 is just γ_5, G has the form

$$G = (\not{D} - W + M)^{-1} \qquad , \tag{4}$$

(W = Wilson mass term; $(2aM)^{-1}$ = K = hopping parameter) and

$$m = M - M_c \qquad , \tag{5}$$

with M_c the critical value of M. For staggered fermions Γ_5 is a four link operator which reduces to γ_5 in the continuum limit, G has the form

$$G = (\not{D} + m)^{-1} \qquad , \tag{6}$$

and m is the parameter appearing in the fermion action.

The renormalization factor is most easily discussed with the Ward-Takahashi relation for the axial current [3-6]. It can be calculated in the weak coupling perturbation theory, $\kappa_P = 1 + O(g^2)$ (with g^2 the _bare_ gauge coupling), but non-perturbative contributions may be very important. The necessity of the renormalization κ_P in (3) can be understood as follows: the parameter m in (5,6) corresponds to $mS(x) = m\bar{\psi}(x)\psi(x)$ in the fermion action, whereas the $m\Gamma_5$ in (3) corresponds to a pseudoscalar density $mP(x) = m(\bar{\psi}i\Gamma_5\psi)(x)$. Chiral symmetry is broken on the lattice which disturbs the relative normalization of S(x) and P(x). In the scaling region a multiplicative renormalization κ_P is necessary to bring

κ_P P(x) in accordance with S(x). Previous suggestions for using the divergence of the fermion axial current to define an 'anomaly' $\bar{q}$(x) and charge $\bar{Q}$ did not take into account the necessary renormalization κ factors and were only valid in the strict continuum limit $g^2 \rightarrow 0$ where $\kappa \rightarrow 1$ [7–10].

If we latticize a smooth external gauge field and let the lattice distance a approach zero, then by the Atiyah–Singer Index theorem [11],

$$\hat{Q} := \frac{m}{N_f} \; \text{Tr} \; \Gamma_5 \; G \tag{7}$$

$$\sim \frac{m}{N_f} \; \text{Tr} \; \gamma_5 \; G \; \Big|_{\text{continuum}} = \frac{n_+ - n_-}{N_f} = Q \quad , \tag{8}$$

with $n_+(n_-)$ the number of zero modes of the continuum $\not{D}$ operator with positive (negative) chirality, and Q the topological charge of the continuum gauge field.

We have studied the properties of the eigenmodes of the lattice Dirac operators and of $\bar{Q}$ for smooth and 'slightly rough' U(1) gauge fields in two dimensions [12] as well as for compact U(1) configurations generated by the Metropolis algorithm. For <u>staggered fermions</u> ($1/N_f = \frac{1}{2}$ in two dimensions),

$$\hat{Q} = \frac{1}{2} \sum_s \frac{m \; r_s}{i \lambda_s + m} \quad , \tag{9}$$

where s labels the eigenfunctions of $\not{D}$ with eigenvalues $i\lambda_s$ and r_s is the expectation value of Γ_5 in the mode s. In the continuum $r_s = 0$ except for the zero modes ($\lambda_s = 0$, $r_s = \pm 1$). On the lattice we find 'zero modes', characterized by small λ_s and relatively large $|r_s|$ (but smaller than 1). See fig. 1a for a picture of the spectrum. The non-'zero modes' have a suppressed r_s. Taken all together the non-'zero modes' balance the 'zero modes' in the sense that

$$0 = \text{Tr} \; \Gamma_5 = \sum_s r_s \quad . \tag{10}$$

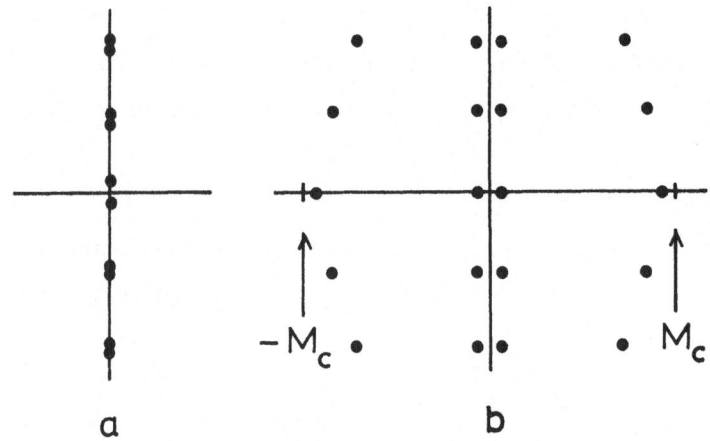

Fig. 1 (a): Qualitative spectrum of the staggered Dirac operator $\not{D}$ in a 'not too rough' configuration with $Q_L = 1$; $i\lambda_s$ and $-i\lambda_s$ are both eigenvalues. The zero modes are situated near the origin. There is approximate flavor symmetry ($N_f = 2$).

(b): Similar for the Wilson-Dirac operator $\not{D}-W$ with $N_f = 1$; μ_s and μ_s^* are both eigenvalues, as well as $-\mu_s$. The deviation of the left most zero mode away from $-M_c$ can be positive or negative.

Hence, the lattice regularization has the following effects:

(a) zero mode shift, i.e. $\lambda_s \neq 0$ for the 'zero modes'

(b) $|r_s| < 1$ for the 'zero modes'

(c) $r_s \neq 0$ for the non-'zero modes'.

These effects cause $\hat{Q}$ to depend (quadratically) on the mass parameter m. In particular, $\hat{Q} = 0$ for $m = 0$ because of (a) and m has to be substantially larger than the zero mode shifts in order that $\partial\hat{Q}/\partial m \approx 0$. The value of $\hat{Q}$ at $\partial\hat{Q}/\partial m \approx 0$ is reduced because of (b) and furthermore because of (c). This has to be compensated by a renormalization factor $\kappa_p > 1$,

$$\bar{Q} = \kappa_P \, \hat{Q} \quad . \tag{11}$$

For $m \to \infty$, $\hat{Q}$ vanishes because of (10).

For <u>Wilson fermions</u> the analogue of (9) can be written as (taking $N_f = 1$)

$$\hat{Q} = \sum_s \frac{m \, r_s}{i\lambda_s + \Delta m_s + m} \quad , \tag{12}$$

where $i\lambda_s + \Delta m_s + m = \mu_s + M = \mu_s + M_c + m$, with μ_s the eigen-values of $\not{D} - W$. The qualitative spectrum of $\not{D} - W$ is shown in fig. 1b. Here we note:

(a) Re $\mu_s \approx -M_c, 0, M_c$. The modes with Re $\mu_s \approx 0, M_c$ are the species doublers; they have $\Delta m_s \approx M_c, 2M_c$. The modes with Re $\mu_s \approx -M_c$ are the Wilson fermions, which have small Δm_s.

(b) there are modes where $\lambda_s = 0$ which we call zero modes

(c) zero mode shifts, $\Delta m_s \neq 0$. Note that M_c is defined by an average over gauge field configurations (e.g. M_c is the value where the quenched pseudoscalar mass vanishes), such that Δm_s fluctuates from configuration to configuration, similar to the zero mode shifts in the staggered case.

(d) $|r_s| \gtrsim 1$ for the zero modes

(e) $r_s \neq 0$ for the non-zero modes.

Because of the Δm_s this $\hat{Q}$ is linearly dependent on m and it vanishes in general as $m \to 0$. However, here $\hat{Q}$ may have poles for $m > 0$ because it can happen that $\Delta m_s < 0$ for a zero mode. As in the staggered case, m has to be sufficiently larger than the zero mode shifts, in order that $\partial\hat{Q}/\partial m \approx 0$. For $m \to \infty$, $\hat{Q} \to 0$. Because of (d) we expect $\kappa_P < 1$.

The zero mode shifts depend on the 'roughness' of the gauge field, which depend on $\beta = 1/a^2g^2$. For $\beta \to \infty$ the zero mode shifts will vanish. One expects similar behavior in QCD. For good con-tinuum behavior the zero mode shifts should vanish fast enough: if m equals the value required by e.g. the experimental pion/nucleon mass ratio, the ratio (zero mode shift)/m should vanish too as $\beta = 6/g^2 \to \infty$.

We now turn to the determination of κ_P. One convenient non-perturbative method for obtaining κ_P is as follows [3,4]: first construct a very smooth U(1) lattice gauge field $V_\mu(x)$ with non-zero topological charge Q^V. For each QCD lattice gauge field $U_\mu(x)$ compute the propagator $G(\hat{U})$ in the field $\hat{U}_\mu(x) = U_\mu(x)V_\mu(x)$. The average over U configurations gives κ_P,

$$N_f \langle \hat{Q}(UV) \rangle_U = m \langle Tr \; \Gamma_5 \; G(UV) \rangle_U = N_f \; N_c \; Q^V \; \kappa_P^{-1} \quad , \qquad (13)$$

where N_c is the number of colors. Note that for Wilson fermions this formula also provides a determination of M_c, since it states that $\langle Tr \; \Gamma_5 \; G(UV) \rangle_U$ has a pole at $m = M - M_c = 0$. This can only be strictly true in the extreme continuum limit where the zero mode shifts are expected to vanish. In the scaling region single pole behavior will hold for $m \gg$ zero mode shifts.

Sufficiently deep in the scaling region κ_P should be independent of m. However, from the discussion of lattice effects in $\hat{Q}$ it is clear that κ_P^{-1} as given by (12) will have an m dependence which is similar to that of a typical $\hat{Q}$. Therefore, even on the edge of the scaling region where $\hat{Q}(U,m)$ and $\kappa_P^{-1}(V,m)$ have still a substantial m dependence, the product $\bar{Q} = \kappa_P \hat{Q}$ may be fairly independent of m, for reasonable Q^V (e.g. $Q^V = 1$) and away from the zero mode shift region.

These remarks were inspired by our preliminary results for the case of two dimensional compact U(1) on a periodic N^2 lattice. For example, for $N = 8$ and $\beta = 4$ fig. 2 shows the susceptibility $a^2 \hat{\chi} = \langle \hat{Q}^2 \rangle / N^2$, $\bar{\chi} = \kappa_P^2 \hat{\chi}$ in units of $a^2 \chi_L = \langle Q_L^2 \rangle / N^2$, with Q_L the topologically motivated charge (see [13] for example):

$$Q_L = \frac{1}{2\pi} \sum_x F_{12}(x) \quad , \qquad (14)$$

$$F_{12}(x) = - i \; \ell n \; U_{12}(x) \in (-\pi, \pi) \quad , \qquad (15)$$

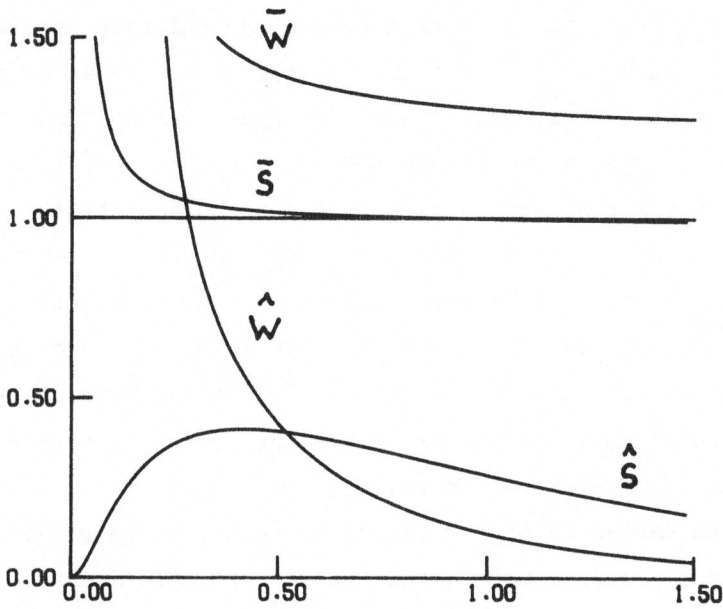

Fig. 2. Susceptibilities $\bar{\chi}/\chi_L$, $\hat{\chi}/\chi_L$ as a function of m/g, for
$\beta = 4$, N = 8 ; S: staggered fermions (512 configurations,
errors $\lesssim$ 15%); W: Wilson fermions (128 configurations,
errors $\lesssim$ 30%).

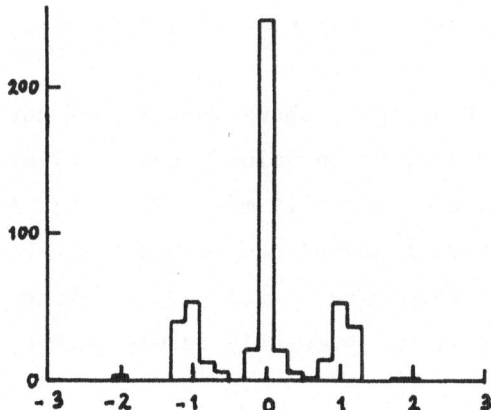

Fig. 3. $\bar{Q}$ distribution for staggered fermions, for $\beta = 4$, N = 8
and m/g = 0.4 .

where $U_{12}(x)$ is the plaquette U. The cancellation in the m dependence in $\bar{Q}$ for am $\gtrsim 0.2$ is striking. (For $\beta = 4$ the zero mode
shifts are of order $|a\lambda| \approx 0.07$ (staggered) and $|a\Delta m| \approx 0.07$
(Wilson).) Fig. 3 shows that the distribution in $\bar{Q}$ is peaked round
integers for β as low as $\beta = 4$ and $m/g = 0.4$. The lattice size
N = 8 may appear to be small, but the correlation length (string
tension)$^{-\frac{1}{2}} = 2.6$ in lattice units at $\beta = 4$, which does not seem
unreasonable. The finite size effects in χ_L can be calculated
exactly and they are of the order of 0.3%. We have checked for
staggered fermions that fig. 2 remains essentially unchanged if the
lattice is doubled to N = 16.

The masses in fig. 2 extend to rather large values. For $m \to \infty$,
$\hat{Q}$ becomes proportional to the 'naive topological charge' $\tilde{Q}$, which
is obtained by using

$$\tilde{F}_{12}(x) = \text{Im } U_{12}(x) \tag{16}$$

in (14). This follows from the hopping expansion representation
of $\hat{Q}$. Hence $\bar{\chi} (m = \infty) = \tilde{\kappa}^2 \tilde{\chi}$, with $\tilde{\kappa}$ and $\tilde{\chi}$ defined by replacing
$\hat{Q}$ by $\tilde{Q}$. We find $\bar{\chi}$ to be approximately independent of m all the way
up to $m = \infty$. If this behavior translates to QCD, then we would be
able to compute χ very simply by computing the naive $\tilde{\chi}$, and,
instead of performing a controversial subtractive renormalization
as in [14], perform a multiplicative renormalization with the just
as easily measurable $\tilde{\kappa}^2$.

For $m/g \lesssim 0.3$ fig. 2 shows that $\bar{\chi}$ does not make sense. This is
somewhat disturbing, since values used in simulations of the
Schwinger model are typically $m/g \lesssim 0.2$ [15]. As mentioned before,
for good continuum behavior the zero mode shifts should vanish
faster than m (in lattice units) upon entering deeper into the
scaling region. We are presently investigating these questions in
QCD.

Acknowledgement

This investigation is supported by the "Stichting voor Fundamenteel Onderzoek der Materie (FOM)".

References

[1] E. Witten, Nucl.Phys. B156(1979)269.

[2] G. Veneziano, Nucl.Phys. B159(1979)213.

[3] J. Smit and J.C. Vink, Neutral Pseudoscalar Masses in Lattice QCD, ITFA-86-18, to be published in Nucl.Phys.B.

[4] J. Smit and J.C. Vink, in preparation.

[5] L.H. Karsten and J. Smit, Nucl.Phys. B183(1981)103.

[6] M. Bochicchio, L. Maiani, G. Martinelli, G.C. Rossi and M Testa, Nucl.Phys. B262(1985)331.

[7] J. Smit, Nucl.Phys. B175(1980)307.

[8] E. Seiler and I.O. Stamatescu, Phys.Rev. D25(1982)2177; D26(1982)534.

[9] M. Bochicchio, G.C. Rossi, M. Testa and K. Yoshida, Phys.Lett. 149B(1984)487.

[10] F. Karsch, E. Seiler and I.O. Stamatescu, Nucl.Phys. B271(1986)349.

[11] A. Atiyah and I. Singer, Ann.Math. 87(1968)484.

[12] J. Smit and J.C. Vink, Remnants of the Index Theorem on the Lattice, ITFA-86-14, to be published in Nucl.Phys.B.

[13] C. Panagiotakopoulos, Nucl.Phys. B251[FS13](1985)61.

[14] P. di Vecchia, K. Fabricius, G.C. Rossi and G. Veneziano, Nucl.Phys. B192(1981)392; Phys.Lett. B108(1982)323.

[15] S.R. Carson and R.D. Kenway, Ann.Phys. (N.Y.) 166(1986)364.

SCALING STUDIES OF QCD ON ASYMMETRIC LATTICES

H. B. Thacker

Fermi National Accelerator Laboratory
P.O. Box 500, Batavia, IL 60510

J. C. Sexton †

Institute for Advanced Study
Princeton, NJ 08540

ABSTRACT

   Using the deconfinement temperature as a probe, the scaling properties of QCD are studied on asymmetric lattices. We find that the measured
asymmetry dependence of the lattice Λ parameter at $\beta \approx 5.7$ agrees precisely
with that predicted by one loop perturbation theory. This result holds on lattices with asymmetry in one spatial direction and on lattices with asymmetry
in all three spatial directions and suggests that the perturbative scaling violations observed on symmetric lattices at these values of β are independent
of asymmetry and therefore unlikely to be lattice artifacts.

† supported by U.S. DOE Contract DE–AC02–76ER02220

An understanding of the scaling properties of lattice QCD is of prime importance in relating the results of numerical calculations to the the continuum physics those calculations hope to investigate. Studies of scaling have therefore received significant attention in the recent literature and some considerable progress has been achieved for the pure (*i.e.* without fermions) QCD model. However, there are still a number of issues pertaining to scaling and universality in pure gauge QCD which have not been clarified.

The scaling region in pure lattice QCD is, by definition, that region of coupling constant space in which all dimensionless ratios of dimensionful physical observables become independent of the coupling constant. More explicitly, consider how a dimensionful physical observable, for example a mass M, is measured on the lattice. We find that

$$M = F_M(g)/a \tag{1}$$

where a is the lattice spacing and $F_M(g)$ is a measurable function of the coupling g. Requiring that M be independent of the lattice spacing now forces us to choose the coupling to be a function of the lattice spacing (*i.e.* $g \equiv g(a)$). For general couplings each different physical observable will require a different choice for this function. In the scaling region, however, there exists a single unique choice (up to an integration constant) for $g(a)$ which leaves all possible low momentum physical observables independent of lattice spacing.

In pure QCD the scaling region includes the point $g = 0$ and close to this point the function $g(a)$ satisfies the perturbative renormalization group equation [1]

$$a\frac{dg}{da} = -\beta(g) = b_0 g^3 + b_1 g^5 + \ldots \tag{2}$$

which has the solution

$$\Lambda a = \exp\left(-\frac{1}{2b_0 g^2} - \frac{b_1}{2b_0^2} \ln\left(b_0 g^2\right)\right) \tag{3}$$

Λ is the constant of integration which defines the physical scale for the theory. Recently Gottlieb *et al.* and Christ and Terrano [2] have shown that the deconfinement temperature scales according to this perturbative equation in the region $6.1 \lesssim 6/g^2 \lesssim 6.5$. This is encouraging and suggests that perturbative or asymptotic scaling sets in at $6/g^2 \gtrsim 6.1$.

For $6/g^2 \lesssim 6.1$ the picture is not so clear. There is evidence from measurements of the deconfinement temperature and the string tension, and from renormalization group calculations [3] that nonperturbative scaling holds in the region $5.6 \lesssim 6/g^2 \lesssim 6.1$. Glueball mass calculations in this region are also consistent with non perturbative scaling [4] but the error bars on the glueball results are large enough that a definitive statement is not possible. If scaling does indeed hold in this region then continuum lattice calculations become possible at $6/g^2 \approx 5.7$ resulting in large savings in the computer resources required to do any given calculation. More importantly even if scaling doesn't hold in this region it would be comforting to have some understanding of exactly what is causing the breakdown in perturbative scaling observed here. Thus lattice studies of pure QCD in this region are of great interest.

There is a another aspect of scaling which also needs further investigation. This arises from the fact that there are many different possible lattice actions which give rise to continuum QCD in the limit $a \to 0$. These actions include the Wilson action, the fundamental-adjoint action [5], the Manton action [6] and the Villain action [7]. Also the lattice on which each of these actions is embedded need not always be the standard hyper-cubic lattice. Hyper-rectangular (asymmetric) [8,9] or indeed random lattices [10] can all be used. Each of these various possible lattice actions will have their own scaling regions which will include the point $g = 0$ together with a region of perturbative scaling about this point. It is important to check that these various choices give rise to the same continuum physics.

Some work has been already be done on this question. Patel *et al.* have made measurements of the ratio of the glueball mass to the string tension for a lattice action which includes 1×1 plaquettes in the fundamental, 6 and 8 representations of $SU(3)$ and also 1×2 plaquettes [11]. The relevant coupling for these calculations was $6/g^2 \approx 5.9$. The results obtained are significantly different from similar results obtained for the Wilson action. This suggests that universality, in the sense that different lattice actions are giving the same physics, has not yet set in at these values of $6/g^2$.

More recently Toussaint and Buendia [12] have looked at the universality of the deconfinement temperature for various choices of fundamental-adjoint lattice actions for values of $6/g^2 \approx 6$. The results of this work suggest that universality may be present at least for a subset of actions close to the Wilson action. However as the action studied varies further from the pure Wilson action universality appears to break down.

Concurrently with the work of Toussaint we have been studying universality on asymmetric lattices [9]. Specifically we have chosen to examine the deconfinement temperature on lattices with spacing $a_\mu = \xi_\mu a$ along the μth axis. The variables ξ_μ introduced here are dimensionless asymmetry parameters. In the perturbative scaling region different choices for the asymmetry parameters represent different lattice regularizations and all asymmetry dependences in the theory should therefore be removable by appropriate redefinitions of the scale parameter Λ.

The pure gauge $SU(N)$ asymmetric action which we have chosen to use is given by [9,10]

$$S(U) = \sum_{x,\mu<\nu} \frac{6}{g^2} \frac{\xi_x \xi_y \xi_z \xi_t}{\xi_\mu^2 \xi_\nu^2} \left(1 - \frac{1}{3}\mathrm{Re\,Tr}\left(U_{\mu\nu}(x)\right)\right) \qquad (4)$$

where $U_{\mu\nu}(x)$ is as usual the path ordered product of 3×3 unitary link matrices about the fundamental plaquette in the μ-ν plane at the point x.

For general choices of the asymmetries ξ_μ there is insufficient symmetry in the lattice action of Eqn.(4) to guarantee that a Lorentz invariant continuum theory is recovered as $a \to 0$. However, when three of the four asymmetry parameters are equal then the resulting three dimensional cubical symmetry on the lattice is sufficient to enforce Lorentz symmetry in the continuum. In our case, since we wish to study the deconfinement temperature, the time direction of the lattice has special significance. There are therefore two distinct classes of asymmetric lattices which we can study. The first of these has asymmetry along the x-axis only. The second has equal asymmetry along each of the three spatial axes. Specifically the two classes are

$$\begin{array}{ll} \text{Class 1}: & \xi_x = \xi_{\text{space}}, \quad \xi_y = \xi_z = \xi_t = 1 \\ \text{Class 2}: & \xi_x = \xi_y = \xi_z = \xi_{\text{time}}, \quad \xi_t = 1 \end{array} \qquad (5)$$

For various choices of ξ_{space} and ξ_{time} we have measured the critical coupling $((6/g^2)_c)$ at which deconfinement occurs on lattices with two and four sites in the time direction.

Before discussing the Monte Carlo results let us briefly mention the behavior that perturbative scaling implies. In one loop lattice perturbation theory it is a relatively simple matter to calculate how $(6/g^2)_c$ should change with asymmetry. One uses standard background field methods to relate the relevant Λ parameters. The result of these calculations is an expression for the difference, $\Delta(6/g^2)$, of the coupling constants at different asymmetries and lattice spacings which takes the form

$$\Delta\left(\frac{6}{g^2}(\xi,\xi_0;a,a_0)\right) = \frac{6}{g^2}(\xi,a) - \frac{6}{g^2}(\xi_0,a_0) = B(a/a_0) + C(\xi,\xi_0) \qquad (6)$$

$B(a/a_0)$ depends only on the relevant lattice spacings and represents the finite difference of infrared divergent Feynmann integrals. $C(\xi,\xi_0)$, on the other hand, is independent of the lattice spacing. For further details we refer the reader to references [9] and [13]. The aim of our Monte Carlo calculations will be to test how well these one loop scaling results hold in the region $6/g^2 \lesssim 6.0$.

In order to determine the critical coupling $(6/g^2)_c$ at deconfinement we used the Polyakov line operator $P(\vec{x})$ which is defined to be

$$P(\vec{x}) = \text{Tr}\left(U_t(\vec{x})U_t(\vec{x}+\hat{t})\ldots U_t(\vec{x}+N_t\hat{t})\right) \tag{7}$$

where the matrices U_t are the time like link matrices and N_t is the number of sites in the time direction. The expectation value of $P(\vec{x})$ is an order parameter for the transition from confinement to deconfinement in pure QCD. In the confined phase $P(\vec{x})$ has expectation value zero. In the deconfined phase $P(\vec{x})$ has finite non zero expectation value. The procedure we have adopted to measure the deconfinement temperature therefore was to measure the expectation value of $|P(\vec{x})|$ at each Monte Carlo sweep. If the parameters of a given Monte Carlo run were close to the transition values then the sweep to sweep values obtained for $|P(\vec{x})|$ clustered in two peaks, one corresponding to confinement and centered at a small value of $|P(\vec{x})|$, and one corresponding to deconfinement and centered at a larger value of $|P(\vec{x})|$. By counting the number of sweeps under each peak we can catalogue the percentage of sweeps of any given run which are confined. Then to determine the critical coupling for any given value of asymmetry and lattice size we generate from 20,000 to 50,000 sweeps at couplings close to the transition and interpolate to that value of $6/g^2$ for which 50% of sweeps are confined.

The results of our Monte Carlo studies are summarized in the three figures included here. In each case the figures show how the critical coupling at deconfinement varies as the asymmetry varies. For comparison each figure also contains the one loop perturbative predictions for the asymmetry dependence of $(6/g^2)_c$. These perturbative results were obtained from the analytic forms given in reference [9]. Let us now consider the three figures in turn.

In Figure 1 we have plotted the results obtained for lattices with two sites in the time direction as we vary the x-axis asymmetry ξ_{space}. Note first that the critical coupling for the symmetric lattice ($\xi_{\text{space}} = 1$) is $(6/g^2)_c = 5.069$. This is a value of coupling in the strong coupling sector of the theory far removed from the perturbative scaling region $6/g^2 \gtrsim 6.1$ and also far removed from the suggested nonperturbative scaling region $5.6 \lesssim 6/g^2 \lesssim 6.1$. Thus it is altogether understandable that the Monte Carlo data and the perturbative scaling predictions are in complete disagreement for this case.

In Figure 2 we show the results obtained for lattices with four sites in the time direction as the x-axis asymmetry changes. The symmetric lattice critical coupling for this case is $(6/g^2)_c = 5.676$ which is in the suggested nonperturbative scaling region but still far from the perturbative scaling sector. However, as the figure shows, we find that for $0.66 < \xi_{\text{space}} < 1.1$ there is surprisingly good agreement between the Monte Carlo data and the one loop perturbative predictions. Also note that the transitions from one loop behavior which occur at $\xi_{\text{space}} = .667$ and $\xi_{\text{space}} = 1.10$ appear to be very abrupt in nature.

Finally Figure 3 shows the results obtained for lattices with four sites in the time direction as ξ_{time} changes. As in Figure 2 we find a region of surprisingly good agreement between the Monte Carlo data and the one loop perturbative predictions. This region of agreement extends from $\xi_{\text{time}} = 0.85$ to $\xi_{\text{time}} = 1.03$. Note also that the transition from one loop behavior which occurs at $\xi_{\text{time}} = 1.03$ appears to be quite abrupt.

The results shown in Figures 2 and 3 pose two interesting questions. First, how can the asymmetry dependent perturbative agreement shown in these figures be consistent with the known nonperturbative behavior of the critical coupling on symmetric lattices when measured as a function of the lattice spacing a? Secondly, what is the origin of the apparently abrupt transitions from scaling which occur in both these figures?

Let us first address the question of the nature of the transition from scaling as the asymmetry changes. Note that it is not at all surprising, as the asymmetry increases, that lattice effects might set in and cause deviations from continuum scaling. Indeed such behavior is expected since increasing the asymmetry in a given direction increases the coarseness of the lattice in that direction. Thus the transitions occurring at $\xi_{\text{space}} = 1.1$ in Figure 2 and at $\xi_{\text{time}} = 1.03$ in Figure 3 are qualitatively reasonable. On the other hand it is initially surprising that similar transitions occur as the asymmetry decreases. One might expect that decreasing the asymmetry could only improve the continuum behavior of the theory since it makes the lattice grid finer in either one or all three spatial directions. (At the critical coupling, the lattice spacing in the temporal direction is by definition a fixed rational fraction of the inverse deconfinement temperature.)

A possible counter-argument arises, however, when one considers that the basic unit of lattice action is the trace of the product of link matrices about fundamental plaquettes. When the asymmetry gets large we find that in certain plaquettes one side gets large relative to the other. When the asymmetry gets small one side gets small relative to the other. Our results then suggest that the important considerations in determining whether a given asymmetry gives rise to continuum behavior are not only the overall size of the elementary lattice cell, but also the relative sizes of the sides in the fundamental plaquettes.

The more fundamental question which we need to address concerns the consistency between our result that the asymmetry dependence of the critical coupling is perturbative and the results of references [2] that the lattice spacing dependence of this same critical coupling is non perturbative at $6/g^2 \approx 5.7$. Our explanation for this phenomenon is based on the result summarized in Eqn.(6). This equation describes how the critical coupling changes both as the lattice spacing changes and as the asymmetry changes. Thus this equation is applicable both to the lattice spacing dependent scaling studies of Gottlieb et al. and of Christ and Terrano [2] and to the asymmetry dependent scaling studies described here. The important point to note, however, is that the two cases decouple. Eqn.(6) tells us that the lattice spacing dependence and the asymmetry dependence are completely independent of each other.

There is a further important point to be made. The term in Eqn.(6) which gives rise to the lattice spacing dependence of the theory $(B(a/a_0))$ receives contributions only from the infrared sector (*i.e.* only from momenta p which satisfy $p << \pi/a$). This explains how this term can be insensitive to the details of the lattice structure since these details are probed only by momenta on the order $p \approx \pi/a$. However this implies that the lattice spacing dependent term in Eqn.(6) is common to all lattice actions within the range for which perturbative asymmetry dependence is observed.

The scaling behavior exhibited in Figures 2 and 3 strongly suggests that the asymmetry dependence of the critical coupling at $6/g^2 = 5.7$ is given exactly by the term $C(\xi, \xi_0)$ in Eqn.(6). Thus our result suggests that the nonperturbative behavior observed at this value of coupling in reference [2] is infrared in nature and is confined to the term $B(a/a_0)$ only.

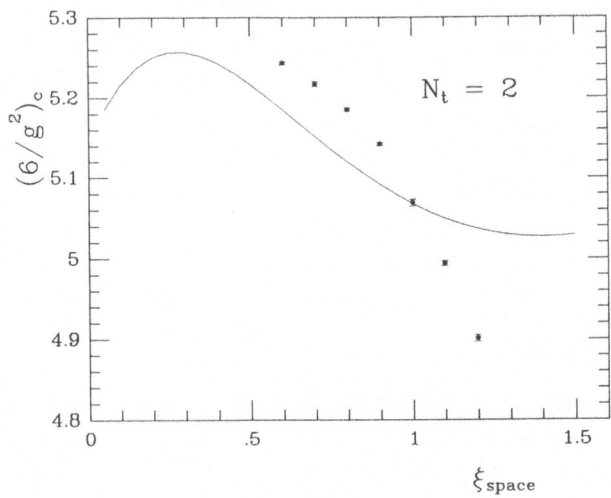

[1] Plot of the critical coupling $(6/g^2)_c$ as a function of asymmetry along the x-axis (ξ_{space}) for a lattice with two sites in the time direction. The solid line shown represents the prediction of one loop perturbation theory. The data points represent the Monte Carlo Results.

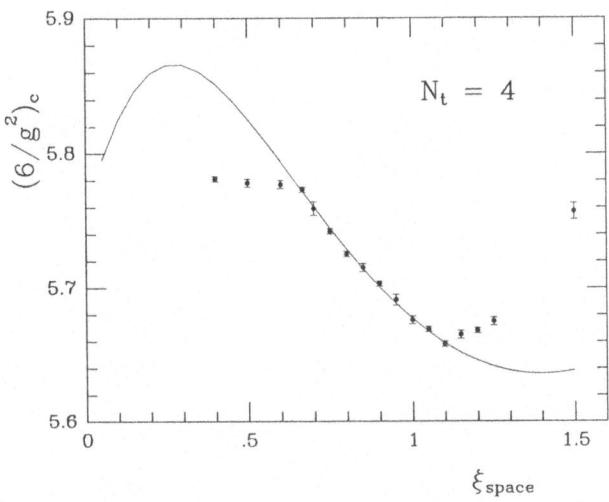

[2] Plot of the critical coupling $(6/g^2)_c$ as a function of asymmetry along the x-axis (ξ_{space}) for a lattice with four sites in the time direction.

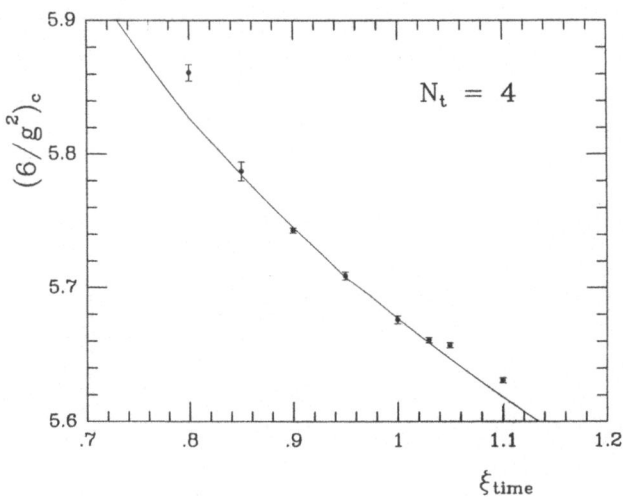

[3] Plot of the critical coupling $(6/g^2)_c$ as a function of asymmetry along the three spatial directions (ξ_{time}) for a lattice with four sites in the time direction.

REFERENCES

[1] D. J. Gross and F. Wilczek, Phys. Rev. Lett. 56 (1973) 3633

[2] S. A. Gottlieb, J. Kuti, D. Toussaint, A. D. Kennedy, S. Meyer, B. J. Pendleton, and R. L. Sugar, Phys. Rev. Lett. 55 (1985) 1958

   N. H. Christ and T. E. Terrano, Phys. Rev. Lett. 56 (1986) 111

[3] R. Gupta , G. Guralnik, A. Patel, T. Warnock, C. Zemach, Phys. Lett. 161B (1985) 352

[4] Ph. De Forcrand, G. Schierholz, H. Schneider, M. Teper, Phys. Lett. 152 (1985) 107

[5] G. Bhanot and M. Creutz, Phys. Rev. D24 (1981) 3212

[6] N. S. Manton, Phys. Lett. 96B (1980) 328

[7] P. Menotti and E. Onofri, Nucl. Phys. B190 [FS3] (1981) 288

   M. Nauenberg and D. Toussaint, Nucl. Phys. B190 [FS3] (1981) 217

[8] F. Karsch, Nucl. Phys. B205 [FS5] (1982) 285

[9] J. C. Sexton and H. B. Thacker, Phys. Rev. Lett. 57 (1986) 2131

[10] N. H. Christ, R. Friedberg, and T. D. Lee, Nucl. Phys B202 (1982) 89, Nucl. Phys. B210 (1982) 310, Nucl. Phys. B210 (1982) 337

[11] A. Patel, R. Gupta, G. Guralnik, G. W. Kilcup, and S. R. Sharpe, Phys. Rev. Lett. 57 (1986) 1288

[12] D. Toussaint and G. Buendia, San Diego Preprint USCD-10P10-263

[13] R. K. Ellis and G. Martinelli Nucl. Phys. B235 [FS11] (1984) 93

MONTE CARLO INVESTIGATIONS OF ASYMPTOTIC

SCALING IN QCD

D. Toussaint
with S.A. Gottlieb, A.D. Kennedy, J. Kuti, S. Meyer,
B.J. Pendleton, and R.L. Sugar

University of California at San Diego
La Jolla, CA 92093

ABSTRACT

We discuss a Monte Carlo study of the phase transition in pure gauge lattice QCD. The effects of the criterion for transition temperature on a finite lattice are discussed and an estimate for the latent heat of the transition is obtained.

About a year ago we presented results for the deconfinement temperature from a Monte Carlo simulation of pure gauge QCD, using lattices with 8, 10, 12 and 14 spacings in the Euclidean time direction.[1] The object of this work was to discover at what value of β physical quantities start to scale according to the perturbative β function. Of course, any physical quantity could be used for this test and in principle they should all be checked. It is not logically impossible for one quantity to scale with the predicted form before another one does. We chose to work with the deconfinement temperature because of our feeling that thermodynamic quantities could be studied on a smaller lattice than particle masses at the same value of β. Previous work by Kennedy et al. on smaller lattices had failed to reach a lattice spacing small enough to see perturbative scaling.[2] It was therefore gratifying to find the results in figure 1, where we plot the critical temperature in units of the lattice Λ. On this plot the constancy of the last three points indicates that as the

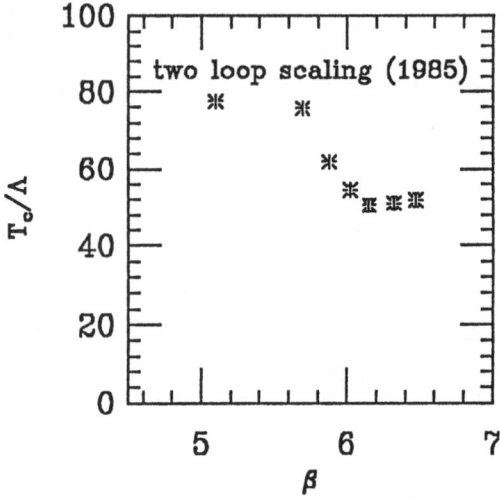

Figure 1: T_c/Λ *versus* β

lattice spacing is varied (from $(10\ T_c)^{-1}$ to $(14\ T_c)^{-1})$ the change in g, defined at the scale of the lattice spacing, is approximately the amount predicted by the two loop perturbative β function. At the same time, a group from Columbia also reported T_c measurements on lattices with ten, twelve, and fourteen time slices.[3] While we had quoted 6.15(03), 6.32(03), and 6.475(030) for 10, 12 and 14 time slices, the Columbia group found 6.06(02), 6.26(02), and 6.35(03) respectively. Largely in an effort to understand this discrepancy, we have taken a little more data and reanalyzed our old data, trying to make our analysis less subjective. We have also considered more carefully the effects of the small spatial size of our lattices. These effects account for the differences in the reported numbers for twelve and fourteen time slices, but the results for ten time slices remain in clear disagreement. We have also extracted an estimate of the latent heat from our data. As I will discuss, the latent heat and the effect of finite spatial size are closely connected, so they must be discussed together.

On a finite size system, what would be a first order phase transition in the bulk is manifested by the appearance of two populations of configurations with the characters of the two phases. During a Monte Carlo simulation the system stays in one "phase" for what can be a very long time, then "tunnels" into the other phase. This behavior is seen over a range of temperature (or coupling constant) proportional to 1/volume. In the transition studied here, this volume is the spatial (three dimensional) volume of the lattice. The size of the lattice in the Euclidean time direction is fixed at $1/T$. The temperature in physical units can be varied by changing g^2 for fixed N_T, so we will see the coexistence of two phases over a range of $\beta = 6/g^2$ for fixed N_T. The assumption that we can divide the configurations encountered during a Monte Carlo run into confined and deconfined configurations is the central assumption of our analysis. Obviously this is an idealization, since we have some configurations that are in between the two phases. It is not clear to us whether our small systems tunnel by nucleation of a domain which spreads or by a more or less simultaneous and homogeneous movement of the system. We ignore the existence of such configurations in our analysis.

Before discussing how we divide a run into confined and deconfined parts, let us discuss the rounding of the transition. Let $n_1(S)$ and $n_2(S)$ be the density of configurations in the two phases. That is, the number of possible configurations of the finite system which we classify as phase 1 and which have action between S and $S + dS$ is $n_1(S)dS$. The average action in phases 1 and 2 is

$$<S>_{1,2} = \frac{\int dS \; S \; n_{1,2}(S) \; e^{-\beta S}}{\int dS \; n_{1,2}(S) \; e^{-\beta S}} \tag{1}$$

The fraction of the time spent in each phase in a long Monte Carlo run is

$$f_{1,2} = \frac{\int dS \; \dfrac{n_{1,2}(S)}{n(S)} \; n(S) e^{-\beta(S)}}{\int dS \; n(S) \; e^{-\beta S} \; dS} \tag{2}$$

where $n(S) = n_1(S) + n_2(S)$. Differentiating the above expression for $f_{1,2}$ and substituting in the expressions for $<S>_{1,2}$ and $f_{1,2}$, plus the obvious $f_1 = 1 - f_2$, leads to

$$\frac{\partial f_1}{\partial \beta} = f_1(1 - f_1) \, (<S>_2 - <S>_1) \tag{3}$$

Up to this point the derivation is completely general — it doesn't even require that the system has a phase transition. Since S is extensive, or proportional to volume, this leads to the well known rounding of the transition by order $1/V$.

If the gap $G = <E>_2 - <E>_1$ is approximately constant across the relevant range of β, the solution to this differential equation is a hyperbolic tangent plus one half:

$$f_1(\beta) = \frac{e^{(G/2)(\beta-\beta_{1/2})}}{e^{(G/2)(\beta-\beta_{1/2})} + e^{-(G/2)(\beta-\beta_{1/2})}} \qquad (4)$$

where $\beta_{1/2}$, the value at which $f_1 = 1/2$, appears as a constant of integration. For a large enough system, the gap will always be approximately constant since the relevant range of β is falling as $1/V$. For the QCD deconfinement transition the gap is constant for much smaller lattice sizes than required for the above argument. In this transition the gap can be constant over the relevant range of β even though $<E>_1$ and $<E>_2$ each change by many times the gap over the range. The reason is that the deconfinement transition is driven by the periodicity of the lattice in the time direction, or by phenomena taking place at N_T lattice spacings. However, most of the dependence of $<S>_{1,2}$ on β comes from the ultraviolet modes, or a characteristic length of one lattice spacing. Also, G is proportional to the latent heat of the transition.[4] (This is not a tautology - G is defined as the jump in the action, and the latent heat is the jump in the energy.)

Our problem now is how to estimate $\beta_{critical}$ from the fraction $f_1(\beta)$ measured on a finite lattice. Figure 2 shows a plot of f_1 on two different sizes of lattice. The two curves cross at some value of f_1, and this is the best estimate we can make of $\beta_{critical}$. If we choose any other value of f_1, we will see a shift in the "critical β" proportional to $1/V$. The earlier work of Kennedy et al.[2] looked for $f_1 = 1/2$ for several spatial sizes, then extrapolated using the $1/V$ law. For our large values of N_T we cannot do enough spatial sizes to have this luxury.

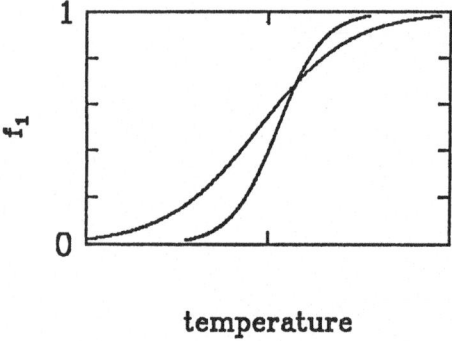

temperature

Figure 2: transition rounding on two lattice sizes

What is the best value of the deconfinement fraction to look for in SU(3)? The Columbia group looks for 1/2 confined and 1/2 deconfined, while we look for 3/4 deconfined, or equal populations in all four vacua. This is the origin of a systematic difference in the results, whose size I will estimate shortly. First let's ask if there is any real evidence for the best value of f_d, the fraction deconfined.

The answer is there is some, but not much. In the old work of Kennedy et al. there is data on $5^3 \times 2$, $7^3 \times 2$, and $9^3 \times 2$ lattices. Figure 3 shows a histogram of the magnitude of the Polyakov loop (averaged over the lattice) encountered during one run. For these lattices, you really can separate the phases! If we divide the histograms at 0.5 and plot the fraction deconfined for the three lattice sizes we find figure 4. (The information necessary to compute error bars has been lost - you will have to estimate them from the unevenness of the lines.) The deconfinement fraction at which these lines cross might be around 3/4 - it is certainly larger than 1/2. Does this extend to larger lattices? We would like to have more evidence. One can also observe that the values of $\beta_{1/2}$ for $N_T > 2$ reported by Kennedy et al. increase as the spatial size increases, which shows that the best estimate of β_c has $f_{deconfined}$ something greater than 1/2. However, the statistical errors on the slopes are large.

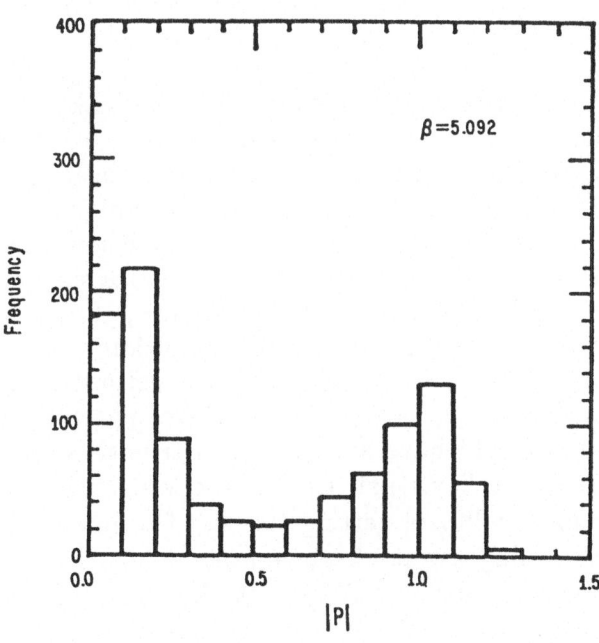

Figure 3: histogram of the Polyakov loop magnitude[2]

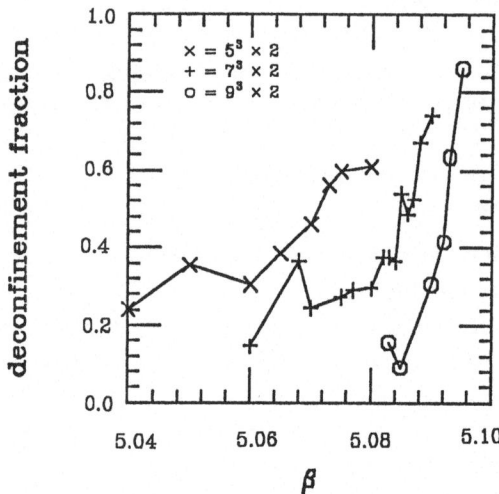

Figure 4: deconfinement fractions for two time slices

The next question is how to divide a run into confined and deconfined parts. We have used two methods. In one method we study the magnitude of the Polyakov loop and in the other the phase. (The second method was developed by the Columbia group, and is described in their publications.[3])

Figure 5 shows the time history of the Polyakov loop magnitude in one of our nicer runs. In order to estimate the fraction of the time spent in each phase, we first smooth the history, essentially by convolution with a Gaussian of width $\sqrt{8/3}$. This results in the time history shown in figure 6. We then count measurements above and below a threshold, in this case .04. The threshold is different for each lattice size, although the same for all values of β at a given size. Of course, if you set the threshold wrong you will get wrong answers. In addition to an estimate of the deconfinement fraction, we require an error estimate. Because our runs have few tunnelings, it is not clear that if we repeated similar runs many times the results would have a Gaussian distribution. Therefore we use a maximum likelihood fit;[5] that is, we ask what values of $\beta_{1/2}$ and G maximize the probability of getting the set of results we have. Our model is that the tunneling events have a Poisson distribution, with different lifetimes in the two phases. We count the tunnelings and compute the likelihood of finding a particular f_d/f_c in a run containing this number of tunnelings. This is actually an approximation, since our runs were made for a predetermined number of sweeps rather than for a predetermined number of tunnelings. Also, the number of tunnelings is

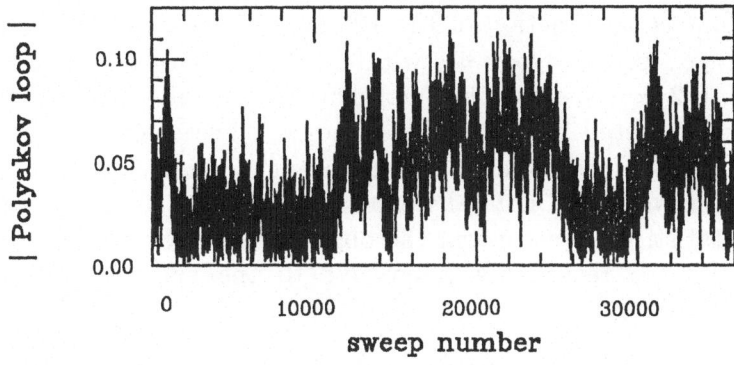

Figure 5: time history of the Polyakov loop magnitude

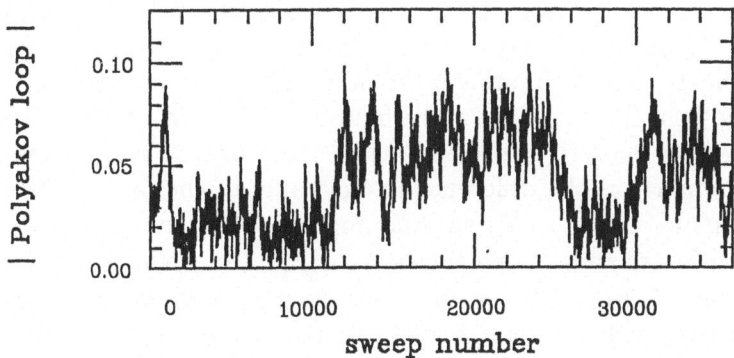

Figure 6: the same time history smoothed

ambiguous since we see many short excursions that may or may not represent tunnelings into the other phase. However, the tunneling count affects the statistical weight of the run, but not the deconfinement fraction for the run.

We have also repeated the analysis of the Columbia group on our data. In this analysis, which has been described in print,[3] one looks at the phase of the Polyakov loop rather than the magnitude. In particular, the Polyakov loop is averaged over 100 sweeps and the fraction deconfined is taken to be

$$f_d = \frac{3}{2} f_{20^\circ} - \frac{1}{2} \tag{5}$$

where f_{20° is the fraction of the 100 sweep blocks in which the phase of the average Polyakov loop lies within 20° of one of the cube roots of one, and the 1/2 corrects for the fact that 1/3 of the points in the confined phase are expected to fall in these wedges. The error estimates for f_d in this method are corrected for the apparent autocorrelations in the sample, but the analysis does not take explicit account of the effect of tunnelings on the statistics. We confirm the observation of Christ and Terrano that the results are reasonably insensitive to the exact numbers of sweeps blocked together.[3]

Despite all the caveats affecting these analyses, we were gratified to find that the results of the two procedures applied to our data were very similar, and we conclude that the differences in analysis procedure are not the cause of the discrepancies in reported values of β_c.

In figure 7a we show the deconfinement fractions estimated by cutting the smeared Polyakov loop at magnitude 0.04. At each point we also indicate our estimate of the number of tunnelings in the run, and show the best fit to the data. In figure 7b we show the result of the Christ and Terrano analysis[3] applied to the same data. What does account for the discrepancy is the difference in the value of the finite lattice f_d used in the estimate of β_c. Using our measured latent heat, which will be discussed later, and the perturbative β function, which we claim describes the scaling of T_c with N_T in this region, as well as the relation of Δ_{plaq} to the latent heat and equation 4, we find an approximate formula for the difference of the two criteria for our range of lattice sizes

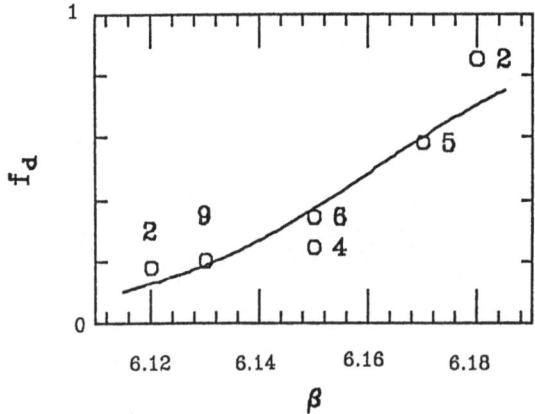

Figure 7a: maximum likelihood fit using magnitudes

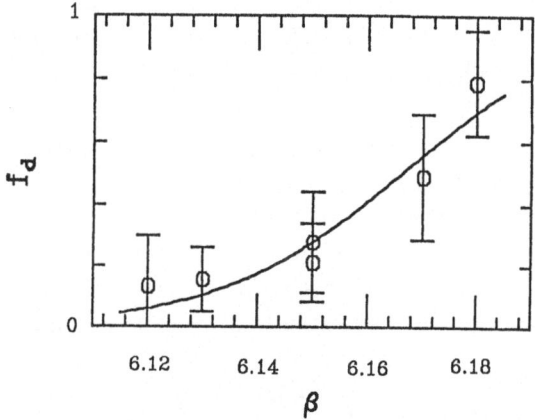

Figure 7b: least squares fit using phases

$$\beta_{3/4} - \beta_{1/2} \approx 0.20 \left(\frac{N_T}{N_S} \right)^3 \qquad (6)$$

which we tabulate for our lattices with $N_T = 10$, 12, and 14.

Effect of T_c criterion

n_S	n_T	$\beta_{3/4} - \beta_{1/2}$
17	10	.04
19	12	.05
19	14	.08
21	14	.06

We can now make another table of β_c. In this table we show the β at which we find $f_d = 3/4$ as well as $\beta_{1/2}$, or $f_d = 1/2$. The numbers differ slightly from our 1985 numbers because we took some more data and because we have gone through the analyses I have described.

Revised β_c Results

N_T	$\beta_{3/4}$	$\beta_{1/2}$	Columbia ($\beta_{1/2}$)
10	6.18 (2)	6.14 (1)	6.06 (2)
12	6.33 (2)	6.28 (1)	6.26 (2)
14	6.45 (2)	6.39 (2)	6.35 (3)

The systematic effect of the choice of criterion is larger than the statistical errors. For $N_T = 12$ and 14 this explains the discrepancy. (The Columbia lattices were $16^3 \times 12$ and $16^3 \times 14$, so in fact we expect their $\beta_{1/2}$ to be lower than ours.) There remains a clear discrepancy at $N_T = 10$.*

We have also tried to extract the latent heat of the transition from our data. The theory here is in the literature.[4,7] The important point for our purposes is that the latent heat is related to the jump in the average plaquette across the transition and the β function of the theory. Plugging the two loop perturbative beta function and the relation between Λ (lattice) and $\beta = 6/g^2$ into the result of Svetitsky and Fucito, we find

$$\frac{\Delta \epsilon}{\Lambda^4} = 12 \left(\frac{16\pi^2}{66} \beta \right)^{-204/121} \exp \left(\frac{32\pi^2}{66} \beta \right) \tag{7}$$

$$\times \left(\frac{22}{(4\pi)^2} + \frac{204}{(4\pi)^4} \frac{6}{\beta} \right) \Delta_{plaquette}$$

where $\Delta_{plaquette}$ is the jump in the average plaquette across the transition. As we have seen $\Delta_{plaquette}$ is related to the rounding of the transition from the finite spatial size. Our procedure is to divide our Monte Carlo runs into confined and deconfined portions, and average the plaquettes separately in the two phases. To be able to do this, we need a long run containing one or more tunnelings, and for which we kept a complete record of the plaquettes. In our data we have seven useful runs, with $N_T = 10$, 12, and 14 and β ranging from 6.13 to 6.45. Four of these runs were made on the ST100 and three on the Cyber 205. The jump in the plaquette is observable, although the statistical errors are large. It is reassuring that the measured gap on a particular run is reasonably independent of the Polyakov loop magnitude at which we separate the phases. Also, if we divide a run "by eye" we get about the same gap. Finally, $\Delta_{plaquette}$ is consistent with the rounding of the transition (G in eq. 4), although the statistical errors on G are always rather large when we fit the deconfinement fraction to eq. 4. There are undoubtedly large systematic errors in Δ_{plaq}. The largest of these arises from the fact that we have assigned all our configurations to one of two phases. In reality our samples contain transitional or mixed configurations. We do not know if these configurations are best characterized as two regions separated by domain walls or by the entire lattice moving from one phase to the other

---

* The Columbia group is doing more running at this point, and indications are that this number may change as a result.[6]

as a whole, but it is certain that with our small spatial sizes the separation of the phases is not clear cut. With these caveats in mind, we show the seven measured gaps together with their weighted average in figure 8.

The error bars were computed by blocking the plaquette lists for each of the two phases into blocks of 100 measurements (400 to 1000 sweeps) and taking the standard deviation of the mean of the partial averages, and these errors were propagated through the subtraction. Although the points seem to fall all over the map, the χ^2 is reasonable — 7.2 for six degrees of freedom. The average latent heat, under all our assumptions, is

$$\frac{\Delta\epsilon}{\Lambda^4} = 3.17 \times 10^7 \pm 0.35 \times 10^7 \pm \text{systematic error} \quad , \tag{8}$$

or, to get a more manageable number

$$\frac{\Delta\epsilon^{1/4}}{\Lambda} = 75.0 \pm 2.1 \pm \text{systematic error} \tag{9}$$

We reiterate that the systematic errors are probably large.

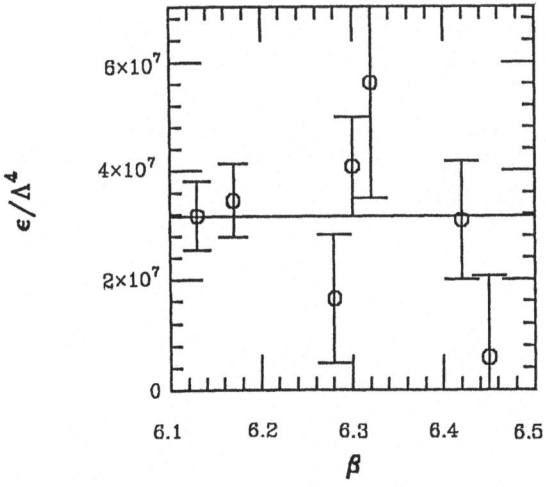

Figure 8: latent heat measurements

ACKNOWLEDGMENTS

We thank the Rechenzentrun der Universitat Karlsruhe, Argonne National Laboratory, and the Florida State Supercomputer Research Institute for the use of their machines. We thank J. Cavallini, K. Fengler, R. Hagstrom, P. Hanauer, M. Rushton, D. Sandee, and A. Schreiner for their advice and assistance. We are especially indebted to Norman Christ for discussing his results and data analysis with us. We would like to thank the Deutsche Forschungsgemeinschaft for their generous support. We gratefully acknowledge Control Data Corporation, E.I. Dupont de Nemours and Company and Xerox Corporation for their support. D.T. is an Alfred P. Sloan Foundation fellow. This work was supported by NSF grants DMR83-20423, PHY83-13324, and PHY82-17853, supplemented by funds from the U.S. National Aeronautics and Space Administration. We were also supported by DOE grant DE-AT03-81-ER40029 and by a DOE grant of supercomputer time. We also thank NATO and the United Kingdom SERC for their support.

REFERENCES

1. S.A. Gottlieb, A.D. Kennedy, J. Kuti, S. Meyer, B.J. Pendleton, R.L. Sugar, D. Toussaint, *Phys. Rev. Lett.* **55**, 1958 (1986).
2. A.D. Kennedy, J. Kuti, S. Meyer, B.J. Pendleton, *Phys. Rev. Lett.* **54**, 87 (1985).
3. N. Christ and A. Terrano, *Phys. Rev. Lett.* **56**, 111 (1986).
4. B. Svetitsky and F. Fucito, *Phys. Lett.* **131B**, 165 (1983).
5. A.G. Frodesen, O. Skjeggestad and H. Tøfte, *Probability and Statistics in Particle Physics* (Universitetsvorlaget, Bergen, 1979).
6. N. Christ, private communication.
7. J. Engels, F. Karsch, I. Montvay and H. Satz, *Phys. Lett.* **101B**, 89 (1981), *Nucl. Phys.* **B205**, 545 (1982); F. Karsch, *Nucl. Phys.* **B205**, 285 (1982); T. Celik, J. Engels and H. Satz, *Phys. Lett.* **129B**, 323 (1983).

RESULTS ON SU(3) GAUGE THEORY IN A FINITE VOLUME

Pierre van Baal

Institute for Theoretical Physics
State University of New York at Stony Brook
Stony Brook, NY 11974

ABSTRACT

The semiclassical evaluation of the energy of 't Hooft type electric flux for $SU(N)$ gauge theories on a torus is given. Using the perturbative energy eigenvalues for $SU(3)$, recently published by Weisz and Ziemann, we predict that a crossover in the energy of electric flux occurs when the length of the cubic volume equals 1.6 times the inverse 0^{++} glueball mass. A contradiction with previous Monte Carlo data by Berg e.a. is resolved by their new data, presented at this conference.

INTRODUCTION

By formulating $SU(N)$ gauge theories on a torus, 't Hooft[1] originated a new and promising approach to non-perturbative features of non-abelian gauge theories. Hardly appreciated until recently, his formulation is the exact context in comparing with Monte Carlo calculations on a finite lattice[2]. What I present here is new analytic results[3]. However, I will emphasise the comparison with lattice gauge theory, the topic of this conference.

't Hooft introduced the notion of electric and magnetic flux by using so-called twisted boundary conditions on a space-time box. In the Hamiltonian formulation, pioneered by Lüscher in a perturbative context[4], one formulates the theory on a spatial torus. Choosing the gauge fields periodic one singles out zero magnetic flux, but electric flux can still be defined, employing the symmetries of the Hamiltonian[1,5]. In lattice gauge theory, using the transfer matrix as the equivalent of the Hamiltonian, one can similarly define electric flux on a lattice[6], which nevertheless has purely periodic boundary conditions. However, the time extent (T) should be sufficiently large if we want to suppress finite temperature effects and compare with the analytic Hamiltonian approach. As with a spectrometer, how small the temperature needs to be is determined by the resolution in energy one wants to achieve. This will play an important role further on.

THE EFFECTIVE HAMILTONIAN

The analytic results are valid for a small volume, where due to the asymptotic freedom of non-abelian gauge theories, the effective coupling constant is small. In perturbation theory, integrating out non-zero momentum modes yields Lüscher's effective Hamiltonian[4], in lowest order:

$$H_{eff} = \frac{-g^2}{2L} \frac{\partial^2}{\partial c_i^{a2}} - \frac{1}{Lg^2} \mathrm{Tr}\left([c_i, c_j]^2\right) \quad , \tag{1}$$

where c_i is a constant element of the $SU(N)$ Lie-algebra related to the zero momentum gauge fields by $A_i = c_i/L$, where the finite volume is a cube with sides L. L also sets the scale for the renormalised coupling constant appearing in eq.(1):

$$g^{-2} = -11N \ln(\Lambda \cdot L) \big/ (24\pi^2) + \cdots \quad . \tag{2}$$

Lüscher's effective Hamiltonian has a <u>discrete</u> spectrum representing the effective interaction of the zero momentum sector of the theory. The 0^{++} glueball mass is therefore the difference in energy between the ground state and the first excited 0^{++} state of this effective Hamiltonian. At this place is is important to observe that the 0^{++} glueball in a small volume obviously does not behave as a particle state. By a simple rescaling of the dynamical variables c_i, one easily deduces that in lowest order the energies are proportional to $g^{2/3}/L$ and as shown by Lüscher[4] one finds the following perturbative expansion for energies and masses: $E = L^{-1} \sum_{k=1}^{\infty} \varepsilon_k g^{2k/3}$. The effective Hamiltonian is non-integrable[7] and the coefficients ε_k need to be calculated numerically. This was done up to one loop for $SU(2)$ by Lüscher and Münster[8] and for $SU(3)$ by Weisz and Ziemann[9]. For details we refer to the literature or to the review by Kuti[25] in these proceedings.

It is the hope that quantities such as mass ratios can be smoothly connected from the short to the long distance regime[4]. There can be only a posteriori justification for such an assumption. Indeed, we will discuss two crossovers, and only a good understanding of these will allow us to make predictions for the large distance domain. The first crossover is associated with <u>zero action</u> tunneling[10] and occurs in a domain claimed to be under analytic control[11]. A discussion of the second crossover will follow later.

THE TUNNELING

The potential of eq.(1) has the special property of a flat direction for abelian configurations c_i ($[c_i, c_j] = 0$). This gives rise to a classical vacuum valley (for $SU(N)$ also called a toron valley[12]). The effective potential along the vacuum valley has a multiple vacuum structure. Different vacua are connected by gauge transformations periodic up to an element of Z_N (the centre of the gauge group $SU(N)$). This gives the appropriate connection with the

412

twisted boundary conditions of 't Hooft in the time direction[1]. These vacua are degenerate to all orders in perturbation theory[4] due to the presence of a quantum induced potential barrier. Tunneling through such an induced barrier was discussed for a toy-model in ref.13 and for $SU(N)$ gauge theories in great detail in ref.14. The energy split (ΔE) of the ground state due to this tunneling is exactly the energy of 't Hooft type electric flux. The relevant effective Lagrangian along the instanton (pinchon[10]) path can be represented by ($\mathbf{x} \in \mathbf{R}^3$)[14]:

$$L_{eff} = \frac{L(N-1)}{Ng^2}\dot{\mathbf{x}}^2 - (N-1)V_1(\mathbf{x}) \quad , \tag{3}$$

with $V_1(\mathbf{x})$ the SU(2) one-loop effective potential[4,10], which has 2π periodicity. Gauge invariance restricts lattice momenta ($\mathbf{p}$) to multiples of $2\pi/N$ (i.e. $\mathbf{p} = 2\pi\mathbf{e}/N$, $\mathbf{e}$ = 't Hooft type electric flux). The following result is obtained for arbitrary N in the steepest descent approximation[3]:

$$\Delta E(L) = 2L^{-1}\sin^2\left(\frac{\pi}{N}\right)|B_N|^2 \lambda_N\, g^{5/3} \times$$
$$\exp\left(-S_N\, g^{-1} + T_N\left(\varepsilon_1(N) - \varepsilon_1(N-1)\right)g^{-1/3}\right) \quad . \tag{4}$$

The action S_N, the tunneling time T_N and a contribution from transverse fluctuations λ_N have simple N-dependence, which can be understood from eq.(3):

$$S_N = \frac{2(N-1)}{\sqrt{2N}}S \quad ; \quad S = 12.4637\ldots$$
$$T_N = \frac{2}{\sqrt{2N}}T \quad ; \quad T = 3.9186\ldots$$
$$\lambda_N = \frac{\sqrt{2N}}{2(N-1)}\lambda \quad ; \quad \lambda = 0.6997\ldots \quad , \tag{5}$$

whereas the first coefficient, $\varepsilon_1(N)$, of the perturbative expansion for the groundstate energy of Lüscher's effective Hamiltonian and the coefficient B_N (which is determined by the asymptotics of the perturbative wavefunction[14]) have no simple N-dependence.

The crossover in ΔE occurs with the onset of tunneling through the quantum induced potential barrier, and is determined to a good approximation by the value of the renormalised coupling constant where the exponent in eq.(4) changes sign:

$$g_c{}^{2/3} = \frac{(N-1)S}{T(\varepsilon_1(N) - \varepsilon_1(N-1))} \quad . \tag{6}$$

Using $\varepsilon_1(2) = 4.116719\cdots$[8] and $\varepsilon_1(3) = 12.5887\cdots$[9] allows us to evaluate the crossover value for SU(3): $g_c{}^2 = 0.423$ (compared to SU(2): $g_c{}^2 = 0.461$).

It is because of tunneling through a quantum induced barrier, that this tunneling sets in at small coupling. This also means that the effective potential itself will hardly change over this crossover, which is therefore not a phase transition and cannot be compared with the deconfining transition.

COMPARISON WITH MONTE CARLO RESULTS

In order to compare with lattice gauge theory, viewed as an alternative regularisation of the continuum theory[2], one should convert to renormalisation group invariant (dimensionless) parameters. A particularly useful set, advocated for the past few years is the universal scale parameter $z = M_L(0^{++})L$ [15] and the energy of electric flux in units of the 0^{++} glueball mass $\mathcal{E}(z) = \Delta E(L)/M_L(0^{++})$ [10]. Using the perturbative expansion of z for SU(3) [9], our prediction of g_c^2 translates into a prediction of the z value at crossover: $z_c \simeq 1.6$ (compared to the earlier SU(2) prediction of $z_c \simeq 1.1 - 1.2$ [11]).

Figure 1 represents the existing Monte Carlo data for SU(3) [16,17], but before we compare with our analytic prediction there are a few things worth pointing out. Berg and Billoire[18] have opened up the finite volume to Monte Carlo calculations by using elongated lattices in the time direction.

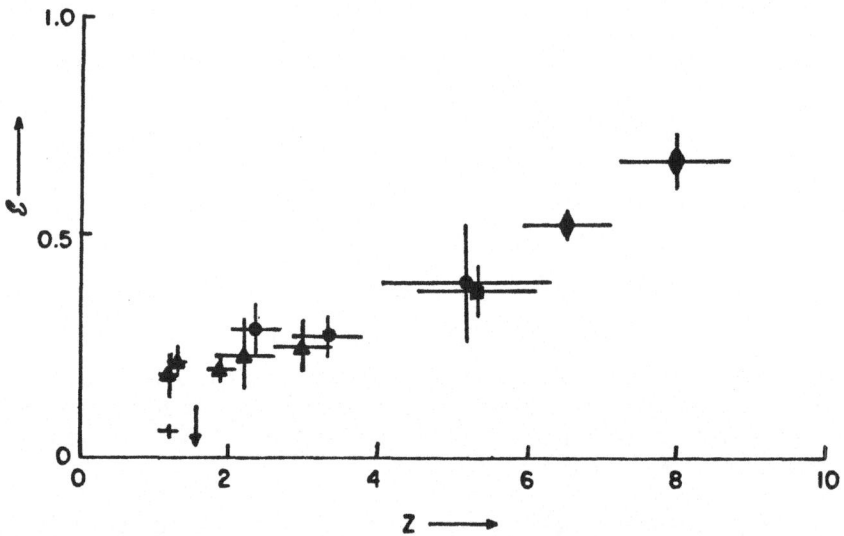

Figure 1: The energy of electric flux versus the size of the cubic volume in dimensionless units. Indicated by an errow is the analytic prediction of z_c, below which $\mathcal{E}$ drops to zero. The Monte Carlo data for a $N_s^3 \times N_t$ lattice were taken from: ▲: $N_s = 4, N_t = 32$, ●: $N_s = 6, N_t = 32$[16],■: $N_s = 6, N_t = 21$,◆: $N_s = 9, N_t = 21$[17], +: $N_s = 4, N_t = 64$[26,27].

414

Equally important was their observation that in a small volume the spatial Polyakov loop in the adjoint representation couples much better to the glueball than does the plaquette. Following Parisi e.a.[19] they also measured the time-time correlation function of the spatial Polyakov loop in the fundamental representation. In attempting to relate our analytic work for SU(2) to Monte Carlo data, we clarified[11] the connection of this with the energy of 't Hooft type electric flux. As was already pointed out by 't Hooft[1], such a Wilson loop creates one unit of electric flux and the correlation function therefore measures the energy of one such a unit of electric flux, which we called $\Delta E(L)$, disregarding whether L is large and disregarding the validity of a string picture (making $\Delta E(L)$ linear in L [1]). The quantity usually considered[16,17,18,19] is $K_L = \Delta E(L)/L$, but it is $\underline{\text{misleading}}$ to call K_L a string tension for $z <$ 5. As we will see, no string has formed for these small volumes. It was also observed in ref.11 how there is a $\underline{\text{dramatic}}$ improvement in universality by considering for SU(2) the ratio $(\mathcal{E}(z)/z)^{1/2} = \sqrt{K_L}/M_L(0^{++})$, instead of $\sqrt{K_L}/\Lambda_{latt}$ and $M_L(0^{++})/\Lambda_{latt}$ separately[18]. The two-loop beta-function apparently deviates considerably from the non-perturbative beta-function[20], thereby causing "artificial lattice artifacts".

FINITE TEMPERATURE PROBLEMS

If we now consider figure 1, we see that there are two data points from ref.16 at z considerably below where we predict tunneling to set in. For these two data points we would expect $\mathcal{E}(z) << 1$. As was pointed out in ref.3, if indeed tunneling is suppressed, the energy resolution necessary to measure the energy of electric flux has to be better than ΔE, which can only be realised by an $\underline{\text{impractically}}$ large time extent: $N_s/N_t << z\mathcal{E}(z) << 1$, where $N_{s(t)}$ is the number of lattice sites in the space (time) direction. Hence we would expect strong temperature dependence for $z < 1.6$. Indeed, the new data presented by Berg in his talk[26,27] for the lowest z value ($z = 1.20 \pm 0.16, N_s = 4, N_t = 32, \beta = 6.6$) showed a substantial $\underline{\text{decrease}}$ for the energy of electric flux when the time extent is doubled to $N_t = 64$. To be specific, at $N_t = 32$ one has $\mathcal{E}(1.2) = 0.18$ and at $N_t = 64$ it is reduced to $\mathcal{E}(1.2) = 0.06$ (we used $\mathcal{E} = (\sqrt{K_L}/M_L(0^{++}))^2 \cdot z$, with $M_L(0^{++})/M_L(2^{++}) = 1.6 \pm 0.2$, $(\sqrt{K_L}/M_L(2^{++}))_{N_t=32} = 0.62 \pm 0.12$ [16] and $(\sqrt{K_L}/M_L(2^{++}))_{N_t=64} = 0.37 \pm 0.06$ [27]). In order to obtain the zero temperature value of $\mathcal{E}(1.2)$, $N_t = 64$ is still not big enough, as we argued above. Upon further increase of N_t, $\mathcal{E}(1.2)$ will continue to decrease. We predict a similar N_t dependence for the other data point below $z \simeq 1.6$ in figure 1, whereas for $z > 1.6$ there should be hardly any N_t dependence. It is important that this will be verified, since all data in ref.16 is at the same value of N_t.

There is another serious problem with the low z data of ref.16. As was pointed out by Weisz and Ziemann[9]. Their analytic value of 1.2 for the $M_L(0^{++})/M_L(2^{++})$ mass ratio contradicts the Monte Carlo value of 1.6 at $z = 1.2$. In our opinion, it is quite conclusive from the new data[26] that $z = 1.2$ is in the region where tunneling is suppressed; this is a strong point in favour of the accuracy of the analytic prediction[9]. Although glueball masses do not significantly suffer from the energy resolution (i.e. temperature) problem for $z < 1.6$, there is a potential problem in estimating the 2^{++} glueball mass from Monte Carlo data in this region, as was pointed out by Lüscher[22] (see ref.9; for a similar effect see ref.23). Due to the degeneracy of the states with one unit of electric flux and their almost degeneracy with the ground state, the 2^{++} correlation function as a function of time (t) behaves likely as $A + B \exp(-M_L(2^{++})t)$, with A proportional to $\exp(-\Delta E \cdot T)$, which can differ appreciably from zero at $z < 1.6$. A two parameter fit to the 2^{++} correlation function with $A = 0$ will underestimate the 2^{++} glueball mass.

THE SECOND CROSSOVER

Let us finally discuss the second crossover. The first argument for expecting such a second crossover was presented in ref.11, where it was observed that within the Nambu-Goto string model $\mathcal{E}(z)$ is exactly calculable, with (due to the tachyon[24]) a squareroot singularity at $z'_c \simeq 5$. Another reason to expect something to happen is that the separation of zero and non-zero momentum states is likely to breakdown when the glueball mass becomes of the order of a typical non-zero momentum energy of $2\pi/L$ [11], leading to a z value of 2π. Two further arguments were presented in ref.21. Namely, using the finite temperature analogue one can expect a crossover at $L \simeq 1/T_c$ (T_c the deconfining temperature) yielding again $z'_c \simeq 5$ [29]. How to relate z'_c to T_c using 't Hooft's duality principle[1] is discussed in some detail by Kuti[25]. Let us emphasise that this crossover, despite its similarity with the deconfining transition, is also not a phase transition, because we are still in a finite volume. The fourth argument[20] for a second crossover is based on the observation that if zero action tunneling leads to a crossover, the onset of non-zero action tunneling should likewise show up as a crossover, necessarily at much larger z. Actually, with the impressive high statistics data on the topological susceptibility in large volumes[28], it should not be difficult to measure precisely where non-zero action tunneling sets in. Unfortunately, existing Monte Carlo data is (deliberately chosen) above $z \simeq 5$. We cannot stress enough the importance of such a calculation. It is clearly suggested that the crossover at $z \simeq 5$ is associated with restoration of rotational symmetry, the formation of strings and the localisation of the glueball. Establishing a connection with the onset of non-zero action tunneling (topology) would clearly be of paramount importance. Finally there is a quite unrelated argument for a crossover at approximately $z = 5$ due to Michael[23]. This however relies heavily on a "pseudo particle or string" picture below $z \simeq 5$, which we find hard to justify.

EPILOGUE

In conclusion we have tried to make plausible the existence of three regions for SU(N) Yang-Mills on a torus: $z < z_c$ where perturbation theory[4,8,9] is adequate; $z_c < z < z'_c$ where a perturbatively defined effective Lagrangian can be used to include the tunneling effects[3,10,11,15]; and $z > z'_c$ the truly non-perturbative strong coupling (string-like?) domain. In this picture we can also understand why the glueball for $z < 5$ couples more strongly to an extended operator (the adjoint Polyakov loop[16,18]), whereas for $z > 5$ it prefers to couple to a localised operator (the plaquette[17]). However, this picture also implies that naive extrapolations of quantities like the $2^{++} - 0^{++}$ mass ratio, from $z < 5$ to $z > 5$, have to wait for a better understanding of the second crossover.

There are two reasons for being interested in relatively small volumes[3]. Lattice artifacts can be substantially reduced by going to smaller values of the coupling constant thereby, however, reducing the physical size of the volume. But for these smaller volumes, analytic results described above, might provide confidence in Monte Carlo results. Having established this confidence, Monte Carlo calculations can subsequently help us asking the right questions in order to <u>understand</u> the dynamics of the longdistance domain of QCD.

ACKNOWLEDGEMENTS

Most of the work described here was done together with Jeffrey Koller, I thank him for a most pleasant collaboration. I also wish to thank the organisers of this conference for giving me the opportunity to present our work. My deep gratitude goes to Peter Weisz for providing me with his SU(3) results, long before they were published. I also greatly appreciate Martin Lüscher's continued support. His criticism and advice have been a continued source of inspiration. I thank Julius Kuti and the other participants of the conference for their interest and discussions. This work was supported in part by NSF grant number DMS-84-0665.

REFERENCES

[1] G. 't Hooft, Nucl. Phys. B153(1979)141.
[2] K. Wilson, Phys. Rev. D10(1974)2445.
[3] P. van Baal, J. Koller, Finite size results for SU(3) gauge theory, Stony Brook preprint, ITP-SB-86-76, September 1986.
[4] M. Lüscher, Nucl. Phys. B219(1983)233.
[5] P. van Baal, Twisted boundary conditions, a non-perturbative probe for pure non-abelian gauge theories, thesis, Utrecht, July 1984.
[6] C. Borgs, E. Seiler, Comm. Math. Phys. 91(1983)329.
[7] G.K. Savvidy, Nucl. Phys. B246(1984)302.
[8] M. Lüscher, G. Münster, Nucl. Phys. B232(1984)445.
[9] P. Weisz, V. Ziemann, Weak coupling expansion of low-lying energy values for SU(3) gauge theory on a torus, Hamburg preprint, September 1986.
[10] P.van Baal, The energy of electric flux on the hypertorus, Utrecht, internal report, march 1984, Stony Brook preprint, ITP-SB-84-72, unpublished; P. van Baal, Nucl. Phys. B264(1986)548.
[11] J. Koller, P. van Baal, Nucl. Phys. B273(1986)387.
[12] A. Gonzalez-Arroyo, J. Jurkiewicz, C.P. Korthals Altes, Proceedings of the 1981 Freiburg Nato Summer Institute, Plenum, New York, 1982.
[13] P. van Baal, A. Auerbach, Nucl. Phys. B275[FS17](1986)93.
[14] P. van Baal, J. Koller, QCD on a torus, and electric flux energies from tunneling, Stony Brook preprint, ITP-SB-86-31, April 1986, to be published in Annals of Physics.
[15] M. Lüscher, Phys. Lett. 118B(1982)391.
[16] B. Berg, A. Billoire, C. Vohwinkel, Phys. Rev. Lett. 57(1986)400.

[17] A. Patel, R. Gupta, G. Guralnik, G. Kilcup, S. Sharpe, Phys. Rev. Lett. 57(1986)1288; the data presented here are those from the preprint version UCSD-10P10-260, april 1986, which differ only slightly from the finally published ones.

[18] B. Berg, A. Billoire, Phys. Lett. 166B(1986)203.

[19] A. Hasenfratz, P. Hasenfratz, U. Heller, F. Karsch, Phys. Lett. 143B (1984)193.

[20] G. Parisi, R. Petronzio, F. Rapuano, Phys. Lett. 128B(1983)418.

[21] T.H. Hansson, P. van Baal, I. Zahed, Chromomagnetic energy of SU(2) gauge fields on a torus, Stony Brook preprint, ITP-SB-86-49, July 1986.

[22] M. Lüscher, private communication.

[23] C. Michael, Urabana preprint, P/86/5/67, June 1986.

[24] P. Olesen, Phys. Lett. 160B(1985)408.

[25] J. Kuti, Hamiltonian spectrum, chromoelectric flux and deconfinement in a small periodic world, these proceedings.

[26] B. Berg, Finite size scaling and lattice gauge theory, these proceedings.

[27] B. Berg, A. Billoire, C. Vohwinkel, Tunneling and deconfinement, Tallahassee preprint, September 1986.

[28] A.S. Kronfeld, M.L. Laursen, G. Schierholz, U.-J. Wiese, High statistics computation of the topological susceptibility of SU(2) gauge theory, DESY preprint, DESY 86-082, July 1986; talk by M. Laursen at this conference.

[29] I owe the seed of this observation to A. Patel, R. Gupta, A. Gocksch and J. Kuti, who first suggested that our analytic prediction for SU(2) could not be correct since $1.2 \neq 5$.

THE GF11 PARALLEL COMPUTER

J. Beetem M. Denneau and D. Weingarten

IBM, T.J. Watson Research Center
Yorktown Heights, NY 10598

ABSTRACT

GF11 is a parallel computer currently nearing completion at the IBM Yorktown Research Center. The machine will have a peak arithmetic rate of 11.4 Gflops and a total memory of 1.14 Gbytes. The computational power and memory are uniformly distributed among 566 floating-point processors which communicate through a switching network. At each machine cycle any of 1024 preselected permutations of data can be realized among the processors. The main intended application of GF11 is a class of calculations arising from quantum chromodynamics. For a detailed discussion of GF11 see:

J. Beetem, M. Denneau, and D. Weingarten, in IEEE Proceedings of the 12th Annual International Symposium on Computer Architecture, IEEE Computer Society, Washington, D.C., 1985.

J. Beetem, M. Denneau, and D. Weingarten, Journal of Statistical Physics 43, 1171 (1986).

J. Beetem, M. Denneau, and D. Weingarten, to appear in Experimental Parallel Computing Architectures, (edited by J.J. Dongarra) North-Holland, Amsterdam (1987).

---

* Present address: Dept. of Electrical and Computer Engineering, University of Wisconsin, Madison, WI 53706.

PARTICIPANTS

I. Angus	*California Inst. of Technology, Pasadena, USA*
Y. Arian	*Boston University, USA*
K. Barad	*Barnard College, New York, NY, USA*
I. Barbour	*The University of Glasgow, Scotland, UK*
G. Batrouni	*Boston University, USA*
B. Berg, SCRI	*Florida State University, USA*
C. Bernard	*University of California/Los Angeles, USA*
G. Bhanot	*SCRI, Florida State University, USA*
A. Billoire	*DPHT, CEN-Saclay, Gif-sur-Yvette, France*
K.M. Bitar	*SCRI, Florida State University, USA*
S.C. Black	*Florida State University, USA*
K.C. Bowler	*University of Edinburgh, Scotland, UK*
R.C. Brower	*Boston University, USA*
A.N. Burkitt	*University of Liverpool, UK*
K. Cahill	*University of New Mexico, USA*
D.G. Caldi	*University of Connecticut, USA*
C. Carlson	*College of William and Mary, USA*
T. Çelik	*Hacettepe University, Turkey*
N.H. Christ	*Columbia University, New York, NY, USA*
A. Coste	*CNRS, Marseille, France*
M. Creutz	*Brookhaven National Laboratory, USA*
E. Dagotto	*University of Illinois, USA*
C. Davies	*University of Glasgow, Scotland, UK*
K. Decker	*Universität Bern, Switzerland*
Ph. deForcrand	*CRAY Research, Chippawa Falls, WI, USA*
T. DeGrand	*University of Colorado, USA*
A. Di Giacomo	*University of Pisa, Italy*
T. Draper	*TRIUMF, Vancouver, Canada*
H.A. Duncan	*University of Pittsburgh, USA*
A. El-Khadra	*Freie Universität Berlin, West Germany*
S. Eubank	*University of Texas/Austin, USA*
R. Fiebig	*Florida International University, USA*

F. Fucito	*University of Rome, Italy*
M. Fukugita	*RIFM, Kyoto University, Japan*
R.V. Gavai	*Tata Institute for Fundamental Research, India*
P. Gibbs	*The University of Glasgow, Scotland, UK*
A. Gocksch	*University of California/San Diego, USA*
A. Gonzalez-Arroyo	*University of Madrid, Spain*
S. Gottlieb	*Indiana University, USA*
M. Gross	*Newman Lab., Cornell University, USA*
R. Gupta	*Los Alamos National Laboratory, USA*
G. Guralnik	*Los Alamos National Laboratory, USA*
O. Haan	*Siemens AG, München, West Germany*
A. Hasenfratz	*SCRI, Florida State University, USA*
P. Hasenfratz	*Universität Bern, Switzerland*
U. Heinz	*Brookhaven National Laboratory, USA*
U.M. Heller	*University of California/Santa Barbara, USA*
D. Heys	*DAMTP, University of Liverpool, England*
A. Irbäck	*University of Goteberg, Sweden*
Y. Iwasaki	*University of Tsukuba, Japan*
J. Jersàk	*Technische Hochschule Aachen, West Germany*
S. Kahana	*Brookhaven National Laboratory, USA*
K. Kanaya	*Technische Hochschule Aachen, West Germany*
F. Karsch	*CERN, Switzerland*
W. Kerler	*Philipps Universität Marburg, West Germany*
G. Kilcup	*Newman Lab., Cornell University, USA*
J. Kiskis	*University of California/Davis, USA*
F.R. Klinkhamer	*Lawrence Berkeley Laboratory, USA*
C. Korthals-Altes	*CNRS, Marseille, France*
M. Kremer	*Johannes Gutenberg Universität, Mainz, West Germany*
H. Kröger	*Université Laval, Quebec, Canada*
J. Kuti	*University of California/San Diego, USA*
E. Laermann	*Universität Wuppertal, West Germany*
C.B. Lang	*Universität Graz, Austria*
G. Lasher	*Watson Research Center, IBM, Yorktown Heights, USA*
M. Laursen	*Nordita, Copenhagen, Denmark*
I-H. Lee	*Brookhaven National Laboratory, USA*
K.F. Liu	*University of Kentucky, USA*
P. MacKenzie	*Fermilab, USA*
J. Mandula	*DOE, Washington, DC, USA*
E. Manousakis	*Massachusetts Institute of Technology, USA*
M. Marcu	*Universität Hamburg, West Germany*
H. Markum	*Technische Universität Wien, Austria*
E. Mendel	*Technion, Haifa, Israel*
H. Meyer-Ortmanns	*Max-Planck Institut, West Germany*

D.E. Miller   Pennsylvania State University, USA
I. Montvay   DESY, Hamburg, West Germany
A. Morel   Saclay, Gif-sur-Yvette, France
A. Moreo   University of Illinois, USA
K.H. Mütter   Gesamthochschule Wuppertal, West Germany
E.A. Myers   Dalhousie University, Halifax, Canada
S. Nadkarni   Rutgers University, USA
A. Nakamura   Freie Universität Berlin, West Germany
T. Neuhaus   SCRI, Florida State University, USA
P. Olesen   Niels Bohr Institute, Copenhagen, Denmark
M. Ogilvie   Washington University/St. Louis, USA
K. Olynyk   Fermilab, USA
P. Orland   Massachusetts Institute of Technology, USA
A. Patel   Lyman Lab., Harvard University, USA
B. Petersson   Universität Bielefeld, West Germany
R. Petronzio   Ist. Naz. di Fisica Nucleare, Roma, Italy
A. Phillips   State University of New York/Stony Brook, USA
J. Polonyi   Massachusetts Institute of Technology, USA
J. Potvin   Brookhaven National Laboratory, USA
P. Rakow   Carnegie-Mellon University, Pittsburgh, USA
C. Rebbi   Brookhaven National Laboratory, USA
K. Redlich   Universität Bielefeld, West Germany
R.Z. Roskies   University of Pittsburgh, USA
P. Rossi   University of California/San Diego, USA
W. Rühl   Universität Kaiserslautern, West Germany
R. Rusack   EP Div., CERN, Switzerland
R. Salvador   SCRI, Florida State University, USA
S. Sanielevici   Brookhaven National Laboratory, USA
H. Satz   Universität Bielefeld, West Germany
G. Schierholz   DESY, Hamburg, West Germany
K. Schilling   Gesamthochschule Wuppertal, West Germany
J.C. Sexton   Fermilab, USA
S. Sharpe   University of Washington/Seattle, USA
J. Shigemitsu   Ohio State University, USA
R. Shrock   ITP, State University of New York/Stony Brook, USA
D.K. Sinclair   Argonne National Laboratory, USA
J. Smit   University of Amsterdam, The Netherlands
R. Sommer   Gesamthochschule Wuppertal, West Germany
A. Soni   University of California/Los Angeles, USA
J.D. Stack   Loomis Lab., University of Illinois, USA
I.O. Stamatescu   Freie Universität Berlin, West Germany
P. Suranyi   University of Cincinnati, USA
B. Svetitsky   Massachusetts Institute of Technology, USA

H.B. Thacker	*Fermilab, USA*
D. Toussaint	*University of California/San Diego, USA*
A. Trivedi	*University of Chicago, USA*
P. van Baal	*State University of New York/Stony Brook, USA*
J.C. Vink	*University of Amsterdam, The Netherlands*
A. Vladikas	*The University of Glasgow, Scotland, UK*
K. Vohwinkel	*SCRI, Florida State University, USA*
D. Weingarten	*IBM Research, Yorktown Heights, NY, USA*
W. Wetzel	*Universität Heidelberg, West Germany*
U.J. Wiese	*Universität Hannover, West Germany*
K. Wilson	*Cornell University, USA*
W. Wilcox	*Baylor University, USA*
U. Wolff	*Universität Kiel, West Germany*
R. Woloshyn	*TRIUMF, Vancouver, Canada*
J. Wosiek	*Louisiana State University, USA*
H.W. Wyld, Jr.	*Loomis Lab., University of Illinois, USA*
D. Zwanziger	*New York University, USA*

Round Table discussion - C. Rebbi, K. Wilson, M. Fukugita, D. Weingarten, R. Petronzio, K. Schilling, and J. Kuti.

INDEX

Dynamical staggered quarks, 347

Effective Hamiltonian, 414
Electric flux, 135, 413, 417
Electroweak interaction, 237
Equilibration time, 311
Ergodicity, 100
Exact algorithm for fermions, 145
Exceptional configurations, 16, 17
Excited hadron, 109
External charge, 184
Eye graphs, 13, 14, 22, 23, 24, 25

Fake loop, 104, 111
Field theory, 29
Figure-eight graphs, 13, 17, 22, 25
Finite baryonic density, 125
Finite size effect, 104, 107
Finite temperature QCD, 75, 99, 100, 110, 111, 413, 417, 418
Finite volume, 413
Finite-size effects, 40, 42, 43, 44
Flavour interpretation, 247
Fokker-Planck equation, 101
Fourier acceleration, 63
Fractal, 29
Fundamental adjoint actions, 389

Glueball masses, 135, 389
Glueball, 39, 414, 417
Gluon condensate, 91
Gluon internal energy, 110
Gluon propagator, 213
Gluon thermodynamics, 80
Gluon vacuum polarization, 270
Gross-Neveu model, 247

Hadron mass, 106, 107, 108, 109
Hadron spectrum, 171, 259, 311

Higgs mechanism, 275
Higgs models, 179
Higgs phase transition, 179, order of, 184,
Higgs phenomenon, 244
Higgs-boson mass, 244
Homotopy group, 295
Hopping parameter, 103, 104, 105, 110
Hot QCD, 275
Hybrid method, 100, 369

Importance sampling, 6
Improved action, 171, 357
Improved ratio method, 113
Index theorem, 379

Jackknife method, 16

KAM theorem, 100
Kaon decays, 21, 25, 357
Kogut-Susskind fermion, 100, 105, 107, 110, 111, 145
$K^0 - \bar{K}^0$ mixing, 17, 21

Lanczos, 2
Landau gauge, 213
Langevin, 369
  equation, 100, 101, 102
  updating, 347
Large scale computing, 259
Latent heat, 55, 402
Lattice fermions, 247
Lattice QED, 201
Linear potential, 109

Magnetic monopole, 297
Mass ratio, 417, 419
Matrix element, 47, 357
  method of extraction, 15,
  weak decay, 13,
Maximum likelihood fit, 406
MCRG, 35, 39, 46
Mean field analysis, 333
Microcanonical, 369
Molecular dynamics, 100
Monopoles, 201

Nucleation, 405
Numerical gauge-fixing, 63

Obstruction theory, 295
Occam, 46
Operator mixing, 357
Order parameters, 225

Parallel computers, 55
Perturbation expansion, 414, 418
Perturbation theory (weak coupling), 19, 23, 275
Perturbative β function, 405
Pinchon, 413
Polyakov line, 110, 389
Polyakov loop, 225, 275, 305, 417, 418
Principal bundle, 294
Projective space, 294
Propagator,
  matrix, 125
  baryon, 41, 43, 44,
  meson, 40, 41,
  quark, 40, 43,
  time-slice, 40, 41, 42, 43, 44,
Pseudofermion method, 113, 136
Pseudofermion,
  Langevin scheme, 101,
  Monte Carlo (Metropolis), 100,
  variable, 100, 101,
Pseudofermions, 311

QCD thermodynamics, 369
Quark confinement, 55
Quark deconfinement, 55, 75
Quark-gluon plasma, 75
Quenched approximation, 15, 35, 40, 99, 107, 109, 110, 125, 259, 325

Renormalisation group, 171, 414, 414
Runge-Kutta, 101

Scalar field condensate, 184

Scale factor, 39
Scaling, 389, 405
Screening energy, 184
Screening, 311
Simplex, 295
Simplicial lattice, 291
Source method, 136
Special purpose computers, 55
Spin dependent potential, 325
Staggered fermions, 40, 113, 247, 357
Static potential, 109
Steepest descent, 413
Stochastic equation, 100, 101
Stochastic quantisation, 100
String tension, 39, 417
String theory, 281
Strong interaction, 297
Supercomputers, 55
Surfaces, 281
Symmetry restoration, 247
Systematic errors, 347

Tangent bundle, 294
Topological charge, 291
Topological susceptibility, 379
Torelon, 137
Toron, 137
Transfer matrix, 413
Transient effects, 40
Transition function, 291
Transition rounding, 405
Transputer, 46
Triangulation, 295
Tricritical point, 207, 209
Tunneling, 414, 413

Unitarity gauge, 275
Universal expansion parameter, 414
Universality, 417

Vacuum condensate, 275
Vacuum saturation, 357
Vacuum valley, 414

Weak decay, 47